Presented to
Blairsville High School Library
by
The Pennsylvania Senior and Junior Academies
of Science in Honor of
Russell Whitfield
In Recognition of receiving a Special Award
May 17, 1999
at the Sixty-Fifth Annual PJAS State Meeting

DATE DUE

SCHOOL

#47-0108 Peel Off Pressure Sensitive

SOLID AND LIQUID WASTES:
Management, Methods and Socioeconomic Considerations

The Pennsylvania Academy of Science Publications
Books and Proceedings

Book Editor: Shyamal K. Majumdar
Professor of Biology
Lafayette College
Easton, Pennsylvania 18042

1. *Energy, Environment, and the Economy,* 1981. ISBN: 0-9606670-0-8. Editor: Shyamal K. Majumdar.

2. *Pennsylvania Coal: Resources, Technology, and Utilization,* 1983. ISBN: 0-9606670-1-6. Editors: Shyamal K. Majumdar and E. Willard Miller.

3. *Hazardous and Toxic Wastes: Technology, Management and Health Effects,* 1984. ISBN: 0-9606670-2-4. Editors: Shyamal K. Majumdar and E. Willard Miller.

4. *Solid and Liquid Wastes: Management, Methods and Socioeconomic Considerations,* 1984. ISBN: 0-9606670-3-2. Editors: Shyamal K. Majumdar and E. Willard Miller.

5. *Proceedings* of the Pennsylvania Academy of Science. Two issues per year; current volume (1984) is 58. ISSN: 0096-9222. Editor: Daniel L. Klem, Jr.

A volume on *Radioactive Materials and Wastes* is in preparation.

SOLID AND LIQUID WASTES:
Management, Methods and Socioeconomic Considerations

EDITED BY
SHYAMAL K. MAJUMDAR, Ph.D.
Professor of Biology
Lafayette College
Easton, Pennsylvania 18042
and
E. WILLARD MILLER, Ph.D.
Professor of Geography and
    Associate Dean for Resident
    Instruction (Emeritus)
The Pennsylvania State University
University Park, Pennsylvania 16802

Founded on April 18, 1924

**A Publication of
The Pennsylvania Academy of Science**

Library of Congress Cataloging in Publication Data

Solid and Liquid Wastes: Management, Methods and Socioeconomic
Considerations

Bibliography
Appendix
Includes Index

I. Majumdar, Shyamal K., 1938-   , ed.
II. Miller, E. Willard, 1915-   , co-ed.

Library of Congress Catalog Card No.: 84-061472

ISBN 0-9606670-3-2
Copyright © 1984 By The Pennsylvania Academy of Science

Printed in the United States of America by

Typehouse of Easton
Phillipsburg, New Jersey 08865

COMMONWEALTH OF PENNSYLVANIA
DEPARTMENT OF ENVIRONMENTAL RESOURCES
P.O. Box 2063
Harrisburg, PA 17120

The Secretary

Nicholas DeBenedictis
Secretary, Department of Environmental Resources

FOREWORD

Implementation of comprehensive federal and state laws to regulate solid, hazardous and radioactive wastes has made business, government and the scientific community examine our current waste disposal methods and evaluate alternative techniques for the future. The preservation of our environment, the health of our economy and the maintenance of our quality of life are dependent upon the safe and proper disposal of the waste produced by all members of our society.

Efforts to establish and develop proper disposal sites have been hampered by public opposition, much the result of problems associated with old, abandoned dumps. To gain the public's confidence, government, industry and community leaders must be willing to lead a constructive, open dialogue to site and manage new and innovative waste facilities.

Our immediate task is to identify and clean up those dumps that pose a threat to the public and the environment and to prevent a repeat of past mistakes. Only then can the confidence of the public be gained to develop new technologies and facilities.

With limited space remaining in many landfills across the Commonwealth and opposition to opening new sites, attention is being focused on resource recovery and waste-to-energy projects.

v

Approximately 300 permitted landfills are in operation in Pennsylvania; over 800 inadequate sites have been closed by the state since 1968. The inability to locate new landfills demonstrates the need for more energy-efficient solid waste disposal.

In 1983, Pennsylvania awarded a record $873,000 to 23 communities for solid waste planning grants to study long-range disposal alternatives and evaluate recycling possibilities. This year, more communities than ever before in the program's 14-year history applied for grants, whcih serve as pilot studies for other municipalities as well as guiding immediate help for solid waste problems.

The safe disposal of hazardous wastes will present a difficult challenge in the future, as public concerns over "midnight dumping" and groundwater contamination from illegal and unregulated disposal flourishes. Much of the scientific community is skeptical of the long-term safety of total dependence on landfills for hazardous waste disposal, creating another impetus for alternatives to land disposal.

The nation's industries could reduce the volume of hazardous wastes through reprocessing of used materials and the evaluation of manufacturing methods, but the development of new treatment and disposal methods cannot be overemphasized. Research into high-temperature incineration, biological treatment and detoxification of hazardous wastes may provide the best disposal alternatives available for the future.

The scientific problems presented by the disposal of radioactive wastes are only part of the social and political considerations surrounding the issue that must be faced in the coming years. For too long, band-aid solutions to the disposal of radioactive wastes have been accepted.

The nation, including Pennsylvania, is required by law to establish low-level radioactive waste disposal sites, either alone or in cooperation with other states, by January 1986. With the assistance of community leaders, scientists and elected officials, Pennsylvania can properly locate a suitable site for low-level radioactive wastes.

The Pennsylvania Academy of Science is to be commended for its efforts in the study of waste management, through the expertise of government, industry and the academic communities. Development of new techniques in waste disposal is dependent on new ideas, such as are included in these volumes. The academy will play an important role in the development of solutions essential to a clean environment and a sound economy.

Sincerely,

Nicholas DeBenedictis
Secretary
Pennsylvania Department
of Environmental
Resources

PREFACE

This is one of three books published by the Pennsylvania Academy of Science on Hazardous and Toxic Wastes, SOLID AND LIQUID WASTES, and Radioactive Materials and Wastes. These volumes are comprehensive, authoritative source books describing and analyzing the technology and disposal of waste products, one of the most critical problems of our times. They consist of an assemblage of quality papers contributed by a group of leading experts from industry, government and academia.

The problem of the disposal of the various types of wastes created by an industrial society is not a new one, but one that has reached a critical dimension in our times. The modern industrial wastes can not only devastate the natural environment, but have the potential to affect the health of millions of human beings. The seriousness of the problem becomes evident when waste materials have not been disposed of for many years and entire neighborhoods may potentially be contaminated, or when scores of abandoned hazardous waste sites are discovered in densely populated regions and the total environment is subject to contamination.

Each of the three volumes considers a different type of waste material. This volume on SOLID AND LIQUID WASTES is divided into five parts. Part one considers waste types, sources and management and covers such aspects as classification and properties, waste management in municipalities, hospitals and the pulp and paper industry, and the role of the Pennsylvania Department of Environmental Resources and local government in waste management. Management of solid wastes in India is discussed in chapter seven. Treatment technology as to waste water, microbial destruction of industrial solid wastes, waste control techniques, and treatment systems is discussed in part two. Part three covers environmental and health impacts as to organic compounds, cadmium and other trace elements, fluorides in the environment, oil pollution, health effects related to sewage effluent discharge, and ecological effects of acid rain deposition. Part four includes chapters on soil chemistry, the marine repository, bio-conversion, solid waste recycling, and economic aspects of waste management, and the last part is devoted to laws, regulations and socioeconomic considerations.

The book on Hazardous and Toxic Wastes describes waste types, treatment and disposal technology; describes sites and their distribution and geological considerations of waste management; evaluates types of emergency response and preparation; examines management, regulations and economic considerations and elaborate on environmental and health effects.

The third book of the series is devoted to the issues on Radioactive Materials and Wastes. The book contains chapters on the types of radioactive wastes and detection methodology; management, treatment and transportation of nuclear wastes; handling, storage and disposal problems; socio-political aspects; preparation and planning considerations; and the environmental and health effects of radioactive materials and wastes.

These books will be of value to a wide audience. Such individuals as engineers, scientists, medical doctors, social scientists, and environmentalists will find them useful. They provide a wide perspective so that a specialist in one field can be informed about developments and trends in another branch of the subject. The volumes will also be of value to individuals who want to be informed of some of the most critical problems of the day.

We express our deep appreciation for the excellent cooperation and dedication of the contributors, who recognize the importance of solving the critical problems of waste disposal. For a task of this magnitude many individuals in addition to the authors made contributions, and we are most pleased to acknowledge them. The advice and guidance of the members of the editorial and advisory boards are gratefully acknowledged. Gratitude is extended to Dr. Robert S. Chase, Head, Department of Biology, Lafayette College, and to Dr. C. Gregory Knight, Head, Department of Geography, The Pennsylvania State University for providing facilities for editorial work to the editors of the three volumes. Thanks are due to Caryn Golden of Lafayette College, and Nina McNeal and Joan Summers of The Pennsylvania State University for competent secretarial assistance. S. K. Majumdar and E. W. Miller extend heartful thanks to their wives Jhorna and Ruby, respectively, who graciously shared weekends and evenings with the preparation of the series and provided help and encouragement.

Shyamal K. Majumdar, Ph.D.
Lafayette College
Easton, Pennsylvania
and
E. Willard Miller, Ph.D.
The Pennsylvania State University
University Park, Pennsylvania
Editors
October 1984

SOLID AND LIQUID WASTES:
Management, Methods and Socioeconomic Considerations

Table of Contents

CONTRIBUTORS

R. P. Albertson, (Chapter 21), 3315 Simmons Street, Oakland, CA 94619

Dale E. Baker, (Chapters 13 and 19), Professor of Soil Chemistry, 221 Tyson Bldg., The Pennsylvania State Univ., University Park, PA 16802.

John A. Bartone, (Chapter 22), President, Organic Processing Systems, Inc., 409 East 26th Street, Erie, PA 16504.

R. D. Bartusiak, (Chapter 9), Engineer, Homer Laboratories, Bethlehem Steel Corporation, Bethlehem, PA 18016.

Robert C. Bealer, (Chapter 29), Professor of Sociology, Weaver Bldg., The Pennsylvania State University, University Park, PA 16802.

Patricia T. Bradt, (Chapter 18), Adjunct Prof. of Biology, Lehigh University, Bethlehem, PA 18015.

J.B. Bundock, (Chapter 17), Scientific Advisor for the Minister of Environment for Quebec, 2549 Carre Pijart, Sainte-Foy, Quebec, Canada, G1V1#8.

Karin W. Carter, Esq., (Chapter 26), Asst. Counsel, Pennsylvania Dept. Environmental Resources, 505 Executive House, P.O. Box 2357, 1015 Second Street, Harrisburg, PA 17120.

Donald Crider, (Chapter 29), Professor of Sociology, Weaver Bldg., The Pennsylvania State University, University Park, PA 16802.

Robert F. Denoncourt, (Chapter 15), Professor of Biology, York College of Pennsylvania, York, PA 17405.

Roger H. Downing, (Chapter 28), Research Assistant, Institute for Research on Land and Water Resources, The Pennsylvania State University, University Park, PA 16802.

Judith Dudley, (Chapter 18), Research Technician, Lehigh University, Bethlehem, PA 18015.

Al Dufour, (Chapter 16), Chief, Health Effects Research Division, Cincinnati, OH 45268.

William A. Eberhardt, (Chapter 3), Environmental Manager, Procter and Gamble Paper Products Company, P.O. Box 32, Mehoopany, PA 18629.

J.S. Evans, (Chapter 10), Senior Project Engineer, Woodward-Clyde Consultants, Plymouth Meeting, PA 19462.

H.Y. Fang, (Chapter 10), Professor, Department of Civil Engineering, Fritz Engineering Lab 13, Lehigh University, Bethlehem, PA 18015.

Hays B. Gamble, (Chapter 28), Associate Director, Institute for Research on Land and Water Resources, The Pennsylvania State University, University Park, PA 16802.

Richard A. Gammon, (Chapter 16), Prof. of Microbiology, Gannon Univ., Erie, PA 16541.

J.R. Graham, (Chapter 17), Member of the Minnesota Bar and Scientific Consultant to the Ministry of the Environment, Province of Quebec, Canada.

David C. Hubinger, (Chapter 8), Marketing and Training Manager, Chemicals and Pigment Treatment, Technical Service Laboratory, Chestnut Run, E.I. Dupont De Nemours & Co., Wilmington, DE 19898.

Teh-Wei-Hu, (Chapter 23), Prof. of Economics, The Pennsylvania State University, University Park, PA 16802.

Prem Shankar Jha, (Chapter 25), Senior Asst. Editor, The Times of India, 7, Bahadurshah Zafar Marg, New Delhi - 110002, India.

Gerald A. Kraus, (Chapter 16), Assoc. Prof. of Microbiology, Gannon Univ., Erie, PA 16541.

I.J. Kugelman, (Chapter 10), Professor of Civil Engineering, Lehigh University, Bethlehem, PA 18015.

Donald A. Lazarchik, (Chapter 5), Director, Bureau of Solid Waste Management, Pennsylvania Dept. of Environmental Resources, Harrisburg, PA 17120.

Robert Lewis, (Chapter 2), Assistant Director of Planning for the Environmental Services Department, City of Pittsburgh, Pittsburgh, PA 15201.

William C. Livingood, (Chapter 27), Professor of Health Education, Health Department, East Stroudsburg University, East Stroudsburg, PA 18301.

Brian K. Mathias, (Chapter 4), Adminstrative Resident, The Western Pennsylvania Hospital, 4800 Friendship Avenue, Pittsburgh, PA 15224.

James J. McKeown, (Chapter 3), Regional Manager, NCASI, Tufts University, Medford, MA

John R. McNamara, (Chapter 24), Professor of Economics, Lehigh University, Bethlehem, PA 18015.

E. Willard Miller, (Chapter 14), Professor of Geography (Emeritus), 318 Walker Bldg., The Pennsylvania State University, University Park, PA 16802.

P.J. Morin, (Chapter 17), Scientific Consultant to the Ministry of the Environment, Province of Quebec, Canada.

Patrick F. Mutch, (Chapter 4), Asst. Executive Driector, The Western Pennsylvania Hospital, 4800 Friendship Avenue, Pittsburgh, PA 15224.

Allen R. O'Dell, (Chapter 6), Joint Planning Commission, Lehigh-Northampton Counties, ABE Airport Govt., Build, Allentown, PA 18103.

Michael R. Overcash, (Chapters 11 and 12), Professor of Chemical Engineering, North Carolina State University, Raleigh, NC 27650-5035.

Raymond Regan, (Chapter 1), Associate Professor of Civil Engineering, The Pennsylvania State University, University Park, PA 16802.

Robert F. Schmalz, (Chapter 20), Professor of Geoscience, 503 Deike Bldg., The Pennsylvania State University, University Park, PA 16802.

A.K. Sharma, (Chapter 7), President, Indian National Academy Science, New Delhi, and Professor, Department of Botany, University of Calcutta, Calcutta, 700019, India.

William R. Sierks, Esq., (Chapter 26), Asst. Counsel, Pennsylvania Dept. of Environmental Resouces, 505 Executive House, P.O. Box 2357, 1015 Second Street, Harrisburg, PA 17120.

Olev Taremäe, (Chapter 6), Joint Planning Commission, Lehigh-Northampton Counties, ABE Airport Govt. Build, Allentown, PA 18103.

James Walker, (Chapter 2), Director of Environmental Services, City of Pittsburgh, Pittsburgh, PA 15201.

Ann M. Wolf, (Chapters 13 and 19), Research Asst. Dept. of Agronomy, 221 Tyson Bldg., The Pennsylvania State University, University Park, PA 16802.

James A. Weaver, (Chapter 23), Manager, Economics Studies, Engineering Science, Inc., Fairfax, Virginia.

Stanley J. Zagorski, (Chapter 16), Prof. of Biology, Gannon Univ., Erie, PA 16541.

Solid and Liquid Wastes: Management, Methods and Socioeconomic Considerations

Editors: Dr. Shyamal K. Majumdar,
Professor of Biology, Lafayette College
Easton, Pennsylvania 18042

Dr. E. Willard Miller,
Professor of Geography and Associate Dean for Resident
Instruction (Emeritus), The Pennsylvania State University,
University Park, Pennsylvania 16802

EDITORIAL BOARD

Dr. Al P. Dufour, Chief Bacteriology Section, Microbiology Branch, HERL, Environmental Protection Agency, 26 West St., Clair St., Cincinnati, Ohio 45268

Sister M. Gabrielle, PAS Past-President, Grove and McRobert Road, Pittsburgh, PA 15234

Donald A. Lazarchik, P.E. Director, Bureau of Solid Waste Management, Pennsylvania Department of Environmental Resources, P.O. Box 2063, Harrisburg, PA 17120

Mark M. McClellan, Executive Director, Citizen Advisory Council, PA. DER, 8th Floor, Executive House Apartment, 2nd and Chestnut Streets, Harrisburg, PA 17120

Dr. Bruce D. Martin, PAS Past-President, Associate Vice President for Academic Affairs, Duquesne University, Pittsburgh, PA 15219

Dr. Heinz G. Pfeiffer, Manager, Technology & Energy Assessment, Pennsylvania Power and Light Co., Two North Ninth St., Allentown, PA 18101

Dr. Arun K. Sharma, Professor, Calcutta University, President, Indian National Science Academy, New Delhi, India.

Daniel E. Wiley, Manager, R/D, PPG Industries, Inc., Industrial Chemical Division, Pittsburgh, PA 15222

Dr. J.B. Yasinsky, General Manager, Advanced Power Systems Divisions, Nuclear Center, Westinghouse Electric Corporation, P.O. Box 355, Pittsburgh, PA 14230

Prof. Stanley J. Zagorski, PAS President, Associate Dean, College of Science and Engineering, Gannon University, Perry Square, Erie, PA 16541

SYMPOSIA

I *Hazardous and Radioactive Wastes.* Holiday Inn, Grantville, Pennsylvania, October 29, 1982.
Chairman: Donald Zappa, President, Vector Corporation, Pittsburgh, PA

II *Solid and Hazardous Wastes.* Host Corral, Lancaster, Pennsylvania, April 10, 1983.
Chairman: Justice John P. Flaherty, Supreme Court of Pennsylvania. Chairperson of The Pennsylvania Academy of Sciences' Advisory Council

III *Radioactive Materials and Wastes.* Marriott Hotel, Monroeville, Pennsylvania, October 27, 1983.
Chairman: Dr. George C. Shoffstall, PAS President Elect. Director of Education and Organizational Development, Western Pennsylvania Hospital, Pittsburgh, PA

ACKNOWLEDGMENTS

The Pennsylvania Academy of Science published this book in association with the Pennsylvania Department of Environmental Resources (DER). Any opinions, findings, conclusions, or recommendations expressed are those of the author(s) and do not necessarily reflect the views of the DER, or The Pennsylvania Academy of Science.

The publication of this book was aided by contributions from The Pennsylvania Power and Light Company, Allentown, Pennsylvania, U.S. Ecology, Louisville, Kentucky and other companies.

OFFICERS OF THE PENNSYLVANIA ACADEMY OF SCIENCE

Message from

JOHN P. FLAHERTY
Justice
Supreme Court of Pennsylvania
Chairman, Advisory Board
Pennsylvania Academy of Science

A society advances, indeed survives, measured by its control and disposition of societal waste. As is quite evident by even a cursory look at history, nothing is more destructive to life than uncontrolled human, industrial and, a fortiori, radioactive waste. During the German blitz of London in 1941-42, for example, Winston Churchill's greatest fear was a breakdown of the sewer system! No amount of explosive then known could have caused the human destruction of which such event was capable. If we are, thus, to accommodate an increasing population on our now highly urbanized and industrialized planet, it is essential that the scientific community devote itself *with priority* to neutralizing the inundating waste which, unabated, will cause catastrophe to mankind, unparalleled in history.

The Pennsylvania Academy of Science, recognizing the importance of the subject, has endeavored by this publication to stimulate the scientific reader to further innovation, as well as to provide an anthology of present methodology.

Highly industrialized, with a large urban population, Pennsylvania is particularly an appropriate situs for this work, as within its borders occurred an event of stark terror, presaging potential future disaster — *Three Mile Island!*

The President of the Pennsylvania Academy of Science

Dr. George C. Shoffstall, Jr.
Director of Education and Organizational Development
The Western Pennsylvania Hospital
Pittsburgh, Pennsylvania

Dr. George C. Shoffstall, Jr., President of the Pennsylvania Academy of Science (1976-78 and 1984-86) came to The Western Pennsylvania Hospital from The Pennsylvania State University where he was Assistant to the Dean in the College of Science (1966-80). Dr. Shoffstall served as President (1981-82) of The National Association of Academies of Science being elected at its Annual Meeting in Toronto, Canada. He was also named a Fellow in the American Association for the Advancement of Science and is currently serving on the AAAS Council (1984-86).

Active at the national and state levels, Dr. Shoffstall holds a doctorate in radiation biology from The Pennsylvania State University. Named an Outstanding Educator in 1975, Dr. Shoffstall is listed in *American Men and Women in Science, Who's Who in the East, Who's Who in Training and Development, Community Leaders of America, 11th ed.,* and the *1981 Directory of Distinguished Americans.*

Along with the American Association for the Advancement of Science and the Pennsylania Academy of Science, his memberships include the Association for the Advancement of Medical Instrumentation, American Society for Training and Development, National League for Nursing, Pennsylvania League for Nursing, Pennsylvania Association for Medical Education and Smithsonian Associates.

Message from the President of the Pennsylvania Academy of Science

Dr. George C. Shoffstall, Jr.

Since its founding in 1924, The Pennsylvania Academy of Science has responded to the growing importance and demands placed upon it by basic and applied sciences. Thus, it was inevitable that the Academy assumed a role in the preparation and publication of a textbook on *Solid and Liquid Wastes*. This book represents the second of a three volume series, the other two are on *Hazardous and Toxic Wastes* and *Radioactive Materials and Wastes*.

With a constantly growing volume of research and technology in the areas of solid and liquid wastes, one problem was to insure the timeliness of this multi-faceted textbook. A major task of the authors was to be selective—they attempted to integrate materials so that the sequence of presentation is ordered, rather than encyclopedic. At the same time they included pertinent new material, eliminated dated material and brought to the reader a variety of the ominous problems and achievements of research in the specialized field of study to which each chapter is devoted.

If we are to preserve our ecosystem and to enhance our quality of life, then it is obvious that we must act in a more efficient, rational way and on a much larger scale to manage solid and liquid wastes produced by human settlements. For many years, conservationists and the scientific community have been warning that resource depletion and global pollution must quickly be slowed and halted because the resources of our ecosystem, earth, are in general finite, and its capacity for absorbing wastes is limited. Yet man persists in challenging these parameters. On any given day, read any newspaper or magazine; watch the newscasts on television and reflect on how many items relate to pollution. Certainly, these are not isolated nor merely local incidents—they are part of a deepening and accelerating global pattern.

The problems involved are very difficult and complex; we are in the middle of a wastes crisis which is getting worse given our pattern of consumption—today's material resources become tomorrow's garbage. We are not only dealing with chemical compounds but with humans. Our human potential is increasing exponentially, in numbers, education and abilities. Therefore, it is incumbent upon society to insure that this raping of our environment through mismanagement of wastes be controlled sufficiently to assure passive health and happiness—for present and future generations.

Unfortunately, society as a rule, is afraid to foresee and to predict—it simply is not accustomed to the idea. Obviously, predictions are not certainty, but only represent what will probably happen. If society is made aware of this difference—predictions present no danger. Indeed, the very earth upon which we build a settlement changes in a more or less predictable way, vis-a-vis, the elimination of hills or forests, quantifiable air and soil contamination, concomitant surface and ground water pollution, are the results of human activity and can be foreseen.

Society's caveat is that the validity of predictions in relation to time will diminish in value the further the predictions venture into the future—styles change, themes evolve, approaches persist. However, tradition should not be the adversary of innovation.

In our ecosystem, societies will change in needs, dimensions and in structure. Ergo, it is imperative that we be concerned with the future because we will live in it, and we and our descendents will be committed by actions taken today.

The Pennsylvania Academy of Science continues to exercise its traditional professional responsibilities by serving society's interest in addition to those in the sciences and technologies. To this end, this book should be regarded as an anthology of holistic knowledge in the sciences and the ethnography of environmental conditions related to . . . *Solid and Liquid Wastes: Management, Methods and Socioeconomic Considerations.*

Part 1
Waste Types, Sources and Management

An industrial society creates a tremendous amount of solid and liquid wastes. The disposal of these materials is now recognized as a major health and environmental problem. Part One provides fundamental chapters for the discussion and analysis of these wastes. The initial chapter presents a classification, sources and types of waste generated nationwide. This is followed by a model for predicting the generation of waste with an application to Centre County, Pennsylvania.

Because of the complexity of handling solid and liquid wastes, management considerations become paramount. Chapters two through four consider some aspects of this problem. Chapter two treats municipal solid waste management as to categories of municipal and residential solid waste management and disposal methods.

Many industries now have problems of unwanted solid and liquid wastes. Of these, the pulp and paper industry is utilized to discuss types, characteristics, and quantities of residual solid wastes produced and the techniques and technologies used to manage them.

Hospitals in their specialized services, such as automated laboratories, radiology treatments and surgical units, generate tremendous amounts of waste. For the welfare of the community, it is essential that a system of waste disposal be devised that is safe, functional and effective.

The importance of government in waste management is increasing at all levels. Chapters five and six consider the Pennsylvania Department of Environmental Resources' role, and local government activities in the Lehigh Valley in solid waste management. The evolution of legislation controlling solid wastes is presented, culminating with Act 97, one of the most far reaching and powerful pieces of environmental legislation in the nation. In implementing Act 97, Pennsylvania's DER has two objectives: to assist in the development of treatment and disposal facilities in order to comply with the law and to enforce the law against those who attempt to circumvent it.

Waste management planning in the Lehigh Valley began in 1971 when the Joint Planning Commission of Lehigh and Northampton counties developed the initial solid waste plan. The major challenges for the future are the need for feasibility studies to determine areas for future waste sites and the feasibility of using waste to generate steam and/or electricity. The solving of waste disposal problems is among the most pressing health and environmental considerations of our times.

Physical and chemical characteristics of agricultural, industrial and city wastes in India are described in Chapter seven and various options to manage these wastes are discussed.

Solid and Liquid Wastes: Management, Methods and Socioeconomic Considerations. Edited by S. K. Majumdar and E. Willard Miller. © 1984, The Pennsylvania Academy of Science.

Chapter One

Classification and Properties of Solid and Liquid Wastes

Raymond W. Regan, Ph.D, P.E.
Associate Professor of Civil Engineering
The Pennsylvania State University
University Park, PA 16802

In an age of expanding technology and scientific innovation, one of mankind's oldest problems, the collection and disposal of waste products, remains as a significant concern. It appears that rapid technology changes and increased affluences have often stimulated waste production by reinforcing a "throwaway" philosophy in many modern communities[1].

Congress, by the passage of the Resources Recovery Act of 1970 (P.L. 91-512), focused attention from the sanitary landfill (SLF) technique to the development of technology necessary to recover and utilize the materials discarded by U.S. communities. When in 1973, energy availability became a concern, the focus of attention for resource recovery (R/R) technology included materials and energy recovery systems.

The objective of this chapter was to summarize recent information concerning the classification and properties of waste materials produced within U.S. communities, categorized as municipal solid waste (MSW) and wastewater treatment plant (WWTP) sludges. The specific purposes include: (a) summarizing the gross quantities of MSW and indicating the limitation of the data for projecting the order of magnitude potential for materials and energy R/R, (b) categorizing MSW into nine sub-categories to provide an approach for predicting total MSW and R/R quantities based on population and physical characteristics of the community and (c) illustrating an application of the MSW quantities for various R/R processing and disposal options for the Centre Region and Centre County, Pennsylvania.

The primary sources used for this chapter include Regan[2], McKinney[3] and Tchobanoglous et al.[4].

Sources and Types Generated Nationwide

Some confusion has been caused by the definition of "solid waste," which has led to disagreements on the estimated quantities and composition of MSW[5]. This chapter has dealt with two main sources, (a) the solid waste generated in individual households, office buildings, commercial and service establishments and local industries collected commonly by municipal or private sanitation vehicles and disposed collectively in a regional SLF (i.e., MSW) and (b) the residual sludge solids following treatment at a municipal WWTP. Garden waste and bulky items collected by the sanitation vehicle or separate trash collection truck was included. This definition excludes many other types of solid waste, including those resulting from mining, agricultural and industrial sources not collected as MSW.

The Environmental Protection Agency (EPA) has projected the MSW based on 1975 data (as cited by Vesilind and Rimer[5]) (Table 1). The national average MSW production was shown to increase from 3.4 to 5.0 lbs/capita-day (1.5 to 2.3 kg/cap-d) between 1975 and 1990.

Components of a typical MSW, their relative distribution and potential energy content were prepared by Regan[2] based on data reported by Tchobanoglous et al.[3] (Table 2). The data indicate a composite energy content of 4,762 Btu/lb (11,076 kJ/kg) for each unit of MSW.

From the information on gross or potential quantities (Tables 1 and 2), misleading projections might be determined. Cost effective R/R technologies would most likely become utilized only in the more populated urban centers, where markets for the recovered materials were available[2]. Communities isolated from R/R markets and rural areas would most likely continue to utilize land disposal methods such as SLF for the MSW produced.

Therefore, national averages of MSW were of limited value because the data cannot be used with any degree of precision as an indicator of local or regional MSW quantities and R/R potentials[5].

TABLE 1

U.S. Solid Waste Generation Projections, Based on 1975 Data.

Year	Million tons per year	Pounds per capita per day
1975	136	3.40
1980	175	4.28
1985	201	4.67
1990	225	5.00

Predicting MSW Generation

A method for predicting MSW generation based on the population of a local community, and physical characteristics, developed by McKinney[3] has been summarized in this sub-section.

TABLE 2

Materials and Energy Content of a Typical Waste Stream.

Component	Pounds of Individual Fraction per 100 pound Waste Stream Solid	Energy Content BTU's/pound Individual Waste
Food Waste*	15	2,000
Paper*	40	7,200
Newspaper	10	7,200
Cardboard	4	7,000
Plastics*	3	14,000
Textiles*	2	7,500
Rubber*	0.5	10,000
Leather*	0.5	7,500
Garden Trimming*	12	2,800
Wood*	2	8,000
Glass	8	60
Tin Cans	6	300
Non-Ferrous Metals	1	—
Ferrous Metals	4	300
Inerts	4	3,000
TOTAL	100	4,762

*NOTE: Indicates combustible fraction suited for energy recovery.

The overall MSW generated has been subdivided into nine categories including, residential; commerial; industrial; construction and demolition; street sweepings; water and wastewater sludges; automobiles; dead trees and bulky wastes. The generation of MSW for each category as a function of population has been summarized. (Table 3). The data indicate the categories providing the major contribution to the MSW load include residential, commercial and industrial sources, (74 to 83 percent of the total MSW). The increasing per capita generation rate for the commercial and industrial categories indicates the stimulating effect of these two factors on the community MSW production. Therefore, it was proposed that the increased MSW generation factor observed from the National average predictions (Table 1) is in fact more closely associated with the commercial and industrial development of a given community, rather than an increase with time attributed to an improved standard-of-living. At the present time the annual MSW production rate for many major urban cities has exceeded the projections indicated (Table 3). Secondly, the generation factors for the remaining six categories, indicate a solid waste fraction (17 to 26 percent) that most likely would not be easily processed for R/R and thereby may require disposal in a SLF or other land based processing facility.

Solid Waste Materials Available for Resource Recovery in Centre County, Pennsylvania.

The Centre Region and Centre County, Pennsylvania has examined the possibilities for R/R, as reported by Regan[2]. The potential amounts of MSW

TABLE 3

Average Municipal Solid Waste Generation Rates

Category	Population	Physical Characteristic	Generation Factor (lb/cap-d)
1. residential	—	suburban	2.4
		urban	2.0
2. commercial	< 1000	—	1.5
	1 to 10,000	—	2.0
	10 to 100,000	—	2.5
	> 100,000	—	3.5
3. industrial	< = 1000	—	0
	1 to 10,000	—	0.5
	10 to 100,000	—	3.0
4. construction and demolition	—	increased with major activity	0.3-0.4
5. street sweepings	—	increased with dust areas	0.1
6. water and waste— water sludge	—	dry solids basis	0.15-0.5
7. automobiles	—	—	0.25
8. dead trees	—	increased with emergencies	0.1
9. bulky wastes	—	—	0.3-0.4
SUBTOTAL CATEGORIES 1-3			3.5-8.9
SUBTOTAL CATEGORIES 4-9			1.2-1.8
COMBINED TOTAL			4.7-10.7

to be processed for R/R were established for five options, namely,
- Option 1, source separation of clean newspaper
- Option 2, source separation of clean paper, glass and metallic containers
- Option 3, energy recovery from unsegregated MSW
- Option 4, combination of Options 2 and 3
- Option 5, continue the use of a SLF

The information presented in this section was limited to the classification of the MSW for R/R. The reader is referred to the full report for further information.

Before the feasibility of a R/R option was chosen, however, the quantity, composition, and energy content of the MSW was determined. As a means of making this assessment the solid waste tonnage records maintained by the Centre County Solid Waste Authority was obtained for 1974 through 1979. The amounts of materials that could become available for R/R was estimated by assuming the physical composition of the MSW (Table 2) based on nation averages with quantities projected to 1995 (Table 4).

The projected recoverable quantities estimated using a standard composition values (Table 2) and the total yearly projected production (Table 4) as

TABLE 4

Summary of Projected Solid Waste for Various Centre County Jurisdications (tons/year) to 1995.

Jurisdiction	Correlation Coefficient	Year			
		1980	1985	1990	1995
		----------- thousand of tons -----------			
State College Borough	0.90	15.5	17.0	19.0	20.0
Penn State University	0.90	5.5	5.5	5.5	6.3
COG¹ Townships	—	26.0	34.7	42.8	51.1
Centre Regions (Subtotal)	0.97	47.0	57.2	67.3	77.4
Non-COG Townships	—	11.7	16.1	20.5	24.6
Centre County (Total)²	0.98	58.7	73.3	87.8	102.0

NOTE 1: Centre Regional Council of Governments including the Borough of State College,
College, Ferguson, Harris and Patton townships.
       2: As received at Snow Shoe SLF.

TABLE 5

Projected recyclables available for Center County and the Center Region in tons per year*

Component	----------- Centre Region -----------				---------- Centre County ----------			
	1980	1985	1990	1995	1980	1985	1990	1995
Newspaper	4,701	5,716	6,730	7.744	5,867	7,325	8,783	10,240
Cardboard	1,881	2,286	2,692	3,098	2,347	2,920	3,513	4,096
Glass	3,762	4,572	5,384	6,196	4,694	5,860	7,026	8,192
Aluminum	470	572	673	774	587	732	878	1,025
Tin Cans	2,820	3,429	4,038	4,646	3,520	4,395	5,270	6,144
Ferrous Metal	1,881	2,286	2,692	3,098	2,347	2,930	3,513	4,096
Energy	448	544	641	738	559	698	836	975
975 (billion BTU's/yr)								

calculated (Table 5). The amounts of recoverable materials shown in Table 5 should not be confused with the practical amounts of materials that could be obtained in practice. The projections presented were the amounts available if all the components could be separated. The degree of community participation and removal efficiencies of any R/R technology determine the actual amount of materials recovered.

The proejcted residual quantities of non recyclable items were estimated using the standard composition values (Table 2) and the total yearly projected production (Table 4) was calculated (Table 6). The amounts of residual materials were asumed to be 12, 2 and 4 percent garden trimmings, wood and inerts, respectively. Possibly some of the garden trimmings and wood, could be recovered, rather than landfilled. Therefore, the residual quantities reported represent an upper value.

Overall, the residual fraction requiring ultimate disposal in Centre County was approximately 18 percent of the overall MSW. The residual non-recyclable MSW fraction was believed to be representative of most American Communities.

Promoters of various R/R technologies have at time advertised that their system might replace the SLF for MSW disposal. These promoters have claim-

TABLE 6

Summary of Projected Solid Waste Residuals for Various Centre County Jurisdications (tons/year) to 1995.

Jurisdiction	Year			
	1980	1985	1990	1995
	------------ thousand of tons ------------			
State College	2.79	3.06	3.42	3.60
Penn State University	0.99	0.99	0.99	1.13
COG Townships	4.68	6.25	7.70	9.20
Centre Region (Subtotal)	8.46	10.03	12.1	13.9
Non-COG Townships	2.11	2.90	3.69	4.43
Centre County (Total)[1]	10.6	13.6	15.8	18.3

NOTE 1: As received at Snow Shoe SLF.

ed 95 or more percent recovery, with "minimum" inert fraction. This claim may be based on an erroneous conclusion from information such as that presented (Table 2).

DISCUSSION

In the following discussion the initial decision making steps for implementing R/R were presented. As outlined by the National Solid Waste Management Association[6] the implementation of a resource recovery program involves three phases, namely initiation, development and procurement (Figure 1). The initiation phase involves the following considerations:

1. *Determine community and local government interest in studying the possibilities of a new plan.* The advantages of R/R include:

   a. Reduction of hauling and landfill costs
   b. Generation of revenues from the sale of recovered material
   c. The saving of fossil fuels by burning wastes for energy
   d. Reduction in the demand for virgin materials.

Some of the options to be considered require a high percentage of cooperation from the community or a large capital outlay from the local government for the plan to work.

2. *Determine the present and future quantity, composition and energy content of the available wastes.* The quantity of wastes generated and possible trends in generation rates can be determined. Population prediction tables for the specific community can aid in estimating future waste production. Waste composition based on National averages and provide a range of expected quantities for the different recoverable components. The energy content of a waste stream can also be estimated.

3. *Choose a recovery option.* The amount of materials available, potential purchasers, resale values, equipment needs, location of markets and the amount

FIGURE 1. Three Phase R/R Implementation Process, as Suggested by the NSWMA (1980).

of community participation must be identified to determine the most suitable recovery plan.

4. *Identify potential energy customers and their energy requirements.* Potential customers must be centrally located, have good access roads and large enough energy demands to exceed that which can be supplied by the wastes. Continuous energy demands, an available site for a recovery land that would be aesthetic to the neighborhood and the amount of interest in purchasing the energy are also criteria for identifying customers or users.

5. *Examine the available technologies for materials or energy recovery.* Several types of recovery technologies are advertised in the technical literature. Many technologies may be too sophisticated or costly to be practical. Based on user experience, which technologies are reliable?

6. *User or market for recovered materials.* Is there a market for recovered materials? This is the essential question! All the steam, all the metals and glass recovered will not be of any value if a viable, long-term user is not available. The monitary incentive still turns the wheels of progress.

7. *Find out the public opinion of the various recovery options.* Some of the options require considerable community participation to make recovery a feasible alternative. Public education and feedback can be obtained by way of presentations at local government meetings and public hearings. As the planning proceeds potential site locations and nuisance problems can also be addressed using these methods.

CONCLUSIONS

The information presented in this chapter has attempted to describe the quantites of MSW and WWTP sludge available for R/R within smaller and medium size communities. The MSW classification system offers to provide a rational framework for projecting MSW quantity as related to increasing community population. The fractions available for recovery by various R/R options was identified. This system also suggests that the non-recyclable residual MSW represents a significant fraction for continued disposal by SLF.

The actual implementation of a R/R program was presented as a three phase effort involving initiation, development and procurement. The classification of MSW for potential R/R represents only one step in the initiation phase.

REFERENCES

1. Pavoni, J. L., J. E. Heer and D. J. Hagerty. 1975. *Handbook of Solid Waste Disposal. Materials and Energy Recovery;* Van Nostrand Reinhold Co., New York.
2. Regan, R. W. 1981. *Resource Recovery Plan for the Centre Region and Centre County, Pa.* Institute for Research on Land and Water Resources. The Pennsylvania State University, University Park, PA 16802.
3. McKinney, R. E. 1973. *Solid Waste Management.* Unpublished manuscript. Kansas University, Lawrence, KS 66044.
4. Tchobanoglous, G. 1977. *Solid Wastes; Engineering Principles and Management Issues;* McGraw Hill Book Company, New York.
5 Vesilind, P. A., A. E. Rimer. 1981. *Unit Operations in Resource Recovery Engineering,* Prentice Hall Inc., Englewood Cliffs, NJ 07632.
6. National Solid Waste Management Association. 1980. *Resource Recovery Decision Makers Guide.* Washington, DC.

Solid and Liquid Wastes: Management, Methods and Socioeconomic Considerations. Edited by S. K. Majumdar and E. Willard Miller. © 1984, The Pennsylvania Academy of Science.

Chapter Two

MUNICIPAL SOLID WASTE MANAGEMENT

James N. Walker[1] and Robert A. Lewis[2]

[1]Director, Department of Environmental Services
[2]Assistant Director of Planning for the Environmental Services Department
City of Pittsburgh
51-29th Street
Pittsburgh, PA 15201

The importance of municipal solid waste management can not be overemphasized, it is a function that can be ranked on the same priority level with police and fire protection, a potable water supply and sanitary sewer system. Consider the following problems that can occur if municipal solid waste is mismanaged.

Public Health Problems
The most obvious problem would be that of the risk of infectious diseases or other health related problems that occur when refuse collections are delayed or stopped for any reason. The potential for major public health hazards can be blamed on the disease transmittal abilities of flies, mosquitoes and rodents that breed and thrive on solid wastes. Table 1 list the various diseases that can be transmitted by vector, that breed in residential refuse.

Aesthetic Considerations
The aesthetics of a municipality suffers when there are problems with the solid waste management system. Wastes not properly managed, or allowed to accumulate can become a blight on the community. One of the first things a visitor to a city notices is the cleanliness or lack of it in a city. In fact bad impressions made on a visitor in this area can be an inducement not to return. Whenever cities become concerned about the images they are projecting, solid wastes become a matter of priority.

Citizen Concern

Citizen ire can be tremendous when they experience or perceive problems with solid wastes. There are very few other municipality related problems that will initiate a citizen to register a complaint as quickly.

Political Repercussions

Elected officials may also expect to receive political repercussions from problems associated with municipal solid wastes, especially residential wastes. This can be attributed to the reasoning that the public views refuse collection as a personalized service; if a city is slow in repairing a bridge or patching potholes it is an affront to no one particular person, however the resident perceives problems with refuse as a personal display of neglect or disdain.

Implied Pressure

Another factor in the importance of proper solid waste management is the "pressure effect" that it produces. This pressure effect is produced by the consistency of the municipal solid waste problem, especially with residential waste. Just about any Director or Manager of a Sanitation or Refuse Department can testify to the "implied" pressure that they are constantly under; that is they are aware that every day their city generates a certain amount of solid waste and while other conditions may change, such as the condition of the collection vehicles, weather, personnel problems, or budget factors; the waste being generated is constant and must be managed in spite of any other problems being encountered.

This author recalls his duties as a complaint clerk for the City of Pittsburgh's Refuse Department, during the extremely severe Winter of January 1977; when temperatures were in the teens and did not go above freezing for over a month, snow, ice and severe cold had disabled a large portion of the City's collection vehicles and made many of the City's steep winding streets impassable along with freezing the refuse solid inside the resident's containers. All of which naturally delayed collections throughout the City. In spite of this, hundreds of residents jammed the phone lines daily angrily demanding that their refuse be picked up. This demonstrated how much the public values refuse collection services. Proper solid wastes management is a must; the task, therefore, is to maintain or improve your service level in spite of the common problems that will be encountered such as:

— budget problems or cuts
— inflation
— citizen discontent with rising taxes

This chapter offers a detailed definition of municipal solid wastes management and then focuses on the ways that municipalities are managing their residential solid waste, which is the primary category of municipal solid wastes management.

TABLE 1

Vector Associated Diseases

1.	*Fly-Borne Diseases*		
	Typhoid	Myiasis	African Sleeping Sickness
	Bacillary Dysentery	Loiasis	Tularemia
	Amoebic Dysentry	Onchocerciasis	Bartonellosis
	Diarrheas	Ozzards Filariasis	Cararrhal Conjunctivitis
	Asiatic Cholera	Leishmaniasis	Sandfly Fever
	Heliminth/Infections	Yaws	
2.	*Mosquito-Borne Diseases*		
	Dengue	Encephalitis	Filariasis
	Malaria	Yellow Fever	Tularemia
	Lymphocytic Choriomeningitis	Melioidosis	Rift-Valley Fever
3.	*Rodent-Borne Diseases*		
	Echinostomiasis	Rocky Mountain	Swine Erysipelas
	Hemorrhagic Septicemia	Spotted Fever	Trichinosis
	Histoplasmosis	Salivary-Gland Virus	Leptospirosis
	Lymphocytic Choriomeningitis	Infection	Leishmaniasis
	Plague	Salmonellosis	Relapsing Fever
	Rat-Bite Fever	Schistosomiasis	Tularemia
	Rat-Mite Dermatitis	Bilharziasis	Rickettsial Pox
	Rat-Tapeworm Infection	Sporotrichosis	Murine Typhus

It would perhaps be beneficial to offer a definition of the phrase "Municipal Solid Wastes Management," while the meaning will actually vary by municipality, generally it can be defined as "the administration for the collection, transportation, treatment and disposal of various categories of solid wastes that are subject to being generated within the boundaries of a municipality." There are eleven categories of wastes that apply. The involvemnet of the municipality with each category will vary from complete responsibility to little or none. All eleven categories were included because they form the overwhelming majority of solid wastes generated within municipalities. The involvement of each municipality with the different categories is usually determined by the following:

A) *The Municipal Charter or Ordinance Codes*
This will usually determine the responsibilities of a municipality and define its powers and authorities.

B) *Budget Limitations*
This can be considered as the underlying determining factor in any municipal endeavor, in short; it's a question of money; the ability of the budget to accommodate the services offered.

C) *Municipality Size*
The size of the municipality is also a determining factor. As a rule, the

larger municipalities will be more involved while the smaller ones will be less involved, along with a tendency to contract for the services where they are involved.

D) *Areas of Specialization*
Management of certain types of solid waste will require areas of expertise and specialization, not always available in the municipalities.

E) *Specialized Equipment and Disposal Needs*
Another factor that often excludes municipalities from managing several categories of solid waste is the need for specialized equipment and disposal sites and facilities.

CATEGORIES OF MUNICIPAL SOLID WASTES

1. *Residential Refuse and Garbage*
This is the primary category of municipal solid waste, and also the primary area of concern for municipalities. The larger cities generate several thousand tons daily. Included in this category is the following:
- household refuse and trash
- garbage, food wastes (Putrescibles)
- garden and lawn clippings
- furniture and appliances
- fixtures (lights, cabinets and lavatories, etc.)
- lumber
- automobile parts, excluding power train, components, frames and bodies
- building materials

Responsibility
It is probably safe to assume that all municipalities have the responsibility for managing residential refuse. The larger ones will usually have their own refuse and sanitation departments, including their own disposal facilities, while the smaller ones will have contracts with disposal firms to provide this service. Some of the smaller communities will levy a fee against the resident for residential refuse collection.

Potential Hazards
- Pathogenicity

Special Handling Required
- None

Disposal Options
- Pyroloysis
- Incineration (Energy Recovery)
- Separation or Shredding Resource Recovery
- Sanitary Landfilling

2. *Institutional and Civic Refuse and Garbage*
   Includes:
   • Refuse and trash
   • Garbage, food wastes (Putrescibles)
   • Furniture and appliances
   • Fixtures (lights, cabinets, lavatories, etc.)
   Responsibility
   The responsibility will vary; some municipalities will assume responsibility while others may require the institution or civic group to contract for a private hauler to dispose of this category of refuse.
   Potential Hazards
   • Pathogenicity
   Special Handling Required
   • None
   Disposal Options
   • Incineration (Energy Recovery)
   • Separation or Shredding (Resource Recovery)
   • Sanitary Landfilling
   • Pyrolysis

3. *Street Refuse and Litter*
   Includes:
   • Street sweepings
   • Dirt
   • Leaves
   • Catch basin debris
   • Contents of street refuse containers
   Responsibility
   The municipality will be responsible for this category of solid waste. Most will handle this responsibility through their sanitation and/or streets department
   Potential Hazards
   • Pathogenicity
   Special Handling Required
   • None
   Disposal Options
   • Sanitary landfill
   • Incineration (Energy Recovery)
   • Separation or Shredding (Resource Recovery)
   • Pyrolysis

4. *Dead Animals*
   Includes:

Cats, dogs, raccoons and other small animals that fall in the following category:
- residential pets
- stray animals
- laboratory animals that die from diseases; natural causes or are accidentally killed.

Responsibility
The municipality assumes responsibility for this category, usually through their sanitary department. The smaller communities may contract for this service.

Potential Hazards
- Pathogenicity

Special Handling Required
- Specialized vehicle required for transportation to disposal site.

Disposal Options
- Sanitary Landfilling
- State and local laws may regulate disposal procedures.
- Rendering operations

5. *Abandoned Vehicles*
   Includes:
   - Cars
   - Vans and trucks
   - Recreational vehicles

   Responsibility
   a)when abandoned on public property this is a municipal responsibility.
   b)when abandoned on private property it is the owners responsibility to remove it, if the municipality has ordinances against this.

   Potential Hazards
   - None

   Specialized Handling
   - Specialized equipment is required for removal of vehicles.
   - Some of the larger municipalities and private vehicle salvage contractors employ compactors to reduce the size and dimensions of the vehicle for transportation to the disposal site.

   Disposal Options
   - Salvaging and Recycling
   - Sanitary landfilling in landfills used exclusively for bulky wastes.

6. *Commerical Refuse*
   This is the refuse, trash and garbage that results from retail operations and office building activities. Includes:
   - refuse and trash

- appliances and furniture
- fixtures
- garbage (putrescibles)

This garbage is generated by the following:
- offices
- shops
- hotels
- service outlets
- restaurants and fast food outlets
- retail and wholesale operations

Responsibility

Responsibility will vary by location.
- Some municipalities will require establishments to contract for refuse removal services while others will provide the service, but usually for a fee.

Potential Hazards
- Pathogenicity
- Flammability (resulting from containers, drums and bottles of chemicals)

Specialized Handling Required
- None

Disposal Options
- Incineration (Energy Recovery)
- Shrdding or separation (Resource Recovery)
- Sanitary landfilling
- Pyrolysis

7. *Parks Debris*

   Includes:
   - refuse and trash
   - vegetation cuttings
   - garbage (Putrescibles)
   - animal manure (if a zoo is included)

   Responsibility

   The municipality will be responsible for management of the solid waste generated in a park.

   Potential Hazards
   - Pathogenicity

   Specialized Handling
   - None

   Disposal Options
   - Incineration (Energy Recovery)
   - Shredding or separation (Resource Recovery)
   - Sanitary Landfilling
   - Mulching or composting for the vegetation portions

- Pyrolysis
- Manure generated at the zoo operations can be applied to fertilizer operations, if local health laws permit.

8. *Excavation Construction and Demolition Wastes*
Includes: wastes, building material and rubble resulting from the construction, remodeling, repair and/or demolition operations on buildings, roads, bridges, pavements and other structures.
Materials included:
- concrete
- glass
- plaster
- brick
- lumber
- metal
- various types of synthetic material
- dirt
Responsiblity
The municipality is usually not responsible for this type of waste, the contractor involved with the project assumes the responsibility for the proper disposal of this type of debris.
Potential Hazards
- None
Specialized Handling
- Specialized vehicles, those able to support extremely heavy loads are usually required to transport this class of solid wastes.
Disposal Options
- Sanitary landfilling in landfills designed to handle bulky inorganic or slowly decomposing material.
- Incineration of the combustible group of waste in this category, for energy recovery.

9. *Industrial Wastes*
This category includes wastes that results from the following activities:
- mining operation
- manufacturing
- transportation and utility operations
- processing operations
- service oriented operations
- refining operations
- printing operations
- fabricating operations
The types of wastes involved in industrial wastes are best classified by the

industry involved and the nature of the materials being processed.
Responsibility
Municipalities generally are not involved in industrial waste management
activities. The handling of this type of waste is either performed by the cor-
poration or turned over to contractors. The municipal tax base and the need
for specialized equipment handling and disposal facilities prohibits
municipalities from being involved in the management of this category of
waste.
Potential Hazards
• Toxicity
• Pathogenicity
• Explosiveness
Specialized Handling
Handling and transportation equipment tailored to the needs of the in-
dustrial waste being managed is often a requirement.
Disposal Options
• Incineration (special)
  • Chemical Treatment or Neutralizing for recycling
  • Sanitary Landfilling Under Some Conditions
  • Pyrolysis
  • Shredding or Separation (Resource Recovery)

10. *Hazardous Wastes*
This category would be the same as the industrial waste category with the
exception that these wastes have been classified as being hazardous, which
is a popular subject now as it is receiving a lot of government and media
attention. One of the prime examples being the "Love Canal"in New York
State. The problem of hazardous waste being improperly managed is probably the most serious environmental problem in the United States today.
The Environmental Protection Agency estimates that in 1969 the United
States generated almost 60 million tons of hazardous wastes and that 50
million tons of this was treated, stored or disposed of in a manner that potentially threatens human health and the environment, which can be attributed
to the indiscriminate dumping or other improper management of hazardous wastes. Generally the following categories of waste are defined as hazardous wastes:
a) Ignitable wastes
   These are wastes that under standard temperature and pressure are capable
   of causing fire through friction, absorption of moisture or spontaneous
   chemical changes and, when ignited burns so vigorously and persistent-
   ly that it creates a hazard.
b) Corrosive wastes
c) Reactive, volatile or explosive wastes

d) Toxic wastes

e) Radioactive wastes

Responsibility

Generally municipalities will not be involved with the disposal or regulation of hazardous wastes. For the most part they are not equipped to do so, nor are they required by law. Hazardous wastes generators will either have a specialized contractor handling their hazardous wastes or they will attempt to manage it themselves, unfortunately though it is often done incorrectly or illegally.

Potential Hazards

- Toxicity
- Explosiveness and flammability
- Radioactivity

Specialized Handling

- Federal Environmental Protection Agency, State environmental agencies, Federal and State Transportation Departments regulations may apply to the storage, handling and transportation of these wastes.

Specialized Equipment

- Also required in the handling and transportation of hazardous wastes.

Disposal Options

The proper disposal of the various types of hazardous waste is probably one of the most topical subject matters in the United States today. The type of waste generated along with the governmental regulations involved will determine the appropriate method of disposal.

11. *Special Wastes*

This category includes hospital and laboratory wastes.

Includes:

- Refuse and trash
- Garbage (Putrescibles)
- Linen and clothing
- Pathogenic wastes

Responsibility

Responsibility will vary by municipality

Potential Hazards

- Pathogenicity
- Toxicity

Specialized Handling

Attention must be given to precautions in the storage, handling and transportation of these wastes to prevent contact with any infectious materials or equipment.

Disposal Options

- Incineration
- Sanitary Landfilling

RESIDENTIAL SOLID WASTE MANAGEMENT

Major changes have been made in two areas of residential solid waste management over the last decade. (Approx. 1973 to 1983) The two areas of emphasis were:

A) Storage and Collection

B) Disposition

In the earlier part of the 1970's the major emphasis was on innovations in the storage and collection of household refuse. The bulk of these innovations were efforts to semi-automate residential collections. Various types of systems were tried out by municipalities. These systems ranged from mechanical arms that picked up platic bags at curbside to devices that lift and empty 300 gallon capacity or more containers. From these trials and errors approximately three basic types or styles of containers have evolved that use three or four types of collection vehicles. At the time fo this writing there was an estimated 175 cities using some form of mechanized collection. Following is an examination of the three basic types of mechanized systems (containers and trucks) that are presently in use.

System A

Cylindrical multi-family sideloading container system

Development:

Developed and used first in Scottsdale, Arizona in 1969

Components of System

90 to 300 gallon cylindrical containers, usually constructed of some form of heavy gauge plastic or vinyl.

*   collection vehicles are sideloading packers with a chain or hydraulic lift. The trend however has been to design these vehicles with a hydraulic lifting device in order to eliminate the complications that often were encountered with the chain lifting devices.

Operational Design:

The containers are placed in alleyways in residential neighborhoods at intervals of every two or three homes depending on the capacity of the containers and the occupancy rate of the homes. The residents then use these containers for the placement of their normal household wastes. The collection vehicles then mechanically lift and empty the contents of the containers. In some systems a disinfectant is applied to the container after it has been emptied. A separate collection has to be arranged for the collection of bulky or large items.

Associated Problems:

The lightweight design allows for containers to be easily tipped over or moved. Corrosive materials, flames or high heat can damage container beyond repair. The biggest problem would be with containers being destroyed by fire, because of this a large inventory of containers must be kept on hand for replacements.

Recommended areas of use:
- areas with single, family owner occupied homes, low population density with numerous continuous alleyways.

Examples of cities using this system:
- Pheoniz, Arizona
- Scottsdale, Arizona
- Beverly Hills, California

System B
"Box" Multifamily Sideloading Container System
Development:
Developed and first used in Texas around 1970.
Components of system:
- $\frac{1}{2}$ to 3 yard containers, usually constructed of steel with plastic or alumnimum lids.
- Collection vehicles are sideloading packers with hopper openings located on the top of the packing body. The earlier models used chaindriven lifts to empty the containers, however the previously mentioned problems with this type of system also necessitated a change to the hydraulic lift system.

Operational Design:
The containers or "Boxes" as they are often referred to are located in alleyways in residential neighborhoods at intervals of every two or four homes, depending on the container use and occupancy rate of the homes. The containers are also used at housing or apartment complexes. The residents in these areas then place their normal household refuse and trash in the containers. The containers are then mechanically lifted by the sideloading packers and emptied. A separate collection usually has to be arranged for bulky or large items.

Associated Problems:
Because of the large size of the "Box" style container, contractors and other unauthorized persons often use them to deposit an assortment of debris. On occasions the weight of the material dumped into the container will exceed the lifting capacity of the truck, thus requiring that the container be emptied manually.

Recommended Areas of Use:
Apartment complexes or multi-family housing developments or residential areas with numerous continuous alleyways. Because of the construction, design and weight of these containers, they are particularly suited for use in high population density areas as they are highly resistant to acts of vandalism and abuse.

Examples of Some Cities Using the System:
- Pittsburgh, Pennsylvania
- Denver, Colorado
- Detroit, Michigan
- St. Louis, Missouri

System C
Curbside Cart, Single Family Container System
Development:
First used in small southern cities around 1973.
The advantages of using a mechanized system are threefold:
1. Greater productivity
2. Reduced personnel requirements
3. Reduced injury rates.
Components of System:
*   Two-wheeled carts of approximately seventy or eighty gallon capacity. The carts are usually constructed of a heavy gauge plastic or vinyl.
*   Collection vehicles are usually the standard rear loading packer with hydraulic lift attachments added on the vehicle to lift and empty these containers.
Operational Design:
Each residence is supplied with a cart at no cost to the residents. The normal household refuse is placed in the cart and set out at the curbline by the residents for servicing. The container is then attached to a rear loading packer by Sanitation Department employees and then mechanically emptied.
Recommended Areas of Use:
Areas with single family homes, level terrain and off street parking.
Associated Problems:
The carts have several areas that are prone to wear and breakage; such as the wheels and axles and lids, theft and misuse of carts may also be a problem.
Examples of Cities Using This System:
*   Washington, D.C.
*   Atlanta, Georgia

PROBLEMS ASSOCIATED WITH MECHANIZED SYSTEMS

However, mechanized systems cannot be utilized in all cities because of a number of obstacles:
*   A lack of sufficient alleyways could eliminate two of the types of mechanized systems.
*   Irregular housing or street types and patterns can also present a problem for a mechanized system. The various systems, in order to be economically feasible either require numerous continuous alleyways or standardized housing patterns.
*   An extremely hilly terrain, narrow streets or lack of off street parking can also present problems for a mechanized system.
*   High capital costs for initial start ups (equipment acquisition) can be a deterrant to some cities in initiating a mechanized system. Lease purchase plans

though can help offset this problem.

The cities that were unable to utilize any of the various mechanized collection systems had to adjust the practices employed in their conventional collection systems, to make them more efficient and economically feasible.

(Please see Table 2 for a comparison of the types of residential refuse storage systems.)

Some of the methods are as follows:

Other Methods of Improvements in Storage and Collection

A) Elimination of back yard collection.

This is probably the most effective step taken in improving conventional collection practices. By requiring residents to set their refuse and bulk items at the curbline for pick-up, the collection service is improved in several ways:

- the time required for pickups is cut considerably, thus requiring fewer routes and personnel.
- the injury and compensation rate is also reduced as the men no longer have to negotiate flights of stairs with loaded bundles of refuse, nor do they have to be concerned with vicious dogs or obstacles in the yard.

B) Type of container used.

Another method of helping to improve conventional collection methods was to the type of containers the public uses and elimination of the fifty-five gallon drum. In most locations, containers were limited to thirty-two gallons in size with an option allowing residents to use securely tied plastic bags. This was another important change, as fifty-five gallon drums contributed heavily to the injury and compensation rate.

C) Combining Services.

Combining bulk item pick-up with normal household refuse pick-up is another cost savings method. This can only be done with the use of collection equipment that is capable of handling both types of wastes.

(Please see Table 3 for a comparison of curbside vs. backyard collection systems.)

CHANGING DISPOSITION METHODS

As was pointed out earlier, there was in the 1970s an emphasis on changes in the methods of storage and collection of residential solid wastes, in the 1980s however, this emphasis has shifted to changes in the methods of disposal of municipal solid wastes. Up until this time sanitary landfilling had been the primary method of disposal for most municipalities. However, there are two chief concerns with this method which caused municipalities to look at other methods of disposal:

- Environmental Regulations

Because of fears of groundwater pollution, along with other environmental concerns, stringent regulations have limited the expansion of some landfills and forced others to close.

TABLE 2

Potential Advantages and Disadvantages of Types of Residential Waste Storage Containers, and Conditions that Favor the Use of Each

Alternative	Potential advantages	Potential disadvantages	Conditions which favor alternative
Bulk containers for mechanized collection	More efficient than manual collection	Residents oppose storage of other people's waste on their property	Alley space available for storage
Drums (55 gal)	None	Lower collection efficiency	Unacceptable alternative
		Excessive weight can result in back injury and muscle strain	
		Difficult to handle	
		Lack of lids allows insects to breed in waste and odors to escape	
		Rust holes at bottom of drum allow rodents to feed on waste	
Stationary storage bins	None	Inefficient-must be emptied manually	Unacceptable alternative
		Lack of proper cover leads to insect and rodent infestation	
		Necessity for hand shoveling of wastes poses health hazard to collectors	
Paper or plastic bags	Easier to handle-no lids to be removed or replaced	Cost per bag	Curbside collection
	Less weight to lift	Bags can fail if overfilled if too thin	
	Reduces spillage and blowing litter when loaded in truck	Susceptible to animal attacks	
	One-way container-no cans left at curb	Not suitable for bulky, heavy, or sharp objects	
	Eliminates odors and necessity to clean dirty cans		
	Prevents fly entrance		
	Increases speed and efficiency of collection		

| Metal or plastic cans (20 to 30 gal) | Reduces contact of collector with waste; Reasonable size for collector to lift; Economical | Must be cleaned regularly when not used with liners | Backyard collection |

TABLE 3

Potential Advantages and Disadvantages of Curbside/Alley and Backyard Collection, and Conditions that Favor Each

Alternative	Potential advantages	Potential disadvantages	Conditions which favor alternative
Curbside/alley	More efficient; Less expensive; Requires less labor; Facilitates use of paper or plastic bags; Reduces collector injuries; Requires less fuel	Cans at curb look messy; Special arrangements must be made for handicapped and elderly; Residents must remember day of collection	High collection; Unwillingness on part of residents to pay higher taxes or user charge
Backyard	No effort required by residents; No mess at curbs	More expensive; High labor turnover; Increases number of collector injuries; Requires more fuel	Quality of service provided more important criterion than economics

- Site Availability
  As a lot of the present landfills reach the end of their life expectancy, some municipalities discovered that there was no site available for additional land-filling or that the sites that were available were at considerable distances away from the operations.
- Loss of Potential Resources
  There is also the argument that depletable resources were being lost and potential energy sources not taken advantage of. It is presently estimated that the United States generates more than 150 million tons of municipal solid waste and if this waste was converted to energy, it could equal 200 million barrels of oil.

In view of these problems the role of the landfill has been diminished and the future emphasis on municipal solid wastes disposal is in resource recovery and energy conversion.

Resource and Energy Conversion "Definitions"

There are several processes that fall into this category, the two major ones being energy conversion and resource recovery. Following is a brief description of each process.

1. Resource Recovery
   In this process metals, glass, paper, minerals, and other material are separated from the waste stream and recycled.
   Processes Used:
   Source separation
   Air classification
   Magnetic separation
   Trommel screening
   Mechanical separation
   Pulping
   Products:
   Metals
   Glass
   Paper
2. Energy Conversion or Recovery
   The conversion of solid wastes with fuel value into a fuel form or the direct incineration of solid wastes to produce energy (steam or heat).
   Processes used:
   Mass burning
   Shredding
   Pyrolysis
   Hydrolysis
   Products:
   Steam

Heat
Refuse Derived Fuel (R-D-F)
Gas
3. Composting
    This process uses the controlled decomposition of organic solid wastes to
    produce soil conditioner
    Process:
    Composting
    Products
    Fertilizer

Progress in Resource and Energy Recovery

The progress of energy and resource recovery processes has been significant
in countries other than the United States. This is especially true in European
countries. In the United States and Canada as of this writing there were forty-
two projects operating and forty-one being developed. Even though there is
a lot of potential for these types of disposal systems in the United States, there
are also numerous barriers, uncertainties and obstacles that exist that are hinder-
ing their development in the North America. Some of these problems are as
follows:

*   Technical Considerations
    There have been problems with the operations of some of these facilities,
    especially the resource recovery operations. The problems seem to center
    around the separation of the materials.
*   Financial Considerations
    The arrangements for financing the construction of a system can prevent
    a system from being constructed.
*   Marketability of Products
    There has to be a constant and readily available market for the product pro-
    duced. Plus it must be economically competitive with other sources of the
    same product.
*   Availability of Waste
    Some systems require more waste than a community generates, thus requir-
    ing a joint effort by several communities, which can be difficult from an
    economic or political standpoint.
*   Economic Infeasibility
    It is still more feasible economically for some communities to landfill their
    solid wastes rather than send it through a resource or energy recovery pro-
    cess. In fact most cities do not go to these processes unless they are forced
    to; due to one of the previously mentioned problems with landfills. In fact,
    in view of the many disposal options that are available it would be misleading
    to think that sanitary landfills are becoming a thing of the past. Most energy
    and resource recovery operations produce a residue that has to be landfilled.

TABLE 4

Potential Advantages and Disadvantages of Solid Waste Processing and Disposal Methods, and Conditions that Favor Each

Alternative	Potential advantages	Potential disadvantages	Conditions which favor alternative
Sanitary landfilling of shredded solid waste	Extends life of landfill Does not require daily cover under some conditions Waste is more easily placed and compacted Vehicles do not become mired in waste in inclement weather Reduces problems with vectors Does not support combustion or lead to blowing litter Shredding at transfer station or at landfills may be first step in implementing a resource recovery system	Jamming and bridging of the feeding equipment can reduce throughput of the mill High level of component wear, especially of hammers Danger to employees from flying objects, explosions, fires within the mills, and noise Leachate may create water pollution Maintenance and repair costs are high	Cover material is difficult to obtain Shortage of landfill sites requires maximum utilization of available land
Incineration	Extends life of landfill May be more economical than hauling unprocessed waste to distant landfill	Large capital investment High operating cost Large expenditures may be required for air pollution control equipment Conventional incinerators generate large quantities of wastewater which must be treated and disposed of	Land available for sanitary landfilling is at a premium Few if any conditions favor conventional incineration

TABLE 4 (continued)

Potential Advantages and Disadvantages of Solid Waste Processing and Disposal Methods, and Conditions that Favor Each

Alternative	Potential advantages	Potential disadvantages	Conditions which favor alternative
Sanitary landfilling	Simple, easy to manage	Proper sanitary landfill standards must be observed or the operation may degenerate into an open dump	All solid waste systems must have a landfill for unprocessed waste or for the residues resulting from processing facilities
	INitial investment and operating costs are low	Difficult to locate new sites because of citizen opposition	
	Can be put into operation in short period of time	Leachate may create water pollution	
	May be used to reclaim land	Production of methane gas can constitute a fire or explosion hazard	
	Can receive most types of solid waste, eliminating the necessity for separation of wastes	Obtaining adequate cover material may be difficult	
Sanitary landfilling of baled solid waste	Extends life of landfill (double that of a fill for unprocessed wastes)	Resource recovery is precluded once bale is formed	Long hauls needed to reach landfill sites
	Lowers operating costs at the disposal site	Leachate may create water pollution	Shortage of landfill sites required maximum utilization of available land
	Reduces hauling costs where distant sites are used		
Materials recovery systems	Less land required for solid waste disposal	Technology for many operations still new, not fully proven	Markets for sufficient quantities of the reclaimed materials are located nearby
	High public acceptance	Requires markets for recovered materials	Land available for sanitary landfilling is at a premium
	Lower disposal costs may result through sale of recovered materials and reduced landfilling requirements	High initial investment required for some techniques	Heavily populated area to ensure a large steady volume of solid waste to achieve economies of scale
		Materials must meet specifications of purchaser	

TABLE 4 (continued)

Potential Advantages and Disadvantages of Solid Waste Processing and Disposal Methods, and Conditions that Favor Each

Alternative	Potential advantages	Potential disadvantages	Conditions which favor alternative
Energy recovery systems	Landfill requirements can be reduced Finding a site for an energy recovery plant may be easier than finding a site for a landfill or conventional incinerator Total pollution is reduced when compared to a system that includes incineration for solid waste disposal and burning fossil fuels for energy High public acceptance As cost of fossil fuel rises, economics become more favorable	Requires markets for energy produced Most systems will not accept all types of wastes Specific needs of the energy market may dictate parameters of the system design Complex process requiring sophisticated management Needs relatively long period for planning and construction between approval of funding and full-capacity operation Technology for many operations still new, not fully proven	Heavily populated area to ensure a large steady volume of solid waste to take advantage of economy of scale Availability of a steady customer for generated energy to provide revenue Desire or need for additional low-sulfur fuel source Land available for sanitary landfilling is at a premium

So, in a way they complement those operations. Also landfills themselves are a form of energy producers as the methane gas generated in them is often recovered and utilized.

(Please see Table 4 for a comparison of the various disposal options that are available.)

REFERENCES

PAPER: May, 1969. Ad Hoc Committee of the Office of Science and Technology, The Executive Office of the President of the United States. *Solid Wastes Management, A Comprehensive Assessment of Solid Wastes Problems, Practices and Need.*

PAPER: September, 1981. Allegheny County Pennsylvania. *Industrial and Hazardous Waste Management Planning Study.* Main report Volume I. pp. 1-1-3 to 1-1-4.

FEDERAL REGISTER: May 19, 1980. Environmental Protection Agency. *Hazardous Waste and Consolidated Permit Regulations.* p. 33122.

JOURNAL: October, 1980. Solid Wastes Management, Special Report. *History of Automated Collection.* p. 70.

JOURNAL: March 29, 1982. City Currents, U.S. Conference of Mayors Special Issue. *Resource Recovery Activities.* pp. 1-18.

Decision Makers Guide in Solid Waste Management. Office of Solid Waste Management Programs, the U.S. Environmental Protection Agency. pp. 14, 15, 19, 88-97.

Handbook of Solid Waste Management. 1977, David Gordon Wilson, Editor. pp. 10-60, 471.

Solid and Liquid Wastes: Management, Methods and Socioeconomic Considerations. Edited by
S. K. Majumdar and E. Willard Miller. © 1984, The Pennsylvania Academy of Science.

Chapter Three

Solid Waste Management in the Pulp and Paper Industry

William A. Eberhardt[1] and James J. McKeown[2]

[1]Environmental Manager,
Procter and Gamble Paper Products Company
P. O. Box 32
Mehoopany, PA 18629
[2]Regional Manager, NCASI
Tufts University
Anderson Hall
Medford, MA 02155

I. INTRODUCTION

According to Post's Directory[1], there are 713 paper and board and 307 pulp plants in the United States having an annual paper production capacity of about 69 million tons. Pennsylvania's annual paper capacity of 3 million tons is produced by a total of 58 paper, board, and pulp manufacturing installations.

The purpose of this chapter is to inform the reader of the types, characteristics, and quantities of residual solid wastes produced at these installations and of the techniques and technologies used to manage the generation and disposition of these wastes. Emphasis is placed on avoidance of generation, on recycling, and on recovery of beneficial uses of the wastes. These areas hold the greatest economic and environmental promise. Techniques for responsibly managing the unavoidable (disposal) are also provided.

II. SOLID WASTE PRODUCTION

A. *Pulp and Paper Manufacturing Processes*—The pulp and paper industry operates in over 700 sites throughout the United States and has been categorized into about 3 dozen manufacturing segments depending on the type of product produced and the process used. Each manufacturing facility is usually supported by various service (power, water treatment, shipping), converting (cutting, coating, printing), and pollution control systems. It is safe to say that no two operating facilities are exactly alike, and it is important to note that most are significantly different. Thus, in order to appreciate the unique integration

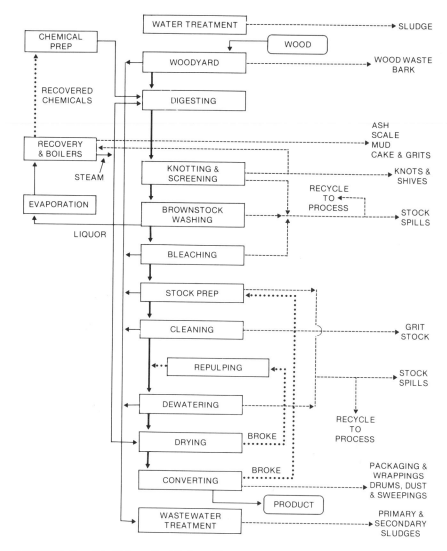

FIGURE 1. Simplified Schematic of Pulp and Paper Manufacturing and Related Residual Waste Generation.

of systems and processes employed in the industry to reduce, utilize, and dispose of residual wastes, it is necessary to have an understanding of the manufacturing and associated service facilities which produce them. For the purposes of this chapter, pulp and paper manufacturing is described in terms of: 1) service, 2) pulpmaking, 3)papermaking, 4) converting, and 5) pollution control systems. Figure 1 provides a simplified schematic of these components.

   1. *Services*—Pulp and paper manufacturing requires large quantities of steam

for digestion of wood and drying of paper. Steam also may be used to produce electric power by employing energy efficient cogeneration schemes. Steam is produced typically by burning a variety of fuels, some of which are generated on site and which otherwise would become solid wastes (i.e. bark, unusable wood, sludges). Significantly, combustion of all these fuels produces residual ashes and the power plant is often the major source of service system residuals.

Power plant ashes are of two general types—1) bottom ash, the inerts and unburned fuel which drops through the grates, and 2) flyash, the ash removed from the flue gases by mechanical separation, wet scrubbers, or dry collection. The quantity of ash produced depends on the composition of the fuel mix, the inert content of each fuel, and combustion efficiency. Systems using gas or oil fueled boilers for power and steam produce significantly less ash than those fueled with manufacturing residuals.

Currently, approximately 20,000 gallons of water are required to make a ton of paper. Conservation practices aimed at reducing water and wastewater treatment costs and energy consumption have decreased fresh water utilization per ton of product in the industry by 60% since 1959.

Water treatment systems usually include screening, sedimentation, and filtration. Chemicals such as alum and lime are used to coagulate the solids and colloids prior to sedimentation and filtration where paper product quality dictates that the water be essentially free of suspended solids and color. The solids removed during treatment become residual wastes.

The amount of residual wastes generated by these water treatment systems depends on the quality of the water supply. Typically, the quantity of sludge generated is greatest where surface waters experiencing high turbidities during run-off events are used. Water treatment sludges are largely inert consisting of soil particles, coagulated colloids, and, where chemicals are employed, inorganic salts. During summers, algae and weeds can be a major component of the sludge. The water treatment residues usually are concentrated by sedimentation in the plant's central waste treatment system or in separate sedimentation basins.

Other water service areas which may generate residuals include boiler water maintenance and cooling water treatment. Cooling water, particularly if recycled, may be treated chemically for slime control and/or filtered for solids removal. Chemical treatment of boiler water for softening, deaeration, and corrosion control is common. Recycled cooling waters and boiler waters are periodically blown-down to purge the systems of dissolved and suspended solids build-ups. The quality and low volume of sludges from these activities is such that they are normally discharged to the general wastewater treatment system where, in fact, they may provide needed nutrients, such as phosphorus for biological treatment. Under certain circumstances, however, they may contain sufficient concentrated metals as to trigger the need for specialized disposal.

Chemical preparation and general plant maintenance activities produce residuals which may contain chemicals, dyes, metals, oils, solvents, or other ingredients which dictate segregation and specialized handling. Discharge to process sewers is the exception.

2. *Pulpmaking*—Chemical pulp is made from digesting wood at elevated temperature and pressure in the presence of chemicals which dissolve the resins and lignins binding the cellulose into wood. Mechanical pulp is made by using pressure, abrasion, and heat to physically free the fibers. In either case, the wood is usually debarked prior to digestion, although whole tree pulping is now practiced in many locations. Following digestion, the pulp is cleaned, refined and often chemically bleached.

The chemical pulping processes usually are named after the chemicals employed—sulfite, sulfate or kraft, neutral sulfite, soda, and others. The mechanical pulping processes are descriptively referred to as groundwood, chemi-groundwood, or thermomechanical.

The bark removed prior to pulping is often used as a source of energy because of its relatively high fuel value. Bark boilers are commonplace. Although burning bark does produce a residual ash, the net effect is a significant reduction in residual waste generation.

Solids removed from the pulp during screening and cleaning are often reclaimed. Screenings, consisting of knots, partially digested wood chips, and other agglomerations of wood which were not initially reduced to pulp, are pressed to reclaim organic chemicals and repulped or burned. Dirt and slivers are centrifically separated from the pulp and usually handled as residual wastes.

The liquors washed from the pulp following digestion are composed of organics and forms of the original pulping chemicals. These liquors are burned for energy and the ash is usually processed to recover most of the chemicals. In the sulfite process, the chemicals are often reovered by scrubbing the combustion flue gases.

Various blowdowns from the pulping and recovery systems produce liquors and sludges suitable for recycle or chemical recovery. Blowdowns created by temporary system imbalances often are stored for later reuse. Cooking liquor or pulp spills can be recycled to the process. Lime mud from the kraft recovery system is a prime example of a reusable sludge. If not reclaimed, these blowdowns or spills enter the general sewer system and contribute to wastewater treatment sludges. The industry generally maintains loss control systems to prevent or minimize such occurrences.

Pulpmaking involving the deinking of recycled paper is worthy of special note. The quality and type of waste paper recycled bears a direct relationship to the quanity and quality of the sludge produced during deinking. In some cases the pulp yield is only 50% of the input with the remainder becoming solid waste. Heavy metals and organics from the original paper may be concentrated in these wastes. Consequently, they may have to receive special handling. Recognition of this is leading to greater selectivity of secondary fiber sources.

3. *Papermaking*—In its simplicity, papermaking typically involved preparation and dilution of pulps, introduction of additives, refining and cleaning, physical dewatering, and, finally, drying. Although the process is designed to minimize losses of fiber, the continual series of unit processes employed affords opportunities for escape of cellulose fines and inerts at various points. When possible, savealls, flotation or filtration processes, return most of the reusable solids to the process, but, on an average, about 0.5 to 2.0% of the product escapes to become wastewater treatment sludge.

The types of paper products made are often a bellweather for the quantity and quality of solid waste requiring management. Lightweight papers, tissue, newsprint, and paperboards will contain a high percentage of fiber and low ash, and so will the sludges. Filled grades, many coated grades, and some specialty grades will contain less fiber and more clay and inerts, as will the sludges.

4. *Converting*—Converting is the conversion of paper to final products, such as cartons, bond paper, facial tissue, and paper towels. The processes are mechanical and are typically dry, printing being a notable exception.

The main solid waste produced during converting, called broke or trim, is usually returned to the papermaking process for reuse. Rejects or off-grade papers also are usually returned for reprocessing. Other residuals trypically result from procurement of raw materials, examples being drums, pallets, cartons, and wrappings. Dust control and clean-up are also sources of waste.

The quantity of solid waste produced during converting is normally small. Excluding broke, it is often segregated and sold or, because it is relatively dry, is blended with trash from the manufacturing operation.

5. *Pollution Control*—Having already mentioned boiler fly ash and water treatment sludge, this section will deal with solids removed during wastewater treatment. Wastewater treatment usually consists of sedimentation or flotation for solids removal and biological oxidation for BOD removal. The latter is often followed by secondary sedimentation.

The primary solids removal processes generate sludges consisting of fiber, clay, and other solids sewered from the various production and service areas. These solids are usually dewatered to 20-30% solids by weight for disposal on land, or 30-45% for use as fuel. In cases where these solids are recycled to manufacturing, the sludge may be pumped directly at 2-8% solids from the clarification unit to the process.

Oxidation of the clarified wastewater generates a biological sludge which is difficult to dewater because of its gelatinous character. The amount of this secondary sludge produced depends on the biological process used and the quantity of BOD removed. Generally, the more compact the biological process, the more sludge is produced. For example, the compact activated sludge process may generate 0.5-0.7 lbs. of settled sludge per pound of BOD removed. However, the more expansive aerated stabilization process may generate little or no sludge because the organisms produced tend to remain dispersed and are self-oxidized

TABLE 1

Types and Distribution of Solid Wastes from Pennsylvania Pulp and Paper Plants

Manufacturing System	Type of Residual	Examples of Residuals	Amount (Tons/Day)
Services	(A) Boiler ash	Bottom ash	131 (a)
	(B) Trash	Office trash, demolition debris, sweepings, scrap parts/machinery, lumber, felts, wires, wrappings	116
Pulpmaking	(C) Wood Waste	Under and over-sized logs, log flume grit and bark, soiled chips	71
	(D) Bark	—	343
	(E) Pulp rejects	Woodroom screenings, knots, uncooked chips, classification wastes, pulp spills, recovery cake and grits	61
Papermaking	(F) Paper rejects	Cleaner, poly, and coating rejects, spills, waste paper	37
Converting	(Typically a part of plant trash)		—
Pollution	(G) Primary sludge	—	177
Control	(H) Secondary sludge	—	42 (b)
	(I) Fly ash	—	278 (a)

(a) Some bottom ash was reported under fly ash
(b) Excludes sludge from publicly owned systems treating plant wastewaters

or carried out in the treated effluent. Often with many compact biological systems, the waste sludge from the secondary clarifier is blended with primary sludge to even out the dewatering properties of the two sludges.

B. *Solid Wastes Produced*—As mentioned earlier, each mill generates a different quantity and type of solid waste which reflects its particular blend of processes and products. With the cooperation of several companies representing a cross-section of the paper industry operating in Pennsylvania, a survey was undertaken to show: a) the variety of solid waste residuals which are being generated, and b) the array of management options being practiced in disposing of these residuals. The survey covered several of the larger mills and a few small mills and is representative of about 50% of the paper production capacity in the Commonwealth; ten mills were surveyed.

Table 1 shows the survey findings with the wastes grouped by point of process origin. Table 2 classifies and quantifies these same wastes by method of disposition.

It is significant to note that pulpmaking produces the largest amount of production related residues and 38% of the total residues reported. Actually, the wastes from this production category are even more significant considering that some were reported under plant trash (woodyard sweepings), and a great deal

TABLE 2

Disposition of Residuals Reported by Selected Papermills in Pennsylvania

Method or Purpose of Disposition	Amount (Tons/Day)	Type of Residual*
Beneficial Uses		
Recycled or Recovered	404	C, D, E, F, G, H
Landapplied	241	A, C, D, G, H, I
Disposal		
Landfilled	595	A, B, C, D, E, F, G, H, I
Lagooned	16	A

* See Table 1 for Residual Code

of the pollution control wastes are pulping-related.

Note also that the power generation facilities produced 32% of the total wastes reported (boiler ash and fly ash)

Finally, it is interesting to observe that pollution control processes generated a very significant amount of the reported residues—by calculation, 40% of the total, although as noted in Table 1, some bottom ash was reported as fly ash.

III. MANAGEMENT OF SOLIDS

Solid waste management in the industry is not left to chance. Chance produces over-generation and unwise handling. Rather, control is exercised at many points starting with restrictions on the use of troublesome chemicals in manufacturing and extending to prevention of material losses, to recycling of wastes back to processes, to recovery of energy and other values from waste materials, and to selection of proper waste disposal endpoints. Also contributing to responsible management are administrative procedures designed to track waste generation and movement and to ensure proper decision-making along the way.

A. *Chemical Management*—The classification of and alternates for disposition of wastes are impacted on by the raw materials and chemicals used in manufacturing. For example, the use of certain slimicides or preservatives could lead to classification of wastes as toxic under the Federal Resource Conservation and Recovery Act. Similarly, the use, without close quality control, of by-product phosphoric acid from metal industries as a nutrient for activated sludge, could lead to unacceptable metal levels in the waste sludge. For these reasons, it follows that a plant should evaluate and control the use of chemicals which might impact unfavorably on waste characteristics. Such control can be integrated with chemical management procedures for personal and product safety.

B. *Avoidance of Waste Production*—It is prudent to question should a particular waste be produced and, if so, in what amount. Waste elimination and reduction considerations should always precede deliberations over disposal alternatives. The authors have found that "why" questioning leads not only to lower waste generation, but also to increased understanding and improvement of manufacturing processes. In one case, abnormally high waste activated sludge production was reduced by improving pulp washer performance. Spin-off process improvements included reductions in bleach usage and pulp quality deviations. Examples of loss control practices which typically reduce wastes include preventive maintenance for chest/tank level controllers, and regular attention to performance of stock screens and centricleaners. On the bonus side, such practices typically increase product yields.

C. *Recycling to Manufacturing*—Characteristics of wastes should be closely examined to evaluate the potential for re-entry into the manufacturing process. It is common practice in the industry to capture and reuse cellulose which has been partially processed. For example, finished product "broke," following repulping and, if necessary, bleaching; and save-all or primary clarifier sludges, directly or following cleaning and bleaching, are recycled to raw material pulp streams. Reject pulp from screenings may be returned to the pulp stream following mechanical action to further release cellulose from incompletely cooked wood chips. Stock spills can be returned to the process following repulping and/or cleaning, rather than being flushed to sewers for disposal. In certain cases, secondary sludges from wastewater treatment are blended into pulp and papermaking processes. Attractive payouts frequently accompany expenditures for such recycling measures.

In reported studies, Brookover[2] presented a review of the approaches and problems encountered in reusing primary and secondary sludge in the manufacture of paperboard. Although the mill had to make major adjustments in its scheduling of production grades, large quantities of residuals were successfully recycled back to the process. Smithe and Abramowski[3] reported on full scale experiments with incinerating secondary sludge with NSSC black liquor in a fluidized bed reactor. The paper reports on the reasons for the trials and the concern for build-up of chemical impurities in the recovery system. Adkins[4] reported that trials on disposing of secondary sludge in batch digesters did not effect pulp quality, but added significant chemical and evaporation costs.

Frederick *et al*[5] reviewed the capabilities of the kraft pulping system to purge itself of the Ca, Mg, P, Cl, and Al contained in recycled secondary sludge. Concern was expressed over the consequences of Al build-up; however, the authors concluded that although secondary sludge use will have some impact on energy recovery and economics, if predigested with white liquor, evaporator fouling should be prevented.

D. *Administration of Residual Wastes Handling*—Residual wastes which cannot be avoided or recycled to manufacturing should be accounted for rigorously and directed to optimum endpoints. This requires support at all organizational levels. Discipline in dealing with wastes is vital to personal safety, the environment, and the economic and legal well-being of the plant.

Recordkeeping and other measures designed to provide control over handling and disposal are essential. Accounting and reporting of waste production serves the company by providing feedback on the performance of those departments and processes producing wastes; focuses on the causes of and prevention measures for spills; helps determine loadings and efficiencies of waste handling equipment; and produces data for cost management and fulfillment of necessary legal obligations.

Formal guidelines should govern the recovery or disposal of residuals. Specific disposal procedures to serve personal safety, environmental, economic, and legal needs should be established on the basis of waste source, characteristics, and volumes. These procedures should be readily available for use and be updated periodically. The management system should provide for verification of waste characteristics so that beneficial changes are identified and necessary procedural modifications are made. Verification of actual recovery or disposal actions, on or off site, is also very important.

E. *Relative Distribution of Disposition Methods*—Table 2 shows the relative distribution of disposition methods for residual wastes reported in the previously mentioned Pennsylvania survey. Before discussing distribution, there are two important observations to make with regard to the disposition methods. First, a distinction between landfilling and landapplication is made. Whereas landfilling employs land to dispose of concentrated residuals, landapplication beneficially uses the waste as a stabilizer or conditioner/fertilizer in a geophysical or nutritional sense. Second, in a collective sense, landapplication, recycling, and recovery all make beneficial use of the residual.

Landfilling, as indicated in Table 2, was reported as both the largest and widest used method of disposition for the Pennsylvania plants. Recycling and recovery is the next largest category. Note that almost every type of residual is recycled or recovered by one or more plants in the Commonwealth. Table 2 also reveals that substantial tonnage is landapplied, demonstrating that Pennsylvania's paper industry is actively developing the technology for beneficial use of residuals on land. Finally, a small portion of the residual ash was reported as lagooned.

In summarizing the distribution of residuals disposition in Pennsylvania, it is observed that 645 tons/day or 51% of the residuals reported were beneficially used. Clearly, the paper industry in Pennsylvania has accomplished a great deal in recycling and recovery of residuals. Based on the experiences of the authors and information in the following section on recovery of wastes, this situation is manifested to a varying degree across the entire United States paper industry.

F. *Recovery of Wastes*—For this review, recovery techniques are classified into four categories: 1) Energy Production, 2) Raw Material Sales, 3) By-Product Sales, and 4) Beneficial Landapplication. On-site recycling, another form of recovery, was presented earlier.

1. *Energy Production*—There is increasing usage of organic residues for energy production. The Pennsylvania survey indicated that the majority of bark produced is combusted for energy. Other organics being used as fuel for boilers include wood residues, pulping reject and liquors, and primary and secondary sludges. Fly ash is finding use to help dewater sludges prior to their use as fuels. In addition, wood pallets and certain combustibles are being reclaimed from the plant trash for use as fuels. At the Procter & Gamble Paper Products Mehoopany, Pennsylvania facility (later reported on in this chapter), more than 94% of the steam consumed in pulping and papermaking is produced from pulping and other woodland wastes.

2. *Raw Material Sales*—Many of the industry wastes are sold for use by others following handling, such as sorting, bundling, and rinsing. These include reject logs, scrap metal and lumber, drums, waste oil, batteries, corrugated materials, and scrap cartons, cores, polyethylene, and plastics. Aggressive in-plant programs of segregation and marketing can produce significant income from these materials. During a recent fiscal year, over 10,000 bone dry tons of these residues were sold by the Procter & Gamble Mehoopany Plant. Another Pennsylvania plant has stripped polyethylene coatings from containers for reprocessing elsewhere.

3. *By-Product Sales*—A growing practice is to produce saleable by-products from waste materials. Typifying the normally non-toxic nature of the industry's wastes, agricultural products are common.

In Pennsylvania, Carter[6] reports on composting of primary, secondary, and filter plant sludges from an integrated pulp and paper plant. Marketing of the compost as a substitute for peat products in farmland improvement has been successful. Eberhardt et al[7] describe the drying and sale as a fertilizer of more than 10 tons/day of activated sludge from sulfite pulping wastewaters. The product, having an average nitrogen content of 8.75%, has been sold without interuption since start-up in 1974 of the commerical flash drying system. Orme[8,9] of the U.S. Bureau of Sports Fisheries and Wildlife studied for more than 4 years the use of this same sludge as a protein source in trout feed. The study concluded that healthy rainbow trout of high eating quality can be grown using the dried sludge as a replacement for both vegetable protein and fish meal in the normal trout diet.

Elsewhere, Church[10] studied the use of single celled protein (SCP) in 6 different dried papermill sludges for use in diets of cattle and sheep. He tentatively concluded that, although lower in palatability and variable in performance, the 6 sludges were non-toxic and could be used to replace some, if not all, supplemental protein sources, such as cottonseed meal. Rogers[11] reports that activated sludge from treatment of sulfite pulping wastewater is dried using the Carver-Greenfield system and is marketed in the state of Washington as an animal feed protein supplement. Sales are sufficient to recover all sludge dewatering costs.

Pulp and paper cellulose residuals also can be used as a source of animal feed ingredients. Pamment *et al*[12] reports that SCP from the fermentation of highly delignified cellulytic wastes by the fungus *Chaetomium cellulolyticum* had sufficiently high *in vitro* rumen digestibility to permit its use as a feedstuff. Baker[13] summarized USDA Forest Products Laboratory studies examining diets containing wood residues for ruminants. For example, a diet of aspen bark, together with corn, oats, soybean meal, and urea, was much less costly than alfalfa hay. Further, sheep fed up to 5 lbs. per day of bark produced normal numbers of healthy offspring. Dinius *et al*[14] fed beef heifers diets including 50 to 75% sulfite pulping fines from the Procter & Gamble Plant in Mehoopany, Pennsylvania. Annual weight gains and the weights of calves were comparable or greater than those involving control hay diets. The cattle consumed the fines readily and there were no observable unusual digestive, metabolic, or calving disorders. On the basis of these and earlier studies, the Mehoopany Plant began a pulp fines give-away program, which later gave way to sales to local Pennsylvania beef and dairy farmers. The program has been reported on frequently by the Pennsylvania State University[15].

Cellulose also offers a source of biomass for the production of synthetic fuels. For example, Emart[16] has developed costs for a process to convert waste pulping cellulose into ethanol and animal feed.

4. *Beneficial Landapplications*—The application of residuals to land for beneficial use is widely practiced by the pulp and paper industry. One Pennsylvania plant is composting sludges and bark for use by nurseries. Another has arranged for bottom ash to be spread on roads during winter. Several mills have applied sludges directly to farmland, and wood and bark have been used for erosion control and landscaping.

NCASI, the National Council of the Paper Industry for Air and Stream Improvement, has published technical bulletins on the physical and chemical properties of the industry's residuals so that management systems can be efficiently engineered. Further, the literature contains guidelines for landapplication of sludges based on their chemical and physical characteristics and local soil composition[17].

Papermill residue use as a soil amendment has been reported extensively. Nolan[18] and Jayne[19] reported on work for Arkansas Kraft Corporation where sludge from an aeration basin was applied to the land. Application rates for subsurface injection into the soil were determined based on sludge nitrogen. Additional work of a similar nature involving Appleton Papers was reported by Nolan[20,21] and Beasom[22].

Cawrse[23] reported on the impact of spray irrigating semi-chemical pulp mill wastewater sludge on the metal and nutrient content of soils. Soil nitrogen and phosphorus increased almost twofold and 20%, respectively, following spraying the sludge. Hill[24] also experimented with using sludge to lower fertilizer costs for corn production.

Buchanan[25] reports on using secondary sludge from the treatment of board mill wastes for application on farmland. Analytical findings on the sludge, wheat, soil, and groundwater led to the conclusion that landapplication is an attractive alternative in which the farming community benefits.

Corey[26] presents an analysis on the use of 3 types of papermill residues on the land. The study concludes that the lime and fiberous and secondary biological sludges could be beneficially used if application rates were matched with soil and crop requirements.

Harkin[27] explored the application of papermill sludges to sandy soils in Wisconsin. Primary effects were a gradual increase in the clay and humus content of the soil, addition of nitrogen and phosphorus, and retention of soil nitrogen in an immobile organic form during fallow periods. Further, secondary beneficial soil effects included increased cation exchange and moisture capacities and tilth. Additionally, leaching of nitrates and water-soluble pesticides was reduced. The author found that these effects are gradual and that all effects are in the direction of improving fertility without degradation of groundwater quality.

Smithe and Morin[28] report on applying semi-chemical waste treatment sludge to 20¼ acre tree farm plots. Tree growth on sludge-applied plots increased 25-30% after 1 year over the control plots. Experiments on corn fields also showed increasing growth with increasing amounts of sludge.

Aspitarte[29] found in U.S. Forest Service Nursery studies involving application of combined primary and secondary papermill sludge to Douglas-Fir seedlings that the seedlings grown in mulched soil produced better growth. Additional experiments with Red Alder and Black Cottonwood grown in riverbed sand plots showed excellent and nominal growth of the two species, respectively. Lime dregs were also added with no detrimental effects.

Gartner and Williams[30] review the use of bark in horticultural systems. Bark was found to be advantageous especially if nitrogen is added, the bark isn't limed and it is aged 60 days prior to use.

G. *Disposal of Residual Wastes*—Despite the industries' successful recycle and recovery applications, disposal requirements are extensive. The primary methods of disposal are landfilling and incineration. A 1979 review of NCASI file data for the industry indicated that landfilling of residuals exceeded incineration by more than seven-fold on a weight basis.

1. *Landfilling*—Since pulp and paper residuals are for the most part non-toxic and are biodegradable, landfilling is typically the responsible choice among disposal techniques. Further, it represents, in many cases, the least expensive alternative available to most plants for disposal of residuals. The non-hazardous, low compaction and consolidation rates, lack of flammability, and lack of disease vector and nuisance characteristics make landfilling of these residuals a relatively straightforward engineering endeavor which does not normally require extensive lining and covering[31]. Gas production and leachate generation

from these wastes is variable, depending greatly on the mixture of wastes involved. For highly delignified cellulose wastes, leachate collection and, possibly, venting, may be necesary. There is a great deal of design information available for landfilling the industries' wastes[32,33,34].

Landfill siting is not usually a difficult choice because plants are often located in close proximity to the sparsely habitated areas supplying wood. Haul distances are frequently well within 16 km of the plant[35]. Reclamation of land by filling in coal strip mines is also practiced. Increasingly, however, plants seeking new or extensions to existing sites are encountering public resistance. Landfill siting is a generic problem today for all types of wastes. Acknowledged information that pulp and paper residuals are generally non-hazardous appears to have little influence on public opinion against landfills. For this, as well as other reasons already expressed, it is prudent that the industry continue its progress in recycle and recovery.

2. *Incineration*—A number of mills have experience with incinerating sludges and other residues. Murdock[36] reports on a costly effort to burn combined primary and secondary sludge. Kekki[37], however, reports that the use of a fluidized bed incineration system to burn a combination of wood and sludge residuals at a sulfite mill in Finland provided an attractive payback. The Mehoopany Plant of Procter & Gamble incinerated, without heat recovery, secondary activated sludge and waste pulping cellulose. The practice has ceased because of the extensive slag build-up from the activated sludge and the development of recovery practices mentioned elsewhere in this chapter.

Increasingly, incineration applications of today include energy recovery, such as that reported by Kekki. Energy recovery boilers are often used in lieu of incinerators. Miner[38] summarized sludge burning practices in combination fuel fired boilers. Although some sludges may contain residues likely to promote clinker formation and cause slagging or tube fouling, these problems have not been too severe to prevent sludge burning in a number of instances. The impact of sludge moisture on net steam generation is another consideration. Additional information on energy recovery is found under the previous section on "Recovery of Wastes—Energy Production."

IV. A CASE HISTORY

A case history involving a large integrated pulp and paper plant is presented to illustrate application of the foregoing waste management concepts. The Procter & Gamble Paper Products Company in Mehoopany, Pennsylvania manufactures products, such as kitchen towels, sanitary tissue, and disposable diapers. Its operations include a Woodyard; an ammonia base sulfite pulping process, somewhat unique in the industry; papermaking, converting, and warehouse facilities; and ancillary power, water treatment, and pollution control facilities. Approximately 300,000 wet tons/year of solid wastes are produced and managed

TABLE 3

Solid Waste Disposition at the Procter & Gamble Paper Products Company, Mehoopany Plant

DISPOSITOIN (Bone Dry Tons/Year Unless Otherwise Noted)

Waste Source	RECYCLE TO Manufacturing	RECYCLE TO Energy Production	RECOVERY THROUGH Raw Mat'l Sales	RECOVERY THROUGH By-Product Sales	RECOVERY THROUGH Beneficial Land Application	DISPOSAL THROUGH Landfill	DISPOSAL THROUGH Incineration
Woodyard—Reject Wood			3,370			2,130	
—Bark		46,400				3,660	
—Clean-Up						18,100	
Pulping—Rejects	5,040					520	
—Fines				1,600		7,750	
—Activated Sludge				4,280	440		
Papermaking—Wastewater Fibers	27,100						
Converting—Broke						610	
—Cartons			430			120	
—Corrugated			4,340				
—Cores			60				
—Dust						420	
—Glues/Inks*						36	
—Drums/Pails			170				
—Raw Materials			820				
Plant Trash—Metals/Wood			1,380				
—Poly			170				
—General						11,500	
—Demolition						1,170	
Miscellaneous Chemicals*						5	33
Oils*			34			1	1
Utilities—Boiler/Fly Ash						11,100	
—Wtr. Trt. Sludge						53	
—Maintenance*						11	
TOTALS	32,140	46,400	10,774	5,880	440	57,186	34
% OF TOTAL	21.0	30.4	7.0	3.8	0.3	37.5	<0.1

(Recycle + Recovery = 62.5)

*As-is weight

toward maximum utilization of their potential value.

Recycle/Recovery—Table 3 summarizes the nature and disposition of these solid wastes. For uniformity, all weights in the table are expressed on a bone dry basis. It is seen that 62.5% of the wastes are recycled or recovered. Key individual waste utilization activities shown with percentage of total waste pro-

Disposal Activities—Approximately 38% of the total solid waste production at the Mehoopany Plant is disposed of at both on-site and contractor landfills. The Plant has two landfills under Pennsylvania permits—one active and the other held in reserve. Surface leachate outcrops at the active site are collected and provided activated sludge treatment. The reserve site is lined and has full leachate collection facilities. Although the 2 sites will provide over 15 years of life for the Plant's total current landfilling demand, efforts continue to increase recycle and recovery of wastes.

Future Recycle/Recovery Projects—Efforts are underway to raise the recycle/recovery of solid wastes to approximately 80%, expressed on a dry weight basis, of the total solid waste production. Planning is underway to incinerate with heat recovery the general plant trash (7.5% of total solid wastes). Further, anaerobic digestion, fuel pelletizing, and fluidized bed incineration with heat recovery of the pulp/paper making waste celluloses currently being landfilled (20% of total solid wastes).

V. SUMMARY

The pulp and paper industry generates a broad array and large quantity of solid wastes. The major types are bark and other wood wastes, steam and power generation ashes, and pollution control sludges. Losses of usable fiber (cellulose) from the manufacturing processes are minimized by well designed and implemented technologies and procedures for recycling and avoiding the generation of wastes.

The wastes generated, with the exception of ash, are largely naturally occurring organics. This readily lends them to recovery practices which derive energy production and agricultural benefits. Further, this property, combined with its generally non-hazardous character, also allows disposal by landfill in the absence of beneficial applications. Some wastes are disposed of by incineration, but this practice is giving way to combustion with energy recovery.

REFERENCES

1. "1982 Post's Pulp and Paper Directory," 1981, Miller Freeman Pub. Inc., San Francisco.

2. Brookover, T.E., May 1977, "A General Review of the History of Processing Primary and Waste Activated Sludge at Sonoco Products Co., Downingtown (PA) Paper Division," Proceedings of the 1976 Northeast Regional Meeting, NCASI Special Report 77]03, p. 9.
3. Smithe, R.J., and Abramowski, H.J., Jan. 1980. "Combination Burning of Secondary Sludge and Black Liquor in a Fluidized Bed Reactor," NCASI Special Report 30-02, p. 51.
4. Adkins, J.C., May 1977, "In-Process Disposal of Secondary Sludges in Batch Digesters," NCASI Special Report 77-03, p. 9.
5. Frederick, W.J., et al., 1981, "Disposal of Secondary Sludge in the Kraft Recovery System," TAPPI, *64*, (1), 59.
6. Carter, N.C. 1983, "Composting Disposes of Sludge, Yields By-Product at Glatfelter," Pulp and Paper, p. 102-104.
7. Eberhardt, W.A., et al, 1978, "Sulfite Pulping BOD Conversion to an Agricultural Product—Dried Activated Sludge," Journal Water Pollution Control Fed. 1893-1904.
8. Orme, L.E., and Lemm, C.A., Dec. 10, 1973, "Use of Dried Sludge from Paper Processing Wastes in Trout Diets," Feedstuffs, 28.
9. Orme, L.E., Aug. 17, 1976, U.S. Dept. of Int. Fish and Wildlife Service, Personal Communication.
10. Church, D.C., June 1, 1979, Dept. of Animal Science, Oregon State Univ., Progress Report, "Nutritional Evaluation of Secondary Pulp Mill Sludge," Personal Communication.
11. Rogers, D., August, 1982, "Summary of Operating Experience and Performance of ITT Rayonier's Deep Tank Aeration/Dissolved Air Flotation Secondary Treatment System, NCASI Special Report 82-05, P. 20-34.
12. Pamment, N., Robinson, C.W., Moo-Yound, M., 1979, "Pulp and Paper Mill Solid Wastes as Substrates for Single-Cell Protein Production," Biotechnology and Bioengineering, XXI, 561, Wiley and Sons, Inc.
13. Baker, A., Dec. 1976, Feedstuffs for Ruminants," NCASI Special Report 76-08, p. 143.
14. Dinius, D.A., and Bond, J., 1975, "Digestibility, Ruminal Parameters, and Growth by Cattle Fed A Wood Waste Pulp," Journ. of Animal Science, 41, (2), 629.
15. Numerous Authors of Unpublished Pennsylvania State University Research Reports, Personal Communications.
16. Emart, G.H., Oct. 1978, "Conversion of Cellulosic Residuals to a Chemical Feedstock," NCASI Special Report 78-06, p. 56.
17. Hermann, Douglas J., Oct. 1981, "Guidelines for the Application of Wastewater Sludge to Land," NCASI Special Report 81-09, p. 84.
18. Nolan, W.E., Oct. 1981, "Land Application of Papermill Sludge at Arkansas Kraft," NCASI Special Report 81-09, p. 83.

19. Jayne, J., Oct. 1981, "Aerated Stabilization Basin Solids Removal at Arkansas Kraft Corporation," NCASI Special Report 81-09, p. 81.
20. Nolan, W.E., Oct. 1981, "Land Application of Papermill Sludge at Appleton Papers," NCASI Special Report 81-09, p. 79.
21. Nolan, W.E., and Jarrett, D.L., Sept. 1981, "Land Application of Paper Mill Sludge," NCASI Special Report, 81-07, p. 194.
22. Beasom, J., Oct. 1981, "The Selection of Farmland Application for the Disposal of Lagooned Sludge," NCASI Special Report 81-09, p. 75.
23. Cawrse, A., Sept. 1981, "Direct Disposal of Paper Mill Sludges on Agricultural Land," NCASI Special Report 81-07, p. 197.
24. Hill, M.E., Sept. 1981, "The Application of Sludge on Agricultural Land," NCASI Special Report 81-07, p. 196.
25. Buchanan, B., Oct. 1978. "Land Application of Secondary Sludge," NCASI Special Report, 78-06, p. 37.
26. Corey, R., March 1977, "Agricultural Application of Pulp and Paper Industry Waste Treatment Sludges," NCASI Special Report 77-02, p. 89.
27. Harkin, J.M., April 2, 1981, "Statement of John M. Harkin Concerning Land Spreading of Papermill Sludges," Univ. of Wisconsin, Dept. of Soil Science, Madison, WI.
28. Smithe, R.J., and Morin, M., March 1977, "The Use of Secondary Treatment Plant Sludge for Fertilizing Intensive Forestry Tree Farms at Packaging Corporation of America, Filer City, Michigan," NCASI Special Report 77-02, p. 92.
29. Aspitarte, R.T., Jan. 1980, "Agricultural Utilization of Wastewater Treatment Sludges," NCASI Special Report 30-02, p. 56.
30. Cartner, J.B., and Williams, D.J., 1978, "Horticultural Uses for Bark," TAPPI, 61, (7), 83.
31. Wardell, R.E., et al., 1978, "Disposal of Papermill Sludge in Landfills," TAPPI, 61 (12), 72.
32. McKeown, J.J., March 23, 1982, "A Summary Review of Issues Involved in Developing State Solid Waste Disposal Regulation and Industry Practices wih Emphasis on Non-Hazardous Paper Industry Residuals," Pre-Publication Paper.
33. Ross, K.S., Jan. 1980. "Promoting Consolidation and Structural Stability with Drainage Blankets," NCASI Special Report 30-02, p. 69.
34. Charlie, W.A., et al., Jan. 1980, "Engineering Properties of Combined Primary and Secondary Papermill Sludges," NCASI Special Report 30-02, p. 59.
35. McKeown, J.J., 1979, "Sludge Dewatering and Disposal—A Review of Practices in the U.S. Paper Industry," TAPPI, 62, (8), 97.
36. Murdock, R.W., Nov. 1981, "History and Operation of Abitibi-Price's Alpena Waste Treatment System," NCASI Special Report 81-13, p. 132.

37. Kekki, R.M., 1981, "Incineration and Heat Recovery of Primary Clarifier Sludge," TAPPI, *64*, (9), 149.
38. Miner, R.A., Nov. 1981, "A Review of Sludge Burning Practices in Combination Fuel-Fired Boilers," NCASI Technical Bulletin No. 360.

Solid and Liquid Wastes: Management, Methods and Socioeconomic Considerations. Edited by S. K. Majumdar and E. Willard Miller. © 1984, The Pennsylvania Academy of Science.

Chapter Four

Solid Waste Handling in Hospitals

Patrick F. Mutch,[1] M.P.H. and Brian K. Mathias,[2] M.P.H.

[1]Assistant Executive Director and [2]Administrative Resident

The Western Pennsylvania Hospital
4800 Friendship Avenue
Pittsburgh, Pennsylvania 15224

The management of a modern hospital represents one of the most unique and complex organizational challenges. The hospital's diverse collection of specialized services, such as critical care units, surgical suites, automated laboratory and radiology departments and patient units, require a vast network of professional and support services. The labor and technology-intensive nature of health care is now a matter of popular awareness and the level of care available is unparalleled in human history.

However, the operation of such a massive technosystem carries with it the unglamorous task of waste disposal. The services described above generate tremendous amounts of waste. This is due in large part to the increased use of disposables in hospitals, and strict standards for sanitation and a sterile environment which far exceed requirements in other facilities. In a large regional referral medical center which offers specialized services—the perspective from which this paper is written—a good rule of the thumb is that approximately ten pounds of waste will be produced per patient day. For a 600 bed hospital operating at 80% occupancy, this would produce nearly two million pounds of solid waste per year and over a million trash can liners to contain it.

Solid waste in hospitals might be described as "routine" (paper, plastic, and similar waste not so different from household rubbbish) and "nonroutine." Routine waste represents the bulk of hospital refuse. Ideally, it should be removed from patient units and other areas at least once a day and preferably several times daily. From there, it is removed to an on-site containment area—usually a trash compactor or dumpster provided by a contractor. These are periodically emptied—several times weekly in a large hospital—and the waste is removed to a sanitary landfill. Incineration is not typically used as a standard means of waste removal except for pathological and other medical or infectious wastes.

It is the "nonroutine" waste, or hazardous medical waste, which represents a more significant problem despite its smaller volume. Removal of these wastes involves special handling considerations specific to the nature of the waste itself. The most common medical refuse in the hospital is something found in virtually every patient area—the hypodermic needle and syringe. In a large hospital, their use will run into the hundreds daily, and proper handling is essential to protect unwary personnel from puncture wounds and possible infection from contaminants, such as tetanus and hepatitis. Direct skin contact with needles—clean or dirty—must be minimized. In Pennsylvania, specific standards govern the storage and disposal of needles. One of the requirements is that needles and syringes must be rendered unusable following their use (Title 28, Pennsylvania Code, Section 109.67).

A most effective means of disposal is a commercial hand-operated needle destruction unit which severs the needle at the syringe tip and drops it into a sealed, disposable bulk storage bin. No hand contact is necessary. Unfortunately, the "human element" is impossible to eliminate and there are occasions when, without thinking, someone drops an unprotected needle into the trash. This can easily result in a puncture wound to a housekeeper with the attendant risk of possible infection.

Another type of hazardous medical waste found on a typical patient unit results from infectious patient conditions. Dressings, treatment trays, dietary trays and other items coming in contact with a contagious patient must be effectively disposed without exposure to other patients or personnel. Refuse must be placed in a contained storage area in the patient's room and when removed, must be placed in specially marked bags. Proper procedure requires incineration of this particular waste.

The hospital's ancillary services, such as laboratories and special treatment areas, also generate hazardous medical wastes. Pathological waste, such as specimens and amputations, must be properly stored to avoid putrefaction; refrigeration is usually the most appropriate short-term storage method. Eventually, pathological waste must be incinerated.

The disposal of nuclear waste is also a concern in larger hospitals. These are produced in the Nuclear Medicine Department and some diagnostic labs. Handling of these wastes must be according to standards set by The Nuclear Regulatory Commission.

In Nuclear Medicine, a pharmaceutical which is "tagged" with a radioactive agent is injected into the bloodstream where it migrates to the specific area of the body under study. Technitium and iodine-125 are among the isotopes used in Nuclear Medicine. Waste which has been contaminated with the isotopes must be placed in lead storage boxes on the unit and removed to a lead-lined containment room. The waste remains in this room until the radiation safety officer, using a Geiger counter, determines that background levels of radiation have been reached. For technitium-contaminated items, this is only a matter of days since its half-life is just six hours. However, iodine-125 has a half-life of eight days and so necessitates longer storage time. While body excretions rarely reach anything approaching hazardous levels, the urine of patients exposed to over thirty millicuries must be stored and patient restrooms in this area are routinely monitored for radiation levels as a precautionary measure.

The last type of hazardous medical waste is generated from the use of drugs to treat cancer. Chemotherapy involves the intravenous administration of chemical agents used in the treatment of malignant tumors. Improperly handled, chemotherapeutic solutions can cause severe burns and might produce mutagenic effect on cells. Preparation and administration require special handling and ventilation. Personnel should wear polyvinyl gloves, cuffed isolation gowns and surgical masks while working under a laminar or vertical flow hood (the latter is preferred). All contaminated items should be placed in a cardboard box and double bagged in some type of distinctive bag alerting those handling the material to use extra caution. The needles used for chemotherapy are placed along with contaminated syringes in an absorbing agent and then disposed with the other chemotherapy waste. Incineration of this hazardous waste type is subsequently necessary.

From the administrative perspective, it becomes apparent that the safe, functional and effective disposal of all hospital waste is essential. It is the hospital's responsibility to occupy the ironic position of insuring that the very agents used in healing its patients do not become instruments of harm to others. Of course, this can be accomplished but only through the proper education of all personnel. From administration and medical staff on down the organizational ladder, waste management is a task shared by the entire hospital staff.

In conclusion, the method of solid waste disposal in the nation's hospitals is one that must receive serious consideration. Both routine and hazardous medical wastes require careful handling and removal. To protect patients, staff, and visitors, Hospital executives must be active participants in the development of local, state and national regulations which relate to solid waste disposal.

Solid and Liquid Wastes: Management, Methods and Socioeconomic Considerations. Edited by
S. K. Majumdar and E. Willard Miller. © 1984, The Pennsylvania Academy of Science.

Chapter Five

Pennsylvania Department of Environmental Resources Role in Solid Waste Management

Donald A. Lazarchik, PE

Director, Bureau of Solid Waste Management
Commonwealth of Pennsylvania
Department of Environmental Resources
P. O. Box 2063
Harrisburg, PA 17120

Pennsylvania, like most other states, largely ignored the management of solid waste for many decades. The State's first statute concerning the management of an increasing volume of solid waste was not adopted until 1968. It was followd by a resource recovery grant and loan program in 1974, and a strong and comprehensive new solid waste management statue to control hazardous and toxic waste in 1980. This chapter deals with the Department of Environmental Resources' role and responsibility under each of these laws.

HISTORY

Pennsylvania's Purity of Waters Act was enacted in 1905 and was followed by the Clean Streams Law in 1937. Air pollution control was addressed in the Air Pollution Control Act in 1960, but not until 1968 was solid waste management considered to be enough of a problem to require legislation. Open dumps were the traditional method of handling trash and garbage in suburban and rural areas. Largely inefficient and overloaded municipal waste incinerators were often utilized in urban areas as a method of volume reduction prior to the dumping of incinerator residues on the surface of the land. Industry was generally free to bury its residual waste wherever it was convenient, so long as these wastes did not cause obvious pollution of surface waters. The health hazards and environmental pollution problems caused by mismanaged solid waste were largely overlooked by everyone except those who were unfortunate enough to reside near an open dump.

Prior to 1968 the only state laws that could be used to correct solid waste problems in Pennsylvania were the State's general public health statutes that prohibited the maintenance of a public nuisance which affected the public health. The water pollution control and air pollution control laws could also be used to abate flagrant abuses of environmental resources, such as the placing of waste directly in a stream or the air polution caused by an open burning dump. In 1965 the Congress enacted the nation's first comprehensive statute on solid waste management. This law was intended to encourage and assist the states in the development of state programs to foster better solid waste management systems. Project 5000, a federal effort to assist the states in closing 5000 open, burning vermin infested dumps across the nation was a part of this initial federal program.

Act 241
The Pennsylvania General Assembly enacted the State's first Solid Waste Management Act in 1968. The law was comprehensive in that it enabled the Department of Health to attack health and environmental problems caused by any type of solid waste, whether it be from a municipality, from agriculture, from mining or from industry. The Act placed responsibility for the planning of effective solid waste management systems on municipal governments with a population density in excess of 300 persons per square mile. 50% state planning grants were authorized to assist municipalities in meeting this planning obligation. All operating solid waste management systems were required to obtain a permit for operation from the Department of Health. Broad rulemaking and enforcement powers were granted to the Department and penalties were established for violations of the act and the Department's rules and regulations. The act was considered to be a model for the nation at that time.

The purposes of the act were:
1. To establish and maintain a cooperative state and local program of planning and technical and financial assistance for comprehensive solid waste management.
2. To utilize wherever feasible and desirable the capabilities of private enterprise in accomplishing the desired objectives of an effective solid waste management program.
3. To require permits for the operation of processing and disposal systems.

The act established the following powers and duties for the Department of Health:
1. To administer the solid waste management program pursuant to provisions of this act.
2. To cooperate with appropriate federal, state, interstate and local units of government and with appropriate private organization in carrying out its duties under the act.

3. To adopt such rules, regulations, standards and procedures as shall be necessary to conserve the air, water and land resources of the Commonwealth, to protect the public health, to prevent public nuisances, and to enable the Department to carry out the purposes and provisions of the Act.
4. To develop a statewide solid waste management plan in cooperation with local government, with the Department of Community Affairs and with the State Planning Board, and when feasible to emphasize areawide planning.
5. To provide technical assistance to municipalities, counties and municipal authorities, including the training of personnel.
6. To report to the Legislature from time to time on further assistance that will be needed to administer a solid waste management program.
7. To initiate, conduct and support research, demonstration projects and investigations and coordinate all state agency research programs pertaining to solid waste management systems.
8. To establish policies for effective solid waste management.
9. To issue such permits and orders and conduct inspections as necessary to implement the provisions of the Act, the Department's rules, regulations and standards.

The Department of Health utilized this new legislation to establish a solid waste regulatory program. Municipal wastes were addressed first and many open dumps were closed in the initial years of this program. Some sites were upgraded to "sanitary landfills." Little was known at the time about the environmental impact of sanitary landfilling beyond the obvious avoidance of health hazards through the compacting, covering and burying of solid waste. Waste from other than municipal sources were largely ignored in the early days of the program, although flagrant abuses were sometimes attacked on a complaint basis. As the program developed much was learned about water pollution, methane gas generation and other problems associated with landfilling. The Department's staff and budget resources grew from a single solid waste coordinator in the Bureau of Community Environmental Control to a staff of 50 full-time employees, first in a Division of Solid Waste Management in 1971 when the Department of Environmental Resources was formed, and finally in a Bureau of Solid Waste Management in 1979.

The role of the Department in solid waste management gradually changed from one of public health education and persuasion to a role of enforcing the environmental requirements of the Act. Case law developed as the program grew. The Department utilized the stiffer penalties of the air and water pollution control laws to correct solid waste management problems and to achieve progress and improvement in operating facilities. Much effort was devoted to the avoidance of future problems in the design and construction of new facilities. Experience dictated a rewrite of regulatory requirements of the Department

in 1977 when rules and regulations were changed from performance-based standards to include some specific design and operational criteria that had proven
necessary. The program gradually expanded to include the disposal of more
and more types of industrial and mining solid wastes as additional environmental
problems were discovered. (Subsequent acts of the Legislature removed coal
mining wastes from the provisions of the Solid Waste Management Act of 1968.)

Public opposition to the land disposal of solid waste became increasingly
important by the mid-1970's. Organized citizen opposition to specific site proposals often resulted in lengthy legal challenges. The Department was placed
in the position of actively closing substandard sites while being unable to
stimulate the acceptance of essential replacement facilites. The municipal planning program resulted in the development of many country and regional plans
that recognized and projected future waste management needs. Unfortunately, few of these plans were implemented or even appoved by the local
municipalities involved when they became aware of the increased costs for
upgraded and impoved dispoal services or the targetting of specific areas within
their municipal jurisdiction for the future establishment of needed new facilities.

Act 198

In the mid-1970's DER actively pursued the need for resource recovery projects to decrease dependence on land disposal and to conserve materials and
energy. The General Assembly enacted the Pennsylvania Resource Recovery
Development Act, Act 198, in 1974 and appropriated a total of $4 million for
75% grants to municipalities to stimulate the construction of new and innovative
resource recovery facilities. The Act also provided for a rotating low interest
loan program, but this part of the Act was never financed or implemented.

The legislative purposes of Act 198 were (1) To promote the construction and
the application of solid waste disposal and processing and the resource recovery
systems which preserve and enhance the quality of air, water and land resources,
and (2) To provide financial assistance to municipalities and development agencies in the planning and development of resource recovery and solid waste
disposal and processing programs. Interest in resource recovery in Pennsylvania
has followed the national pattern. Initially interest was centered in the production of refuse derived fuels and then later moved on to mass burning of municipal
solid waste for energy recovery. Low technology, waste separation and recycling projects have been of interest to some small population areas while the high
technology projects have interested the urban areas that have little space for
new land disposal facilities. The Department also utilized its planning grant
program to stimulate the interest of local government in all aspects of resource
recovery. Numerous and varied projects have been made grant offers under the
provision of Act 198 over the years. Some have succeeded, but many communities
have not been able to raise the additional local funds required to finance these
highly capital intensive ventures.

THE RESOURCE CONSERVATION AND RECOVERY ACT

The enactment of the Resource Conservation and Recovery Act by the U.S. Congress in 1976 provided long overdue national recognition of the nation's very serious and growing solid and hazardous waste problems. State agencies greeted the new Act with enthusiasm as the beginning of a long overdue corrective and preventive program. Federal funding would help many states attack problems caused by less than adequate solid waste management systems and from the residues from air and water pollution control devices, the residuals from the industrial activities and perhaps most importantly from hazardous wastes that affect public health and safety. The initial RCRA program under the federal law involved a three-pronged attack on solid and hazardous waste. The Act required the preparation of statewide solid waste management plans and established minimum national standards for the disposal of non-hazardous solid waste. It also created a complete regulatory system to attack the hazardous waste problem. States like Pennsylvania assisted in the establishment of the minimum national standards and began an inventory of all operating facilities to upgrade them to the minimum standard. Unfortunately, the Federal Government ceased funding of the non-hazardous programs several years ago and at the present time the States are on their own in managing non-hazardous waste.

The hazardous waste management program under RCRA gave the states the opportunity the implement their own hazardous waste management regulatory programs. They must prove that their state programs are equivalent to the national program, adequate in terms of staff resources and in enforcement and consistent with the programs of other states through the Nation. The states have spent a great deal fo time and effort in the years since 1976 attempting to comply with the complex federal requirements and to prove that their programs are eligible for authorization. Another Chapter of this book deals with Pennsylvania's efforts to obtain hazardous waste program authorization from the United States Environmental Protection Agency.

Act 97

After three years of closely monitoring EPA's RCRA regulatory program and in participating in the development of the national requirements, the Department approached the General Assembly and requested the enactment of a new Solid Waste Management Act to enable it to comply with the RCRA requirements. We were able to cite several near disasters that had occurred in Pennsylvania because of the mismanagement of hazardous wastes. The Pittston Borehole was perhaps the classic case of environmental irresponsibility on the part of persons who dumped hazardous waste into underground mine workings. The Bill that became Act 97 was introduced in both the House and Senate in October of 1979. It quickly passed the House, but became the subject of lengthy Senate hearings. A number of amendments were made to the Bill which survived eight separate printings before finally being enacted in July of 1980.

Act 97 was a total rewrite of the original Act 241 and included the concept that hazardous waste is a subset of a much larger solid waste management problem. It recognizes the many interrelationships and interconnections with air, water and other environmental programs that are involved in the management of solid and hazardous wastes. It is a broad, comprehensive legislative authority and while its principal purpose was to enable Pennsylvania to achieve program authorization under the Federal Resource Conservation and Recovery Act, it also cleared up many other problems that had occurred in the initial enforcement of the original Act. Act 97 is a very powerful law with many enforcement tools and with a high degree of public information, public notice and public participation requirements. Responsibility for the ultimate handling and disposal of all wastes is placed on the generators of non-hazardous industrial waste and hazardous waste and on municipal government for municipal wastes generated within a municipal jurisdiction. The Act attempts to close the loop on damage to all phases of the environment by protecting air, water and land from faulty solid waste management methods. The Act also addresses the cleanup of the environment from past waste practices and the full utilization of any federal funds to be made available for those purposes. It also directs implementation of the environmental amendment to the Pennsylvania Constitution and requires consideration of the social and economic impact of waste management activities.

The declaration of policy section of the Act includes the following:
1. Establishment of a cooperative state and local program for comprehensive waste management.
2. Encouragement of the use of resource recovery technology for the conservation of materials and energy resources.
3. Requirements for permits for municipal and residual waste processing and disposal, licenses for the transportation of hazardous waste, and permits for the storage, treatment and disposal of hazardous waste.
4. Protection of the public health, the public safety and the public welfare.
5. Provision of a flexible and effective means of implementation and enforcement of the penalty provisions of the Act.
6. A requirement to adopt a Pennsylvania Hazardous Waste Facilities Plan to assure the availability of the necessary facilities to manage hazardous waste in the future.
7. The development of an inventory of the nature and quantity of hazardous waste generated within the Commonwealth.
8. A projection of the nature and quantity of hazardous waste to be generated in the Commonwealth through the year 2000.
9. A mechanism for the establishment of new hazardous waste treatment processing and disposal sites in the Commonwealth.

10. Implementation of the environmental amendment to the Pennsylvania Constitution by requiring consideration of the social and economic aspects of waste management by the Department in the issuance of permits.
11. The utilization of private enterprise wherever this is advantageous and economical for implementation of the Act.

The Act directs the Department of Environmental Resources, in consultation with the Department of Health on matters of public health significance, to administer the Act; to cooperate with Federal, State, interstate and local units of government and with appropriate private organizations; to develop a statewide solid waste management plan; to provide technical assistance to municipalities; to participate in research demonstration and investigation projects on solid waste management systems; to regulate the storage, collection, transportation, processing and disposal of all wastes; to issue permits, licenses, and enforcement orders where necessary. The Act establishes a fee for the issuance of licenses and permits; it authorizes the Department to receive and expend federal and other public and private agency funds for the development of the solid waste management program; it establishes enforcement authorities; it enables the Department to appoint such advisory committes as are necesary to assist it in the implementation of the Act, and directs the Department to prepare a Pennsylvania Hazardous Waste Facilities Plan in conjunction with a Hazardous Waste Facilities Planning Avisory Committee. It also directs the Department to prepare preliminary environmental, social and economic hazardous waste siting criteria; to establish fees for the inspection of hazardous waste treatment and disposal facilities and finally, authorizes the Department to act as the Commonwealth's agent in the federal Superfund Program.

The Act continues the planning requirements of the original Act 241 for municipal levels of governments, and clarifies the fact that they must plan only for the collection, processing and disposal of municipal wastes generated within their boundaries. It relieves them of the obligation to plan for the handling of residual and hazardous waste. The Act also continues the 50% planning grant mechanism for funding of local solid waste management planning.

The Department is given the authority under Act 97 to order local municipal jurisdictions with less than 300 persons per square mile population density to prepare solid waste management plans when the Department has identified a waste management problem or a potential waste management problem within the municipal jurisdiction. This new provision corrects some of the problems that have occurred with the municipal planning requirements of the original Act.

The Act places a high degree of public notice and public participation on all of the Department's permitting and licensing activities in solid wastes by requiring the notification of local government of the receipt of every application for a solid waste management permit. Local government is given a 60 day period in which to study, review and react to solid waste managemenet permit applications. They may then recommend approval, approval with conditions or denial of the applications. A recommendation for denial must be accompanied by specific cause for that recommendation. If a municipal jurisdiction does not respond within the 60 days provided by the Act, they are considered to have waived their right to comment. The Department is obligated to formally answer the comments of local government by publication of a response in the *Pennsylvania Bulletin*.

The Act, also for the first time, allows the Department to require proof of the financial responsibility of those who manage solid waste within the Commonwealth. A minimum $10,000 bond is required for those who apply for licenses for the transportation of hazardous waste. Everyone who receives a permit from the Department is required to post a bond to assure compliance with the terms and conditions of that permit except for municipal government entities operating solid waste facilities handling solely municipal wastes.

CONCLUSION

Act 97 is a far reaching and powerful piece of environmental legislation. It is certainly one of the best of such laws in the Nation. In implementing the Act, the Department has two objectives; (1) to assist those who wish to comply with the provisions of the law, and (2) to enforce the law against those who would circumvent it. The Department has been utilizing the provisions of Act 97 to encourage the development of needed treatment and disposal capacity to provide for the solid waste management needs of the citizens of the Commonwealth. We have used the enforcement provisions of the law vigorously to curtail illegal disposal activities and hopefully to increase public confidence in our ability to regulate solid waste management activities. We have assigned Department staff to a special Toxic Waste Investigation and Prosecution Unit in the Department of Justice to curb illegal waste disposal activity.

If the Department's implementation of this Act is to be successful we will need the commitment to comply from everyone who generates and handles waste. We need the ingenuity of the private sector to develop ways and means of processing materials and providing for a higher degree of reuse and recycling and resource recovery to eliminate as much as possible of our current waste management problem. We need the support and assistance of everyone to explain the nature of our solid waste problem, the waste management needs of local communities and the importance of effective waste management, not only to the preservation of public health and the environment, but to enhancement of our economy.

While our solid waste problems are enormous, they can be solved and our needs can be met. It will take time.

Solid and Liquid Wastes: Management, Methods and Socioeconomic Considerations. Edited by S. K. Majumdar and E. Willard Miller. © 1984, The Pennsylvania Academy of Science.

Chapter Six

Local Government Response to Solid Waste Management Needs in the Lehigh Valley—A Case Study

Olev Taremäe, B.A.[1] and Allen R. O'Dell, B.S.[2]

[1]Senior Planner and [2]Chief Planner
Joint Planning Commission,
Lehigh-Northampton Counties
ABE Airport, Government Building
Allentown, PA 18103

This chapter presents a regional solid waste management planning program carried out in 1981-1983 in Lehigh and Northampton Counties, located in east central Pennsylvania. The landfill capacity problems facing the region in early 1980's is described along with the extensive landfill siting process which was developed to locate potential new landfill sites. The steps taken to begin consideration of an energy-from-refuse plant are presented including the survey of potential steam customers. The arrangements which were made by the municipalities in the region to guide and finance detailed engineering studies are outlined. Finally, some of the major challenges which will face the region as it begins to implement solutions to the solid waste management problems are discussed.

DESCRIPTION OF THE REGION

Lehigh and Northampton Counties contain approximately 729 square miles (1,888 square kilometers) and comprise the heart of the Allentown-Bethlehem-Easton Standard Metropolitan Statistical Area (SMSA). The counties account for 497,767 of the total SMSA population of 635,481 according to the 1980 Census. In addition to the three cities after which the SMSA is named, the counties include 28 boroughs, 4 townships of the first class and 27 townships of the second class. The two counties are served by a common regional planning body, the Joint Planning Commission of Lehigh-Northampton Counties (JPC). Much of the work described hereafter was undertaken by the JPC and was carried out by the JPC staff. Any opinions exprssed here are the authors', however.

CURRENT SITUATION

The JPC had first prepared a solid waste plan in 1971. The plan included recommendations for the upgrading of many of the landfills then existent. Based on available data, the plan concluded that considerable landfill capacity existed for the two county area. The future feasibility of resource recovery operations was foreseen. The economic non-competitiveness of resource recovery vis a vis landfills was seen as the short term obstacle to the construction of such plants. Between 1971 and 1981, the JPC issued annual solid waste plan supplements tracking the solid waste situation.

In 1981, the JPC started a major update to the Solid Waste Plan in response to the increasingly apparent need to address the area's changing solid waste management needs. The first task of the update was to quantify the existing situation. The availability of recently mandated Department of Environmental Resources' (DER) landfill operations logs provided a picture of where waste was going which was of unprecendented accuracy in the Lehigh Valley. Based on previously completed surveys of municipalities and haulers, seven major landfills serving the two counties were identified (see Figure 1). Of these, two (Heleva and Novak) are in Lehigh County, three (Chrin, Bethlehem, and Grand Central) are in Northampton County and two (Christman and Colebrookdale) are in Berks County. The two counties are also served by the Bangor incinerator which processes 12 tons per day. The incinerator has a 50 ton per day capacity. Table 1 reports the waste received by each landfill as surveyed in 1981 and resurveyed in 1982 and 1983.

Of significance in these figures is the closure of the Heleva landfill (by DER order). Area landfill capacity has been decreased by the closure of or the voluntary termination of landfill operations not complying with DER regulations. The Department has increasingly enforced its regulations in recent years. Such landfills include the Heleva, Herceg and Oswald operations. Secondly, the increase in the Colebrookdale volumes exemplifies the influence of outside forces on the area's landfill situation. The increases in this instance, are largely based on long distance hauling to the site from counties lacking in-county landfill capacity.

In order to calculate the remaining landfill life, a number of assumptions are necessary. Because of the critical nature of these assumptions, the JPC developed four scenarios, the best case, the most likely case, the worst case, and a case which solely reflected DER permitted capacity. The assumptions of the most likely case scenario, first developed in 1981, continued to represent the

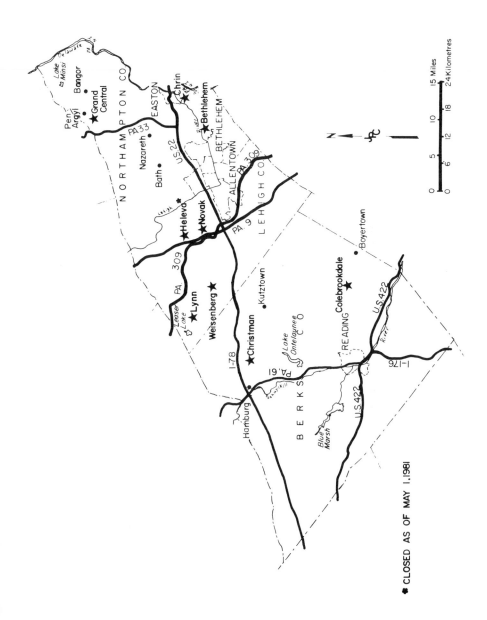

FIGURE 1. Landfills Serving Lehigh & Northampton Counties.

TABLE 1

Tons of Waste Received per Working Day

Name of Facility	1981	1982-3
Bethlehem	244.5	271.1
Chrin	338.0	414.1
Christman	167.5	175.0
Colebrookdale	185.4	1,114.7
Grand Central	258.7	276.7
Heleva	236.8	0.0
Novak	208.4	217.8

TABLE 2

Assumptions and Amounts of Capacity Available per Landfill

Landfill	Assumption	Capacity (tons)
Bethlehem	Permit application for expansion approved	1,586,667
Chrin	Permit application for expansion approved	675,650
Christman	Pending permit application approved	370,762
Colebrookdale	Currently permitted capacity available	3,310,400
Grand Central	Currently permitted capacity available	920,096
Heleva	Closure order appeal denied	0
Novak	Site utilized to full capacity	105,363
TOTAL		6,950,938

best estimate in 1983. By this estimate 6,950,938 tons of capacity exist as of January 1983. Assuming the continued production of waste by the area at current rates, this capacity would last until 1993. The assumptions and amounts of capacity available per landfill are summarized in Table 2.

REGIONAL SOLID WASTE PLANNING PROCESS

Overview

In response to the relatively short term availability of landfill capacity, the JPC compared several basic options for meeting the area's solid waste needs. Four alternatives were constructed for the purposes of the comparison. The first alternative assumed that no additional landfills or resource recovery plants would be constructed. The second alternative assumed that two landfills would be developed. The third alternative assumed that two resource recovery plants, one of 400 tons per day capacity and one of 300 tons per day capacity would be built. The fourth alternative assumed that two landfills would be developed *and* that two resource recovery plants (with the same above listed capacities) would be built. A model was developed and utilized to determine the point at which all area landfill capacity would be exhausted for each alternative. The results of the model are listed below:

Alternative	Description	Year Landfill Capacity Exhausted
1	No action	1993
2	Landfill development	2000
3	Resource Recovery Plant Construction	1996
4	Landfill Development and Resource Recovery Plant Construction	2008

The results of this exercise led the JPC to conclude that a solid waste management plan must provide for both additional landfill capacity and for the construction of resource recovery plants. The other options provided inadequate capacity for long-term needs. Even the amount of capacity gained through the actions outlined in the scenarios may not provide as great an amount of capacity as may be desirable so greater new capacity and additional resource recovery actions will be considered in future steps. To start the implementation of the plan, the JPC began a process by which suitable sites for landfills would be picked and an effort to promote the development of a resource recovery plant. The need for the JPC, as a governmental agency to identify suitable landfill sites was based on a perception that the private sector would be unwilling or unable to meet the needs. This unwillingness results from the difficulties involved in achieving the required DER permit and the problems of dealing with widespread public opposition. The involvement in resource recovery plant facilitation was based on the greater likelihood of plant construction if government involvement existed.

Sanitary Landfill Siting

The landfill siting work involved the examination of the entirety of the two counties relative to a series of factors developed by the JPC and accepted by DER. The factors included economic, societal values, technical adequacy areas, and government regulations. The factors were rated in terms of relative importance. The factors were applied in several groups to progressively eliminate the less desirable sites until the desired number of sites remained.

The initial step involved mapping eight factors. These factors were:
1. Excessive distance to collection areas to permit cost-effective disposal;
2. Contiguity to major concentrations of residential development;
3. Present use of area for urban use;
4. Presence of high water table soils;
5. Presence of shallow depth to bedrock soils;
6. Proximity to airport runways as specified by DER;
7. Recommendations of the JPC Transportation and Recreation Plans for site acquisition; and
8. Presence of flood plains.

Areas characterized by any of the above factors were considered to be poorly suited for landfill uses. They were eliminated from further consideration.

The remaining criteria applied in the selection process were not considered as critical as the initial eight factors. Failure to comply with remaining criteria was not intended to indicate the area's lack of suitability for a landfill. These criteria were used to distinguish better areas from good areas and more acceptable sites from less acceptable sites. From a technical standpoint, it would be possible to develop a landfill in most places which had survived the initial factors.

In the second step, the following areas were mapped:

1. Steep (greater than 15%) slopes;
2. Presence within a DER prescribed distance to a water system intake;
3. Presence within a high quality watershed where service by public sewers is not available; and
4. Smaller sizes of contiguous usable area than required for desired site operation.

These areas were also eliminated from further consideration. At the end of this step, a total of 169 areas ranging in size from 100 acres to several thousand acres remained in continued consideration.

In the third step, the areas were evaluated relative to the following factors:

1. Ease of site accessibility as measured by frontage along roads classified by the Federal Aid Highway Classification Maps as collector or arterial roads. (This factor is also believed to assist in the reduction of neighborhood disruption.);
2. Presence of woodlands;
3. Knowledge of active subdivision proposals;
4. Availability of a permanent stream within a 1/4 mile of the area (for the discharge of effluent from a treatment plant for the leachate). This factor was applied only to areas where public sewers were unavailable.
5. Presence of underground mining;
6. Inadequate contiguous usable area for site operation.

Areas not meeting any of the above factors were eliminated from further consideration. At the end of this step, a total of 92 areas ranging in size from 100 acres to several thousand acres remained in continued consideration.

At this point, the haul distance factor was reconsidered. Given the cost advantages of using closer areas, sites involving round trips of 25 miles or more from the centers of the collections areas were eliminated from further consideration.

The remaining 27 sites were then field checked relative to on-site and adjacent land use, site visibility, and the condition of the nearby access roads. Of the 27, the ten most promising areas were selected for further consideration and evaluation.

After the compilation of additional data and the undertaking of additional studies, two sites were recommended as first priorities for landfill siting consideration. These sites were a new site in Whitehall Township, Lehigh County entitled the "Whitehall-Giant" site and the expansion of the City of Bethlehem Landfill (Lower Saucon Township, Northampton County) to its ultimate capacity.

In November 1982 after the completion of the siting work, the Department issued the first part of the second draft of the revised solid waste management regulations. This draft included one previously unexamined criteria which is of particular relevance to landfill siting in the Lehigh Valley. Proposed Section 275.300 flatly prohibits the placement of landfills on carbonate geology. This regulation is not included in any form in the current regulations, in the first draft of the proposed regulations, or in the DER approved siting criteria. A significant portion of the two counties is underlain by carbonate geology. Eight of the ten areas identified for possible landfill use including the Whitehall-Giant site are underlain by carbonate geology. (The concentration of the selected sites in the carbonate geology areas is not coincidental. These areas are where the bulk of existing urban development has occurred. Therefore, these areas rated favorably on minimum haul distance, availability of public sewers, and ease of access factors. Further, the carbonate geology areas, on average, have deeper soils thus providing more cover material and have lesser slopes.) As such, the JPC was faced with a possibility that eight of the ten selected areas *might* be considered unacceptable for landfill use by future regulations. An assessment as to whether or not the carbonate geology ban would be retained in the adopted regulations could not be confidently made. In response, the JPC conservatively chose to rework the site selection process.

A number of sites generally suitable for landfill use, not located in carbonate geology, were reexamined. Data were accumulated for these sites which had been previously discarded in favor of the ten selected areas. More than 25 of these sites were subjects of field inspections. Staff evaluation and weighting of the data yielded five new areas, where the problems associated with landfill siting would be minimized. These areas have been selected along with the two non-carbonate geology sites previously selected, and two previously selected carbonate geology sites (retained for comparative purposes and in event that the proposed carbonate geology ban regulations are not enacted.) The JPC has not recommended top priority sites within this group.

Resource Recovery

Throughout the 1970's, the Joint Planning Commission monitored experiences at resource recovery projects around the country so that we could learn from both their successes and failures. The JPC staff also visited seven operating resource recovery facilities in Maryland, Pennsylvania and Virginia to gain first hand experience. Although there are still some technical and economic problems, large resource recovery projects that produce steam, electricity or a refuse-derived fuel (RDF) have become practical ways to substantially reduce the amount of solid waste requiring disposal. Plants of this type can generally reduce disposal needs by 70-80 percent by weight with even greater reductions by volume. However, capital costs are very high. For example, a large plant with a capacity to handle all the trash generated in the two counties would typically cost in the range of $50 to $100 million. Tipping fees including debt service and operating costs not covered by the sale of products are typically in the range of $12 to more than $20 per ton. The planning, development, financing, construction and shakedown of such facilities is a lengthy process. It will take at least five years and possibly as much as ten years between the time a project is first conceived until it is in full scale operation.

Judging from the experiences to date, the following are the key attributes of

1. *Guaranteed long-term energy customer*—All of the successful projects have customers for their energy project. On the other hand, lack of a near-by steam customer or customers to purchase the RDF product produced by the plant were key factors in the financial failures of dozens of plants around the country some of which involved losses on the order of tens of millions of dollars.

2. *Adequate supply of waste*—All resource recovery projects need some minimum volume of waste in order to operate relatively efficiently. It is important that the facility has assurance that it will receive at least that much refuse. In areas with very high landfill costs, a few resource recovery facilities have been able to attract waste simply because their dumping fees were competitive. In most cases, however, waste flow needs to be assured through public control of refuse collection or through legislation which requires private haulers to use designated facilities.

3. *Relatively simple process*—The more successful projects have been ones with a relatively simple process such as burning the refuse as received or production of a simple RDF with removal of ferrous metals. Many of the projects with more elaborate processing or more exotic processes such as wet pulping and pyrolysis have had cost or technical problems.

4. *Long-term support by the sponsor*—Virtually all the successful projects have had to persevere through initial problems before the success was realized. Most plants have needed extended shakedown periods and additional capital improvements before they reach full scale efficient operation.

Since an energy customer is a key attribute of a successful resource recovery plant, a survey of 48 large industrial and institutional energy users in the region was undertaken. The survey was conducted with the assistance of an Industrial Work Group which the JPC set up with the assistance of the Pennsylvania Power and Light Company. The survey asked whether the industry would be willing to purchase steam or RDF at a cost which represented a discount from fossil fuels. In addition, background data on the existing boilers were requested. The industry was asked for its base and peak steam needs currently and for any planned expansion. Finally, an optional question on the customer's annual fuel usage and costs was asked.

Responses by mail or phone were received from 37 of the 48 facilities, a response rate of 77 percent. Twenty-seven of the 37 respondents (73 percent) expressed an interest in purchasing steam. Most of those who were not interested gave no reason, but a few expressed concern over cost, quality of steam or reliability. Ten of the 37 (27 percent) expressed interest in RDF; however, most respondents qualified their answer with concerns about conversion cost, storage and handling problems, odors, etc.

Based on an analysis of the data provided on steam demand, it was determined that 16 of the respondents were both willing to consider the possibility of buying steam from a resource recovery plant and were large enough to support a resource recovery project of meaningful size in this region. A siting analysis was done in the vicinity of each of these customers to determine if there is an area of five acres or more which is near their steam production facilities, is suitable for a resource recovery plant, has good road access for heavy trucks, is reasonably isolated from incompatible land uses, and is near a large power line so that electricity could be also generated for sale back to an electric utility company. Also, a personal interview was held with the customers to obtain more detailed information on the potential customer's steam demand variations from day/night, week/weekend and summer/winter. The most consistent year round steam demands in the study area were identified at companies in the chemical, paper manufacturing and food processing industries.

INTERMUNICIPAL COORDINATION

Pennsylvania Act 97 of 1980 (35 P.S. Section 6018.101 et seq.) known as the Pennsylvania Solid Waste Management Act requires all municipalities meeting certain population density and planning criteria to prepare and implement a solid waste management plan. The Act also provides that a municipality may request the county to prepare the required plan on its behalf. Since the draft solid waste plan prepared by the Joint Planning Commission[1] could serve as the basis for the required plan, Lehigh and Northampton Counties sponsored a series of meetings with local municipalities to explain the Act 97 requirements and present the draft plan. Many of those present observed that the planning requirements and the solid waste problems facing the region could be addressed more efficiently by a joint effort rather than by as many as sixty individual municipal efforts. Subsequently, both Lehigh and Northampton Counties sent draft ordinances to each of its municipalities indicating that the county would take a role in making arrangements for the additional studies and contribute half the cost if a substantial number of municipalities agreed to cooperate with the county and provide the additional half of the needed funds prorated on a per capita basis. The ordinance was prepared under the provisions of Act 180 of 1972 relating to intergovernmental cooperation (53 P.S. Section 481 et seq.)

to meet the solid waste planning and implementation requirements of Act 97. By adopting the draft ordinance, each municipality agreed to the following:

1. Participate as an active member in the solid waste management planning process.
2. Designate the county to prepare the Solid Waste Management Plan required by Act 97 on behalf of the municipality.
3. Cooperate with the other participants in approving a budget for the planning process.
4. Agree to share in the costs of the planning process on a per capita basis.
5. Designate an individual to serve on the county/municipal steering committees known as the Lehigh County Solid Waste Association and the Northampton County Solid Waste Management Planning Committee.

Twenty-three of the 25 municipalities in Lehigh County representing 92 percent of the population have agreed to participate in the Association. A final count in Northampton County is not available yet, but it appears that 35 of the 38 municipalities in Northampton County representing 96 percent of the population will be participating in the Planning Committee. Each participating municipality will contribute 17.6 cents per capita for the planning and engineering studies in calendar years 1983 and 1984. The counties will match the municipal contributions. These revenues together with a $102,500 solid waste management planning grant provided by DER will finance the $240,000 engineering and planning studies.

CHALLENGES FOR THE FUTURE

Now that the initial planning work has been accomplished, the task becomes the translation of the plans to concrete actions which will solve the problems which have been identified. The future work will cover both technical and administrative issues. The major technical issues which will be dealt with through engineering consultants are as follows:

1. Prepare a hydrogeologic study and feasibility study to determine the suitability, capacity, capital costs and operation and maintenance costs for a major expansion of the existing City of Bethlehem Landfill and for a new site located in Lehigh County.
2. Evaluate the feasibility of retrofitting the existing Borough of Bangor Incinerator to produce electricity.
3. Prepare a feasibility study of an energy-from-refuse plant to generate steam and/or electricity based on the results of the survey of potential steam customers.

Because of the controversial and emotionally charged nature of solid waste disposal in our society, a great number of challenges must be met. These challenges are summarized in the following text.

1. A mechanism to provide for the needed intermunicipal cooperation must be created. At present, 62 municipalities, two counties, the Joint Planning Commission, a county solid waste authority, and two loosely bound solid waste associations are involved. These groups must agree on a decision making process which can lead to commitments and actions. The solid waste associations are a first step in this direction. A strong leadership advocating, sponsoring, and working for the implementation must be found. Without such leadership, the effort will likely flounder. As decision making processes and leadership are created, the parties that will own, operate, and finance the projects must be selected. Legal agreements and arrangements binding the involved groups together will be necessary. The history of inter-municipal cooperation is fraught with examples of failures. Unless these steps are accomplished, implementation will not be possible.

2. It can be anticipated that whatever project or projects are undertaken, whether with landfill development or resource recovery plant construction, a segment of the population will arise to oppose the project. The recent experience in solid waste management has shown that public opposition whether justified or not by the merits of the project, has often been responsible for the demise of solid waste management projects. The challenge will be to satisfy the legitimate concerns of these groups without permitting the groups to paralyze the process. These efforts will need to deal both at a political level and through the judicial process, the two arenas where the groups will attempt to influence the implementation of the projects. Perseverance through extended legal challenges is often necessary for the success of the project.

3. Technical problems need to be solved. Each landfill site will have its own set of problems whether it be the containment of the leachate or the treatment of the leachate. This mandate is equally important for resource recovery plants. The specific technology of the plant must be workable or, like in too many cases nationwide, the plant is abandoned.

4. A number of challenges relating to the financial aspects of the projects exist. First, the projects which are devised must be financially viable. The revenues accruable thereto must match the expenses incurred. The failure to meet this objective has marked sanitary landfills as well as resource recovery projects. For resource recovery plants, a portion of the picture for assuring financial viability is the securing of a customer for the products of the process. Secondly, the capital improvements proposed must be financed. Finding sources for the financing will be a important task.

5. As an adjunct to the financial viability of the project is the need to institute waste flow control. Unless adequate amounts of waste are brought to the sanitary landfill or resource recovery plant, insufficient revenues needed to cover the financed costs are likely to appear. To forestall this possibility and to assure an adequate supply of waste, many lenders re-

quire the institution of waste control measures. The challenge in instituting such controls is a political one. The attempted introduction of such controls in other areas has often met with controversy. Objections surface from private haulers who perceive their rights being diminished.

6. Whatever projects are undertaken, a number of permits must be obtained. The challenge of negotiating through the regulatory process is a significant one. The annually declining number of landfill permits granted by the Department of Environmental Resources exemplifies the difficulties involved. The reluctance of the private sector to attempt new solid waste mangement projects is in part due to the permit situation. An additional complication is that one must consider both the existing and the pending regulations which deal with the designs. In some areas, the standards and requirement vary widely between the two sets of regulations.

7. The future solid waste mangement needs of the region will be met by some combination of existing public and private facilities, new facilities initiated by the public sector and possibly new privately-initiated facilities. An additional challenge will be to incorporate these various facilities into an efficient system which uses public sector actions when necessary and recognizes the abilities, interests and rights of the private sector while assuring that the long-term disposal needs of the region are met in an environmentally sound and cost effective manner.

REFERENCES

1. Joint Planning Commission, Lehigh-Northampton Counties. 1982. *Solid Waste Management Plan — 1982 Update (Draft)*. Allentown, PA.: JPC

Solid and Liquid Wastes: Management, Methods and Socioeconomic Considerations. Edited by S. K. Majumdar and E. Willard Miller. © 1984, The Pennsylvania Academy of Science.

Chapter Seven

Management of Solid Wastes in India — An Overview

A.K. Sharma

President
Indian National Science Academy
New Delhi
and
Professor and Programme Coordinator, Centre for
Advanced Study in Cell and Chromosome Research
Department of Botany
University of Calcutta
Calcutta 700019

The sophistication of technology in different spheres of science in general, and biological science in particular has indeed paid very high dividends. It is however, ironical that in spite of advances in the different spheres of conversion technology, there is still ample scope for optimum utilization of solid waste from a variety of sources. In addition to the city refuse, solid waste from agricultural products forms a major component of the residues in several parts of the world, including India.

In India, of the agricultural wastes, so far bagasse or sugarcane residues are being treated for conversion into energy source. But in majority of the sugar factories, bagasse is directly used as a fuel and as such the situation poses certain practical problems. The utilization of the city refuse is, as yet, principally as a source of fertilizer for a variety of crops. The optimum exploration of this enormous wealth is still to be achieved.

The mixed city refuse of India has a variety of physical and chemical constituents, which in turn have a high potential for the production of energy (Table 1, Bhide *et al* 1975).

In the disposal of solid wastes, particularly from cities, their use in land filling and composting have been the prevalent practices though incineration, pyrolysis and gasification too have been adopted to a great extent. Methane

TABLE 1

Physical and chemical characteristics of mixed city refuse of India

Particulars	Population range (in lakhs)				
	2	2-5	5-20	20	Average
	Physical characteristics (% by wet weight)				
1 Paper	3.9	4.76	3.80	7.07	4.68
2 Rubber & Leather	0.49	0.63	0.62	0.87	0.65
3 Plastics	0.57	0.59	0.81	0.86	0.71
4 Metals	0.51	0.39	0.64	1.03	0.64
5 Glass	0.29	0.34	0.44	0.76	0.46
6 Ash & Fine earth	46.60	39.97	41.81	31.74	40.03
7 Total compostable matter	33.41	39.76	40.15	41.69	38.75
8 Miscellaneous	15.04	19.56	11.73	15.98	15.58
	Chemical characteristics				
1 Moisture (%)	22.12	25.12	22.45	31.18	25.20
2 pH	8.18	8.16	8.34	7.68	8.09
3 Organic matter (%)	22.02	22.51	21.51	27.57	23.40
4 Carbon (%)	12.26	12.51	11.95	15.32	13.01
5 Nitrogen (%)	0.60	0.61	0.56	0.58	0.59
6 Phosphorus (P_2O_5)(%)	0.71	0.72	0.68	0.59	0.68
7 Potash (K_2O) (%)	0.71	0.74	0.72	0.68	0.71
8 C/N Ratio	20.35	20.47	21.45	26.23	22.13
	Energy content (Kcal/kg)				
Calorific value	801	874	867	1140	920

(Bhide et al. 1975)

production by anaerobic fermentation, waste heat utilization for power production, composting and animal feed from solid wastes and pyrolysis are the different facets of solid waste utilization. In fact, innovations in the material recovery operation are the principal issues to be considered for the optimum utilization of solid wastes.

It has been estimated that in India, nearly 20 million tons of city refuse are generated every year (Vimal and Talashilkar 1984).

A systematic study has been carried out on the availability and characteristics of city garbage (Vidyarthy and Misra 1978) and on mechanical composting, which is its principal use (Table 2). It appears that the ratio of plant intake to garbage is quite high, ranging form 5 to 20%. The acceleration of the capacity of the existing plants as well as their use in different parts of a city need to be emphasized.

The production potential and value as manure of city compost in Delhi and in India as a whole have been worked out (Bhide *et al* 1975, Table 3).

The nature of the waste to a great extent depends on various factors including occupation and diet of the population of the area, in most cases, leading not only to air but water pollution as well and ultimately to serious health hazards. It is well known that the leachates are produced following decomposition of

TABLE 2

Mechanical composting programme in India

Location of plant	Capital cost (in lakhs)	Garbage production/ day (tonnes)	Plant intake/ produced (tonnes)	Plant intake/ garbage produced (in percent)	Rate/ tonne (Rs.)
Baroda	70	700	125	17.9	70
Ahmedabad	60	1000	125	12.5	35
Kanpur	110	7000	350	5.0	60
Calcutta	60	2200	200	9.3	80
Bangalore	89	1000	200	20.0	50
Delhi	75	2000	150	7.5	40
Bombay	100	3500	300	8.6	55

(Vidyarthy and Misra, 1978)

TABLE 3

Production potential and manurial value of city compost in Delhi and on All-India basis during the period 1980-81 to 2000-2001 A.D.

Particulars	Delhi 1980-81	Delhi 2000-2001	All India 1980-81	All India 2000-2001
Population x 10^5	45.49	62.20	1261.19	1861.21
Total quantity of city garbage x 10^5t/yr	6.64	9.08	198.73	271.74
Non-compostable	3.98	6.45	119.24	163.03
Compostable	2.66	3.63	79.49	108.71
Compost	1.33	1.82	39.75	54.36
Nutrient content and organic matter x 10^3t/y.				
Nitrogen (N)	0.93	1.27	27.83	38.05
Phosphorus (P_2O_5)	0.80	1.09	23.85	82.62
Potassium (K_2O)	0.67	0.91	19.88	27.18
Organic Matter (0.M)	53.20	72.80	1590.00	2174.40

(Bhide, *et al.* 1975b)

solid wastes, the biological oxygen demand of which is much higher than that of raw sewage. Some leachates containing both dissolved and suspended matter are the sources of contamination of surface zones and ground water surface and cause a variety of other injurious effects. In several parts of India, where such leachates are formed following waste disposal, stunted growth of crops is often recorded due to the disturbance in the balance of nutrients. The levels of different metals frequently show a heavy increase in grains and in soils.

However notwithstanding such limitations, the sewage and city refuse should still be regarded as potential sources of energy due to a variety of reasons. For other organic or woody biomass, lignocellulose degradation is a major factor in limiting their utilization. With successful degradation process and utilization of suitable microbial techniques agricultural or forest residues can be utilized as feed, fertilizer, chemicals and sources of energy on which investigations have been started in different centres in India. Such optimum utilization of biomass with maximum efficiency is however not commensurate with the cost of the operational processes. The high cost involved in the entire process is a drawback though undoubtedly these technologies lead to the generation of energy with the least disturbance to environment.

In India, in general, the composite municipal wastes have a heating value between 4500 to 6500 BTU/lb (Vimal and Talishilkar 1984), with of course an inherent capacity of heat rise concomitant with drying.

The most simple method of heat recovery as steam is the use of a boiler adjoining the incinerator (Vimal and Talishilkar 1984). Pyrolysis too yields a number of products including carbon monoxide, methane, butane, with high energy content through burning organic matter with anaerobic conditions. These processes should be widely used under Indian conditions.

The most common use of urban waste is the form of compost which essentially is stabilised humus. It has its own self feeding system through thermophilic aerobic processing as well as processing of biodegradable solid organic residues (Gotaas 1956, Bell 1974, Gølucke 1975, Stonehouse 1981).

A survey so far conducted on 33 Indian cities shows that, in general, majority of the cities spend only 10% of budget on waste management (Bhide *et al* 1975). Composting is a multifaceted process in which microbial degradation of varying types plays a crucial role. At lower temperature certain microbial types remain active while with increase in temperature the thermophilic bacteria and other microflora dominate the scene.

In large cities, the disposal of waste poses serious problems leading to unhealthy situation even through a mechanized composting system. It must, however, be admitted that mechanized composting has several advantages, specially in the recovery of nondegradable materials and suitable sanitary control in all reasons. Mechanized composting involves initially the removal of heavy particles through automated scavenging operations, the grinding of comparatively crude components, the separation of glasses and metals and finally mixing and composting the organic residues under regulated moisture and temperature conditions in a closed chamber. The maturation period is mostly 1-3 months. Lastly, the addition of chemicals completes the process (Vimal 1973, Diaz et al 1979, Titus et al 1980). The broad principles adopted at Delhi, a metropolis of India, has been outlined by Jain (1980).

The composting of city refuse, later with sewage sludge, may lead to decrease in heavy metal content (Carlson and Menzies 1971). The most crucial factor in composting is the stability of the organic matter in the waste even when the moisture content is high. Several criteria such as drop in temperature, rise in redox potential, oxygen intake, carbon-nitrogen ratio, toxicity to plant roots, etc. have been used as indicators of stability (Stickelberger 1975, Moller 1968, Chrometzka 1968, Poincelot 1974).

In India, the city compost plays a very important role in agricultural practices. It has been shown to lead to the increase of pH values, level of micronutrients and enrichment of microflora through accelerated nutrient supply. In a variety of plants used in agriculture, horticulture and forestry, garbage compost is of tremendous value. In fact, the use of fertilizer to a great extent has been minimum in India with the utilization of garbage compost. Inspite of the extensive use of garbage compost in India, data available is meagre due to lack of systematic assessment (Prasad 1981). In Nepal, it has been reported that mechanized compost had led to increase in 60% of yield over control (Jugsijinda et al 1978).

The management of solid waste is a crucial issue since dumping leads to pollution of both air and water. Suitable measures for recycling and recovery need to be carried out in the most effective manner so that these become commercially feasible and lead to growth of economy. It is true that metals can be conveniently obtained from industrial wastes but the present system operating in steel processing in India does not allow the input from the commercial standpoint. The use of glass is undoubtedly diverse but it is a fact that glass can be obtained from a variety of sources. As such its recovery through complicated recycling does not have commercial feasibility.

Emphasis has been laid on the high economic value of residues of pyrolysis such as gas, tar and charcoal but there is little commercial acceptance (Vimal and Talishilkar 1984). The gas is utilized for running the process itself; tar needs refinement and charcoal needs fractionation from glass and metals.

All these facts contribute to the low commercial value of products.

So far as urbanization is concerned, except in certain cities, mostly there is horizontal growth. As a result the agricultural land is gradually receding and is becoming restricted to areas far away from cities. This development necessitates a high cost of transportation of compost from urban area to agricultural land. It is indeed a serious problem. Social customs and taboos of viewing the excreta not only as obnoxious but also as untouchable stand in the way of full utilization of compost materials. In some of the neighboring countries in the east, such as Singapore, the entire waste has been converted into wealth through proper management.

With regard to solid waste management, the crucial issues involve adoption of a suitable mechanized method for the separation of components, construction of chambers for proper recycling to meet the multifaceted needs of energy, fertilizer, chemical, and poultry feed. The pyrolytic procedures for energy recovery from waste paper, food and plastics for which technologies have been developed in different parts of the world are yet to be adopted. It is true that some of these processes are not cost effective. Notwithstanding the limitations, the adoption of degradation procedures is essential in view of the tremendous capacity of such wastes for creating hazards. Some of these compounds, especially the high polymer plastics, are normallly non-degradable and outlive in years the civilization of Mohanjodaro and Harappa. These facts, taken in conjunction with the knowledge that the biological organisms have never been exposed during their evolution, to such new chemicals in modern technology, pose a severe threat. Such an irreversible process would destroy the natural ecosystem with the destruction of flora and fauna, not as a transitory phase, but for eternity.

ACKNOWLEDGEMENTS

The author would like to express his thanks to Miss Sarmishta Sen, Junior Research Fellow, University of Calcutta for assistance in preparing the manuscript and to Shri Jagmohan Gupta and Mrs. Madhu Marwah of the Indian National Science Academy for typing it.

REFERENCES

1. Bell, R.G. (1974) *Comp. Sci.* 14(6), 24
2. Bhide, A. D., Titus, S. K., Alone, B. Z., Dixit, R. C., Bhoyer, R. V., Mothghare, L. M., Gautam, S. S.and Patil, A. D.(1975) *Indian J. Environ. Hlth.* 17(3), 223, 315.
3. Carlson, C. W. and Menzies, J. D. (1971) *Bioscience* , 21, 315.
4. Chrometzka, P. (1968) *International Research Group on Refuse Disposal, Information* Bull No. 33
5. Diaz, L. F., Savage, G. M. and Trezek, G. J. (1979) *Comp. Sci. Land Util.* 20(1), 16
6. Golucke, C. G. (1975) *Comp. Sci.* 16(3), 6
7. Gotaas, H. B. (1956) *Composting, Sanitary disposal and reclamation of organic works* WHO monograph 31(World Health Organisation, Geneva)
8. Jain, R. C. (1980) *Compost Technology* Project Field Document No. 13 (FAO, Rome) 171
9. Jugsijinda, A.,Glacwiggram, S. and Takahashi, J. (1978) *Organic recycling in Asia, FAO Soils Bull.* No. 36 (FAO, Rome) 229
10. Moller, F. (1968) *International Research Group on Refuse Disposal* , Information Bull. No. 32
11. Neise, G. (1963) *International Research Group on Refuse Disposal,* Information Bull. No. 17
12. Poincelot, R. P. (1974) *Comp. Sci.* 15(3), 24
13. Prasad, B. (1981)*J. Indian Soil. Sci.* 29(1), 132
14 Stickelberger, D. (1975) *Organic materials as fertilizers,* FAO Soils Bull. No. 27 (FAO, Rome) 394
15. Stonehouse, B. (1981)*Biological Husbandry — a scientific approach to organic farming,* p.351 (Butterworths, London)

16. Titus, S. K., Olaniya, M. S. and Bhide, A. D. (1980)*Indian J. Environ. Hlth.* 22(3), 207
17. Vimal, O. P. (1973)*Wld. Sci. News* 10(1), 40
18 Vidyarthy, G. S. and Misra, R. V. (1978) In: Organic recycling in Asia, *FAO Soils Bull.* No. 36 (FAO, Rome), 417
19. Vimal, O. P. and Talishilkar, Recycling of Urban Solid Waste: The Perspective Ahead, *Jr. Sc. Indust. Res. CSIR* (1984)

Part 2
Treatment Technology

Because of the seriousness of the solid and liquid waste problems, there have been many new technological developments in recent years. Part Two of this volume considers some of these processes. Traditionally, liquid and solid wastes have too often been disposed of with little or no consideration of the environmental hazards. It is now recognized that these practices are not a viable alternative in a modern society.

Four different technologies are presented in this part. The chapter on wastewater treatment describes technologies used by service companies that are commercially treating hazardous and non-hazardous wastewaters. The PACT TM advanced wastewater treatment technology is elaborated. The Du Pont Company's Chamber Works plant in Deepwater, New Jersey provides a practical illustration of how this process handles the complex discharges from a large multiproduct chemical plant.

Microbial technologies are particularly attractive for treating solid industrial wastes, including sludges, because of their ability to process a material in-situ. Further, this technology eliminates the need for large capital expenditures for a processing plant, such as an incinerator. The technical feasibility of microbial destruction of industrial organic wastes has been verified by many practical applications.

The chapter on solid and liquid control techniques presents a systematic engineering approach to the design and construction of passive containment alternatives which can be used to mitigate migration of wastes. The three major control components are top seals (caps), barrier walls, and bottom seals (liners). Each of these is discussed in detail. By practical demonstration it has been shown that passive techniques, properly designed and constructed, can be used to mitigate waste migration.

The final paper considers solid waste land treatment systems. These treatment technologies are primarily centered on sludges and liquid residues. The system design is based primarily upon the limiting constituent or parameter concept for it provides a standard methodology, embodies information from many disciplines, provides a direct relationship between waste characteristics and land assimilative capacity, and the methodology is general so that any waste or site can be investigated.

Solid and Liquid Wastes: Management, Methods and Socioeconomic Considerations. Edited by S. K. Majumdar and E. Willard Miller. © 1984, The Pennsylvania Academy of Science.

Chapter Eight

Wastewater Treatment Technology

D. C. Hubinger

Product Manager—Wastewater Treatment
E. I. du Pont de Nemours & Company, Inc.
Chemicals and Pigments Department
Technical Service Laboratory—Chestnut Run
Wilmington, DE 19898

The subject of wastewater treatment as it applies to hazardous and toxic wastes is a broad one. Each industry has its own characteristic set of contaminants. The chemical industry in terms of quantity and diversity is undoubtedly the most difficult to manage with respect to meeting discharge requirements such as National Pollutant Discharge Elimination System (NPDES) permits. In general, a large manufacturing plant has its own wastewater treatment system that empties to a river or pretreatment facilities that discharge to a municipal treatment plant where organic destruction can take place. Deepwell disposal, which may involve some treatment, is also practiced extensively in some parts of the country such as, but not limited to, Texas and Louisiana.

This chapter will describe wastewater technologies used by service companies that are commercially treating hazardous and non-hazardous wastewater. Major driving forces are the Federal Resource Conservation and Recovery Act (RCRA) of 1976 implemented by the U. S. Environmental Protection Agency (EPA) and individual state laws equal to or more restrictive than RCRA.

Where organics must be removed, the activated sludge process has been the workhorse municipally and industrially. This process, commercialized[1] in Britain in 1914 and the United States in 1916, did not really find widespread application until many years later principally because of patent litigation.

Through the years, refinements in the activated sludge process have been commercialized. In the early 1970's the Du Pont Company made a technology breakthrough with respect to the activated sludge process. Powdered activated carbon was added to the aeration unit of a submerged-culture biological process to enhance bacterial destruction of biodegradable organics and removal of non-biodegradable organics. This process is called PACT™, short for powdered activated carbon treatment. This patented process was developed by

Du Pont for use at one of the Company's largest, most complex plants and has proven to be very reliable and cost effective. A major part of this chapter will be devoted to PACT™ technology.

Historical Perspective

Public awareness of hazardous and toxic wastes and their effects has been increasing over the last decades. Rachel Carson's *Silent Spring* pointed out pesticide hazards. Newspapers kept us aware of oil spills, fish kills, etc. Love Canal publicity pointed to human dangers.

With this background, Congress overrode a presidential veto and approved the Federal Water Pollution Control Act Amendments of 1972 (PL 92-500). The plan was to restore and maintain the chemical, physical, and biological integrity of the Nation's waters. In the decade from 1972 to 1982:[2]

- Congress has appropriated more than $37 billion for grants to aid construction of water pollution control facilities.
- States and local governments have contributed nearly $10 billion for capital improvements.
- Industries have spent an estimated $16 billion in improved treatment facilities at factories.

As of December 1982, 2,960 wastewater treatment projects using federal funds were under construction. More than 1,000 additional were awaiting construction. One interim goal, "fishable" and "swimmable" waters throughout the nation by 1983, was met in many places, but not nationally.

In the early 1970's scientists at the Du Pont Company's Chambers Works Plant, Deepwater, New Jersey, were investigating wastewater treatment processes that could best handle combined discharges from this very large, multiproduct chemical plant. Over 1,500 processes have historically manufactured about 1,000 finished products, mostly organic chemicals. Chemical processes at Chambers Works included manufacture of dyes, rubber chemicals, textile chemicals, intermediates for fibers, Freon® fluorocarbon products, inorganic acids such as sulfuric acid, and specialty chemicals such as Tyzor® organic titanates. Batch and continuous processes were and are in use. Large, highly variable wastewater flows of changing chemical composition were normal. Many of these compounds were chemically structured for stability and so were not biodegradable via conventional technology. It was with this background that Du Pont scientists developed PACT™. The first successful application of PACT™ was documented in 1971[3].

PACT™ is used in conjunction with and in describing systems and processes.
PACT® is a registered U. S. Service Mark, for Wastewater Treatment Service, obtained originally by Du Pont, but now the property of Zimpro, Inc., Division of Sterling Drug Company, Rothschild, Wisconsin.

PACT™Advanced Wastewater Treatment

PACT™ is an efficient and cost effective advanced wastewater treatment technology that uses conventional activated sludge treatment as its base. The process involves the controlled addition of powdered activated carbon to the aeration cell in a biological treatment unit. A synergism develops in the activated carbon/sludge matrix formed by carbon and bacteria in the reactor. The growth of bacteria on the carbon particle surfaces undoubtedly contributes to the enhanced activity observed. The carbon serves as a sorbent for toxic and inhibitory substances and provides a surface for concentration of oxygen, adsorbed organics, micro-organisms and microbial enzymes. This matrix acts to:

- Reduce many more substances in greater amount than biological treatment alone, and
- Adsorb many non-biodegradable pollutants that could not otherwise be removed.

The cumulative result of combining the powdered activated carbon and activated sludge is an effluent lower in dissolved organics and pollutants than with conventional biological treatment alone.

PACT™'s performance has been proven in both industrial and municipal applications. Consider the following:

- PACT™ consistently removes a high degree of BOD, COD and color.
- PACT™ protects the biological unit from upsets.
- PACT™ improves plant capacity, enabling easy handling of overloads.[4]
- PACT™ reduces odor and aerator foaming.
- Addition of powdered activated carbon to a municipal activated sludge process could give a system excellent nitrification capacity.[5]
- Existing biological plants can be converted to PACT™.

PACT™ removes a broad spectrum of pollutants including many EPA listed priority pollutants. Here is a partial list.

Chemical Compound	Influent (parts per billion)	Effluents (parts per billion)	Removal
Naphthalene	30	Nil	99%
Toluene	470	4.7	99%
Benzene	45	0.85	98%
Chlorobenzene	1,500	30	98%
2-Chlorophenol	13	0.6	95%
Carbon Tetrachloride	30	1.4	95%
Phenol	635	38	94%
Tetrachloroethylene	25	1.7	93%
Chloroform	108	20.5	81%
2,4-Dinitrotoluene	695	243	65%

Messrs. David C. Hutton and Frances L. Robertaccio, inventors of the patented PACT™ process, found that this process gave better removal than sequential activated sludge plus tertiary carbon treatment as shown in Table 1.

In the early 1970's Chambers Works case, 40 million gallons per day of

TABLE 1

Sequential Activated Sludge Plus Carbon vs.
Combined Activated Carbon-Activated Sludge (PACT™)

System	Activated Sludge	Activated Sludge plus Tertiary Carbon Treatment	PACT Process
Reaction Time, hr.	2.4	5.0	2.2
Carbon, ppm of feed	—	205	222
BOD_s in Feed, mg/l	167	167	169
BOD_s in Effluent, mg/l	28	—	15
COD in Feed, mg/l	247	247	247
COD in Effluent, mg/l	61	43	31
TOC in Feed, mg/l	93	93	93
TOC in Effluent, mg/l	19	12	10
% BOD Removal	82	—	91
% COD Removal	75	83	88
% TOC Removal	79	86	89

wastewater that was colored, contained a large amount but low concentration of organics having low biodegradability and a variety of priority pollutants needed to be treated. After much study, the Chambers Works' management decided only two advanced treatment techniques were feasible, PACT™ or granular carbon columns following an activated sludge process. Superior performance and substantially lower initial investment helped Chambers Works' management choose PACT™ technology. The results were that PACT™ removed 90% of the BOD, over 80% of the COD while reducing color by over 65%. The treatment consistently removed* over 90% of most of the more than 40 priority pollutants entering the plant. In addition, severe foaming was eliminated, sludge settling was improved, and upsets from shock loads were significantly reduced.

Figure 1 shows the Chambers Works Wastewater Treatment System.

Du Pont in 1982 sold its patent rights to Zimpro, Inc., The Environmental and Energy Group of Sterling Drug, Inc. The U. S. patent and Trademark Office had granted two patents to E. I. du Pont de Nemours & Co., Inc., on what is known as the PACT™ process: U.S. 3,904,518 and U.S. 4,069,148. Both are process patents covering the addition of powdered activated carbon to biological wastewater treatment processes. The first is general in type of waste treated and covers activated carbon dosages between the 50 and 1500 parts of carbon per million parts of wastewater. The second patent is limited to treating industrial wastewater and covers the addition of activated carbon at a rate sufficient to supply between about 4 and $100M^2$ of surface area of carbon per liter of wastewater.

There are three foreign patents on the PACT™ process:

---

*Based on EPA gas chromatography/mass spectroscopy screening protocol.

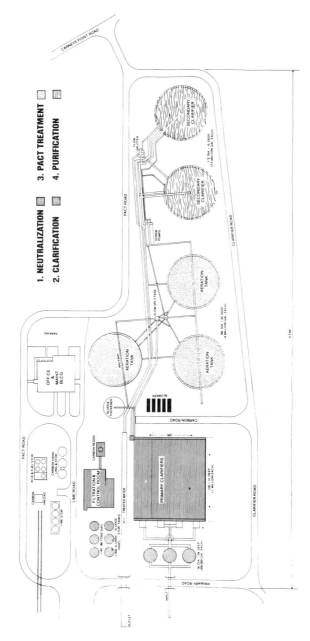

FIGURE 1. Chambers Works Wastewater Treatment System

Canada	954,042
France	71.01176
Great Britain	1,335,464

There are a number of commercially available powdered activated carbons such as Westvaco's Nuchar SA-15, Husky Industries' Husky BC-25, and ICI Americas' various Darco and Hydrodarco grades.

A number of PACT™ process plants have been constructed since the startup of the Chambers Works Wastewater Treatment Plant in November, 1976. PACT™ has been used to treat industrial, municipal, and industrial-municipal wastewater. Following is a list of such plants. All the municipal plants use Zimpro's wet air oxidation process for sludge destruction and carbon regeneration.

Company or Municipal Treatment Plant	Location	Startup	Remarks
Genl. Elect. Co.	Selkirk, NY	1979	1 mgd. Handling wastewater from resin manufacture. Demonstrated reliability in handling wide swings of organics. Odor reduction; foam reduction; improved sludge settling.
Vernon	Vernon, CT	1979	6.5 mgd*. Added powdered carbon 1979.
E. Burlington	Burlington, NC	1981	12 mgd*, nitrification benefit.
Liverpool County Regional	Medina, OH	1981	10 mgd, nitrification benefit.
Mt. Holly Sewage Auth.	Mt. Holly, NJ	1982	5 mgd*, nitrification benefit
Chatsworth	Chatsworth, GA	1982	0.8 mgd, domestic and carpet/organic removal.
Environmental Systems Corp.	Muskegon, MI	1983	1.5 mgd, ground water/organic removal.
Exxon	Clinton, NJ	1983	0.125 mgd, laboratory waste.
S. Burlington	Burlington, NC	1984	9.5 mgd*, nitrification benefit.
Bedford Hts.	Bedford Hts., OH	1984	3.5 mgd, nitrification benefit.
Tenneco	Chalmette, LA	1984	2.3 mgd, refinery/organic removal.
Kalamazoo	Kalamazoo, MI	1985	55 mgd, nitrification benefit. Wastewater includes components from pharmaceutical, paper, polymer manufacturing.
N. Olmsted	N. Olmsted, OH	1985	7 mgd, nitrification benefit.
El Paso	El Paso, TX	1985	10 mgd, aquifer recharge.
American Bottoms	Sauget, IL	1986	27 mgd, chemicals/organic removal reliability.

*Municipal wastewater and large industrial contribution from textile manufacturing and/or textile dyes.

The 40 million gallon per day PACT™ Wastewater Treatment Plant at Du Pont's Chambers Works, Deepwater, New Jersey, was built to handle its internal process needs. However, in the late 1970's, Du Pont withdrew from the dyes business. This took a major wastewater load from the Chambers Works treatment plant. About this time RCRA was being implemented and various external generators of wastewater sought to have their wastewater treated at Chambers Works. After a market survey confirmed that a potential business to treat wastewater from others existed, Du Pont started to market a bulk (tank truck) wastewater treatment service. Later on, the ability to handle tank cars of wastewater was added and a dedicated fleet of tank cars obtained.

Currently, Du Pont handles a large variety of outside wastewaters. A partial list is shown below:

acids; bases; elastomers; heavy metals; inorganics; organics; resins; chemical cleaning residuals; contaminated ground water; dyes wastes; etching, pickling, plating residues; latex residues; paint, pigment residues; pharmaceutical wastes; tank truck and car washings; textile treatment wastes.

Du Pont at Chambers Works has a number of advantages vs. competition:

1. PACT™ process is very effective and reliable. The Wastewater Treatment Plant has never shut down except for very short times for maintenance. At these times and other times such as heavy rains, an excess flow basin acts as a surge tank so that internal processes and outside wastewater receipts can continue.
2. Chambers Works has a 40 million gallon per day treatment plant with significant excess volumetric and treating capacity.
3. Chambers Works is a fully integrated plant that includes its own secure landfill having a double Hypalon™ liner where the leachate is sent to the head of the Wastewater Treatment Plant for processing. The landfill is for neutralization solids formed in primary treatment.
4. Chambers Works is on the Delaware River and access by barge and ship is available. Wastewater unloading facilities at an existing dock have been designed and are planned for 1984 startup and use in treating selected wastewaters.

Additional Wastewater Treatment Technologies

Once again, technologies in this chapter are limited to those that are well known and commercially available to treat wastewater shipped off-site by manufacturers and from other sources such as priority sites. It is beyond the scope of this chapter to specify which technology is the most initial and long term cost effective for various types of streams. One would have to know manufacturing and freight costs, investment dollars, etc., and very importantly, overall responsibility factors including: (1) company reputation and insurance, (2) the degree of destruction or nullification of toxics and (3) the long term safety of stored residues including leachate handling. As usual, decisions are made in the ever changing market place encompassing generators, transporters, and disposal facilities and are influenced by federal, state and local laws, interpretations and enforcement activities.

Broadly speaking, technologies used by disposal companies can be thought of as chemical, physical, biological or a combination of some or all of these. Chemical steps include neutralization of acids and bases. Organic acids in wastewater may be unreacted raw materials or develop as by-products. Many spent inorganic acids such as sulfuric, hydrochloric and nitric come from metal finishing operations including major steel pickling operations. Bases may come from synthesis operations, reactor washings, etc. Those wastewaters which contain ammonia may be a particular problem because of effluent discharge limits. Also caustic and acidic cleaning solutions are used industrially and the spent solutions must be responsibly treated.

Oxidation is another typical chemical operation. Cyanides, for example, can be detoxified by oxidation with hydrogen perodixde such as Du Pont's Tysul® hydrogen peroxide. Organic sulfur compounds can be similary treated.

Wet oxidation refers to the aqueous phase oxidation of dissolved or suspended organic substances at elevated temperatures or pressures. Wet air oxidation is a process employed for those wastes too dilute to economically incinerate yet too toxic to biotreat. Cyanides, sulfides, organic sulfur compounds and most classes of organic can be wet air oxidized. As a general rule, waste in a range from 5 g/l to 300 g/l COD are good candidates for wet oxidation. Wet oxidation is also applied to those wastes which may be reactive when treated. Powdered carbon is also reactivated by the process.

Incineration, in effect, is oxidation at high temperatures. Normally incineration and wastewater treatment are at the opposite ends of the economic spectrum based primarily on organic content. Aqueous wastes containing low concentrations of soluble organics are primarily biologically treated whereas waste organics are typically burned if not economically recovered or blended and reused. There is an economic gray area where either technique has the same total disposal, freight, and handling costs. The water content and the heating value (BTU/lb.) of the organic portion play an important part in the economic evaluation of the stream because of their influence on supplemental fuel requirements. Of course, there are other factors for wastewater treatment and incineration that may preclude use of either technology. For example, bacterial toxicity may preclude use of wastewater treatment. Incineration may be inappropriate because (1) materials may be reactive or explosive (2) fluorine containing compounds could attack refractory brick (3) sulfur compounds may not be adequately removed by scrubber(s) handling incinerator off-gas (4) stack limits such as opacity and particulates may be exceeded, for example, due to heavy metal content.

Reduction, the opposite of oxidation, is another chemical step. Reduction electrochemically is the process of gaining electrons. For example, chromium (which can exist in an oxidation state of $+2$, $+3$, and $+6$) from plating and tanning operations can be reduced from a more dangerous $+6$ to a lower, safer oxidation level such as $+2$ before discharge to a stream or river.

Metal finishing operations may have toxic anions and heavy metals in dilute solution where ion exchange operations can be applied. Here, the undesired pollutant is picked up by the solid form ion exchange resin which can be subsequently regenerated for economic purposes.

One of the simplest physical steps is that of clarification (term used when solids are present in small proportion). Clarifier design is primarily dependent on the settling rate of the sludge. Solids of higher density settle to the bottom of large rectangular or circular clarifier tanks allowing the clear liquid to overflow a weir system. Polymer or ferric chloride can be added to aqueous wastes entering primary and secondary clarifiers to assist in solids agglomeration and settling. Jar tests are useful for screening polymers. Polymer effectiveness is usually site specific requiring tests and economic evaluation.

Filtration is the separation of solids from a liquid and is effected by passing the liquid through a porous medium. The solids retained on the surface of the medium form a cake. Solids pumped from the bottom of a primary clarifier after filtration usually are transported to a landfill. The cake will contain up to 50% solids and the remaining percentage will be water.

Some waste streams can be solidifed or stabilized. Various agents can be added to fix the material. One such is pozzolanic material which contains aluminous and siliceous compounds. Fly ash, a waste product, is the most common artifical form. Calcium from lime or lime based additives reacts with the alumino silicates to form hydrates of calcium aluminates, calcium silicates, and calcium aluminosilicates. These products surround the waste particles and in a month or less develop into a nearly impermeable mass.

As stated earlier, technologies can be grouped in terms of chemical, physical, and biological. The biological system used by Du Pont at Chambers Works was covered in some detail earlier in this chapter. The system at Chambers Works was started up with bacterial "seed" from both a municipal and an industrial wastewater treatment plant. Starting at low flow rates and building rates as bacterial growth occurred allowed the biomass to grow properly without being overwhelmed in quantity (food supply) and quality (toxic compounds). Even today, protection of the biomass is critical to continued operation so as to meet permit limits. At Chambers Works a backup electrical supply system is in place so blowers will continue operating to feed oxygen to the bacteria and suspend the carbon-biomass matrix. The treatment plant vs. operating facilities generally gets first call on electrical power if a shortage develops. Carbon in excess of normal requirements can be added to the aerators (biological reactors) if toxics inadvertently enter the system, a high organic load occurs, or during winter when bacterial activity is at a lower level due to the well-known slowing effect of reduced temperature. With these and other equipment and operating procedures in place, it is not surprising the Du Pont's Wastewater Treatment Plant, utilizing the PACT™ process has never really shut down. When routine maintenance is required, say for eight to ten hours, the excess flow basin at the head of the process is allowed to partially fill. Later, the flow rate to the treatment plant is increased to bring the excess flow basin back to its normal low level.

Another form of biological treatment is land farming which occurs not only in Gulf Coast and West Coast areas, but also on the East Coast and in the Midwest. Basically, aqueous or oily waste material is mixed in or applied on surface soil. The concept is to get the waste in contact with air (oxygen), soil bacteria, and fungi to break down the organics present in the waste. It goes without saying that the soil subsurface must effectively prevent wastes from containing groundwater which could be a source of drinking water. Permeability of the soil must be adequately considered and leachate problems well handled.

Another factor in landfarming is the quality of the waste. Unsuitable wastes include non-biodegradable organics and wastes with high metal content could enter groundwater or get into vegetation consumed by farm animals. Also, there is the consideration of odor of materials directly applied and odors developed during bacterial destruction. Nevertheless, land treating or land farming under strict controls has proved successful in certain applications for many years.

Commercial Treatment Facilities

The purpose of this section is to describe various types of commercial treating facilities generally available. A generator contemplating off-site shipment would be well advised to visit a disposer's facilities and responsible personnel (marketing and production) before making a decision because the generator per RCRA is responsible from "cradle to grave". This is the law for hazardous wastes as defined by RCRA, but so-called "non-hazardous" wastes should also be treated as responsibly by commercial treatment companies. There may be no cost differential to do so.

The most cost effective disposal of wastewater for many industrial concerns because of low transportation (pipeline, etc.) and treatment costs is to utilize a municipal treatment system. Consider, for example, the 300 million gallon per day Passaic Valley Sewerage System, fourth largest in the country. With a secondary unit installed and started up in the fall of 1981[6], this system uses oxygen activated sludge to achieve over 90% removal of high strength, combined domestic and industrial wastes generated in a heavily industrialized Passaic Valley drainage area.

Some disposal companies take advantage of such municipal/industrial wastewater treatment plants, particularly for organic processing. Typical companies, in effect, do partial treating on a batch basis and arrange with others to do organic disposal and sludge acceptance in a landfill. Such disposal companies have tanks or pits to accept various types of wastes such as solvents, oils, oil-water mixtures, acids, bases, cyanides, chromes, etc. After materials are clarified, they are put into mixing or blending tanks. Various physical and chemical treatments as described earlier are used to either recover materials such as oils and solvents or to detoxify the wastes. A typical process involves neutralization using waste acids and waste caustics while charging each customer for bringing his solution to neutral pH of 7. Detoxification may cause; (1) sludges requiring disposal and (2) aqueous effluents suitable for discharge to municipal wastewater treatment facilities. The pretreated aqueous wastes usually do not pass to a municipal system without batch hold tank sampling and approval or automatic monitoring and sampling.

Another type of facility is one that specializes in a certain type of wastes such as from surface finishing operations ranging from small platers to large steel mills that produce large quantities of acidic pickle liquors. Companies in the electronic components industry are included too because of acid and alkaline wastes usually containing heavy metals from processes such as cleaning, stripping, etching, and plating. Typical processes to treat these wastes include neutralization, filtration, chemical detoxification, particularly for cyanides and chromium, stabliization of solids or dewatered solids sent to a landfill, and liquid effluent sent to a municipal sewer. Some companies have obtained from the EPA a partial or total delisting of their sludge as "hazardous" on the basis of tests.

There are some companies that do chemical pretreatment as described earlier. Some then utilize membrane filters under a few atmospheres of pressure to separate solids from the liquids. Other companies, depending on the waste, instead may after primary treatment pump the waste through adsorptive granular activated carbon in a column. Spent carbon must then be landfilled or regenerated.

It should be noted that some companies have mobile equipment which can be set up on site. Each system is designed for the need. Mobile systems include equipment such as clarifiers, filters, centrifuges, stripping units, ion exchange units, carbon adsorption columns, and chemical treatment systems. These mobile systems may be particularly good for one-time cleanups such as abandoned drums or if the site is remote and difficult for bulk transportation.

Deepwell injection of various liquid wastes is practiced to a large extent in the Gulf Coast area, but not exclusively there. Waste types handled include oils, alkalies, inorganic brines, and chlorinated hydrocarbons. If necessary, wastes are neutralized, chemically treated, and filtered to remove undesirable solid contaminants. Materials are pumped from storage into a suitable geological formation through concentric casings which protect the underground water. The waste flows down about a mile and then horizontally through porous strata which lie below thick, fracture resistant strata such as shale, clay and limestone.

SUMMARY

Today's generator of hazardous and non-hazardous wastewater shipped off site has a number of technologies and companies from which to choose. The person or persons responsible must not only consider costs for disposal and freight, but also any long-term costs which may relate directly to the reputation and processes used by the disposal company. Initial on-site inspection of a treatment company by the generator is highly recommended. Understanding how effectively and reliably a generator's waste is treated is essential. Visitation at least once a year after that would be well advised.

REFERENCES

Journal
1. Alleman, J.E. and Prakasam, T.B.S.. May 1983, Reflections on Seven Decades of Activated Sludge History. *Journal Water Pollution Control Federation,* 55:436-443.
2. Jones, Charles H. May 1983, A Decade of Progress, America's Quest for Clean Water, *Journal Water Pollution Control Federation.*

Bulletin
3. "Du Pont PACT™ Process" Bulletin, E. I. du Pont de Nemours & Co., Inc., 1971.

Magazine
4. Adams, A. D. "Powdered Carbon: Is It Really That Good", *Water and Wastes Eng.* 11(3): B-8-B-10, March 1974.
5. Burant, W. Jr. and Vollstadt, T.J. "Full Scale Wastewater Treatment with Powdered Activated Carbon", *Water and Sewage Works,* pp. 42-45, 66, Nov. 1973.
6. New Jersey *Effluents,* November, 1981.

Solid and Liquid Wastes: Management, Methods and Socioeconomic Considerations. Edited by
S. K. Majumdar and E. Willard Miller. © 1984, The Pennsylvania Academy of Science.

Chapter Nine

Microbial Destruction of Industrial Solid Waste

R. Donald Bartusiak, M.S.

Homer Research Laboratory
Bethlehem Steel Corporation
Bethlehem, PA 18016

Process technology for the microbial destruction of wastes has its origins
in the activated sludge process for sewage treatment that began to be used in
Western civilizations around the turn of the twentieth century. As industry
emerged, the chemical structure of the wastes to be treated became increasingly less similar to those found in nature. Nevertheless, microbial process
technology has continued to be effective due to the remarkable ability of life
to adapt to environmental changes. The biochemical "machinery" of the
microbial world has, for the most part, been able to assimilate whatever organic
compounds humanity can concoct, albeit sometimes only slowly and at low
concentrations. Also, microorganisms have demonstrated the capability of
treating inorganic matter either by direct or indirect metabolic processes, or
by comparatively simple absorptive mechanisms. In engineering a better environment for ourselves, we have learned to use the capabilities of various
microorganisms to metabolize many different compounds in domestic and
industrial wastes that are either toxic, carcinogenic, or generally undesirable
when released. Recent advances in molecular biology have greatly increased
our understanding of the fine details of waste treatment biochemistry. With
the advent of genetic engineering, we now have the capability to assist nature
with the process of adaptation by selectively changing small facets of the
biochemical machinery of microorganisms to improve their effectiveness in
treating pollutants.

Our objective in this chapter is to provide a brief overview of the basic science
and some novel technological aspects of microbial processes for industrial solid
waste treatment. We will emphasize degradative processes for organic wastes

characteristic of steel mills, such as lubricating oils and cokemaking residues. Only briefly will we touch upon established technology for wastewater treatment since this topic is extensively documented in the literature.[1,2] The promise of genetic engineering for improving microbial waste treatment technologies will be discussed from a chemical engineering perspective.

BIOCHEMICAL PRINCIPLES

In microbial waste treatment processes, the individual microbial cells can be considered as small chemical reactors capable of converting a waste compound to a less harmful substance. These conversions result in a change in the chemical energy, or free energy ($\triangle$G), in the system of reactants and products. To proceed from reactants to products, the products must have a lower free energy content than the reactants, i.e. a negative $\triangle$G. In microbial treatment processes, wastes are gradually broken down through a series of chemical reactions. The free energy that is released is used in part by the microorganisms for their metabolic needs.

However, a negative free energy change does not guarantee that a reaction will take place at an appreciable rate. If the reaction is slow to proceed, it is because the reactants must first overcome an activation energy barrier before they are converted to products. Increasing the temperature of the reactants is one way to overcome the activation energy barrier. A second way is to lower the activation energy barrier through the use of chemical agents, generally called catalysts, which increase the probability of reactant interaction, thereby speeding up the rate. It is by this second process that microbial waste treatment processes operate. The biochemical machinery within the microorganisms produces catalysts called enzymes which speed up the waste decomposition reactions.

Enzymes are proteins, but some also contain non-protein components. The protein portion of an enzyme is called the apoenzyme and is considered to be responsible for the chemical specificity of catalytic action. The non-protein portion of an enzyme is called the co-factor. Co-factors may be either metal ions or relatively complex molecules. The co-factors are considered to be responsible for effecting the chemical reaction, e.g. oxidation, hydrolysis, transfer, etc. associated with the enzyme.[3]

A simple conceptual model of the enzyme's role in chemical reactions is given below:

$$E + S \rightleftharpoons ES \rightarrow E + P$$

where E is the enzyme, S is the reactant or substrate, ES is a short-lived enzyme-substrate complex, and P is the product. This model illustrates the essential feature of enzymes, and catalysts in general, namely that they participate in but are unchanged by the chemical reaction. In reality, enzymes are slowly deactivated with continued use. Moreover, because they are proteins with delicate, catalytically important three dimensional structures, enzymes are extremely sensitive to heat, extremes of pH, radiation and mechanical stress.

There are two possibilities open to living cells which need to metabolize large insoluble substrates found outside of the cell. The cells must either force the large molecules through the cell membrane so that chemical processes can be performed inside, or they must somehow make their enzymes available outside of the cell. The second mechanism is more common. Enzymes that are available outside of the cell may be fully liberated molecules or they may be bound to the cell membrane. Both liberated and cell-bound external enzymes, exoenzymes, are common. Enzymes responsible for the initial stages of metabolizing a large or insoluble molecule are usually bound to the cell membrane, for example, paraffin oxidases.[4]

The ability of a microorganism to decompose a particular waste compound depends upon its ability to synthesize the necessary enzymes. The enzymes are assembled within microbial cells by the linking of amino acids according to a sequence determined by a DNA molecule. Improving the ability of an organism to effect a desired reaction is a matter of modifying, or adding to, the genetic programming in the cell's DNA.

Artificial intervention into the cell's genetic programming can be by induced mutation or by recombinant DNA techniques. Mutations can be induced by exposure to electromagnetic radiation or a variety of chemical agents, such as phosgene. Following such treatment, the culture is examined by biochemical tests to identify the mutants that exhibit the desired characteristic, for example the increased rate of decomposition of a specific substrate.

While mutation alters the genes within an organism, recombinant DNA techniques involve the transfer of entire genes or parts of genes from one organism to another. It is important to note that microbial DNA carrying the genetic coding can exist either as (1) chromosomal matter which is specific to an individual microbial strain, or (2) as plasmids—extrachromosomal DNA molecules which can replicate independently and, in general, can be transmitted into or out of the cells. A chromosomal gene can be transferred from a donor to a host by process of homologous recombination. The donor gene can replace a gene from the recipient by addition at a homologous (corresponding base sequence) region in the recipient chromosome. Plasmids, because of their ability to move across cell membranes, can be used to carry completely new DNA to the cellular machinery.[5]

AEROBIC HYDROCARBON METABOLISM: REACTION PATHWAYS AND ENZYMOLOGY

Aerobic metabolism of hydrocarbons is an oxidative process. Fatty acids are formed from which smaller organic molecules are taken to be used in the basic cellular metabolic processes, e.g. the tricarboxylic acid (Krebs) cycle. Generally, fatty acids may undergo alpha, beta or omega oxidation pathways. In microbial metabolism, beta oxidation is believed to be the most common. Simply, beta oxidation removes a two-carbon unit acetyl CoA ($CH_3COSCoA$) which is a key metabolic intermediate. The remaining hydrocarbon is oxidized at the terminal position and the oxidation continues.

Degradation of Alkanes

The most common degradative process for aliphatic hydrocarbon chains is the terminal oxidation to the primary alcohol, conversion to the aldehyde, followed by beta oxidation, as shown below:[6]

$$R-CH_2-CH_2-CH_3 \xrightarrow{O_2} R-CH_2-CH_2-CH_2OH \xrightarrow{2H^+} R-CH_2-CH_2\overset{H}{\underset{}{C}}=O$$

$$R-CH_2-CH_2\overset{}{\underset{OH}{C}}=O \xrightarrow{\;H_2O\;,\;2H^+\;}$$

$$RCOOH + CH_3COOH \xleftarrow{\text{beta oxidation}} R-CH_2-CH_2-\overset{}{\underset{OH}{C}}=O$$

Several enzymes are needed in the above process. Oxidation of the alkane to the primary alcohol is accelerated by enzymes variously called hydroxylases, mixed function oxidases or oxygenases. For this reaction, the biological reducing agent NADH is required.[7] Conversion of the alcohol to the fatty acid requires the action of an alcohol and an aldehyde dehydrogenase. Finally, beta oxidation can require several enzymes depending on the length of the hydrocarbon chain. In the simplest case, the conversion of acetate to acetyl CoA is catalysed by acetyl-CoA synthetase.

Degradation of Aromatic Hydrocarbons

Decomposition pathways of aromatic hydrocarbons all require the presence of functional groups on the benzene ring to reduce its resonance stability. Specifically, the benzene nucleus must carry at least two hydroxyl groups, placed ortho or para to each other, before the ring can be opened by a dioxygenase.[8] This requirement makes the compound catechol, (catechol structure), and its homologues central in the degradative pathways for various aromatics. Before discussing the aromatic ring cleaving reactions, we will first examine the processes by which common aromatics are converted to catechol.

Benzene

Microbial conversion, most notably by pseudomonads, of benzene to catechol is known to proceed by way of 3,5 cyclohexadiene-cis-1,2 diol as shown below:

In *Pseudomonas putida*, there are three enzymes involved in the catalysis of the benzene to diol reaction. A single cis-benzenediol dehydrogenase catalyzes catechol production.[9]

Phenol

Metabolic pathways for phenol to catechol are known for *Bacterium NCIB* and *Pseudomonas aeruginosa*. The one step conversion to the diol is catalyzed by phenol 2-monooxygenase.[6]

Napthalene

A composite pathway for the degradation of napthalene to catechol by pseudomonads and other organisms is given below:

It is interesting to note that the ring cleavage of 1,2 dihydroxy-napthalene and of catechol are very similar, as we will see in the next section. However, separate enzymes for these two reactions have been identified.[9]

Catechol

The degradation of catechol can proceed by either of two paths depending upon the location of ring fission, as shown:

Ortho fission is catalysed by catechol 1,2 dioxygenase; meta fission is catalysed by catechol 2,3 dioxygenase. A comparison of the further degradative pathways is given in Figure 1.

The aliphatic and aromatic degradative pathways outlined above are only

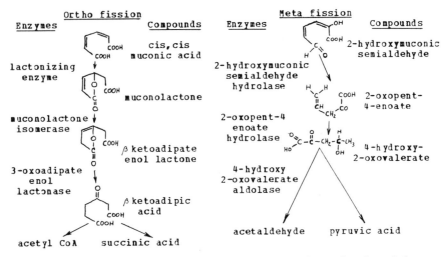

FIGURE 1. A comparison of degradative pathways for ortho and meta cleaved catechol.

a few of those possibly required for steel mill solid waste treatment. Other pathways are documented in the references.[6,7,8,9]

Degradation of Halogenated Organics

As a final note, we should call attention to some current research on the degradative mechanisms of halogenated industrial wastes, particularly polychlorinated biphenyls, PCBs.[10,11] Successful development of microbial treatment of PCBs and other halogenated compounds that are widely spread in the environment could provide a practical alternative to digging up large amounts of contaminated soil for treatment or secure disposal.

TECHNOLOGICAL ASPECTS—SOLID WASTE

Microbial technologies are particularly attractive for treating solid industrial wastes, including sludges, because of their ability to process a material in-situ, for example the remediation of a large mass of soil contaminated by a spill. Also, microbial technologies for solid waste treatment may eliminate the need for large capital expenditures for a processing plant such as an incinerator. The petroleum industry, for example, has used microbial processes for almost 30 years to treat oily sludges by controlled application to topsoil and the rich microbial populations living there.[12] The U.S. EPA has approved this practice, called land treatment, for a variety of wastes subject to certain criteria and operating standards governing unwanted emissions.[13,14]

Bethlehem Steel is currently investigating the use of microbial processes to destroy organic contaminants in steelmaking by-products so that these materials can be recycled to the steelmaking process.[15] Both material resources and land formerly used for fill are conserved. Water treatment sludge consisting of finely divided iron oxide, water and wasted lubricating oils is being treated by a process called composting. Sludge is inoculated with non-recombinant mutant organisms and formed into windrow piles. Nutrients are provided and the windrows are agitated periodically to provide oxygen. The process is similar to the municipal waste treatment process of the same name, with some important distinctions. The oil in the steel plant sludge is less readily biodegradable than the organic compounds in the municipal sludge. Consequently, the process dynamics are quite different.

In steel mill sludge composting, both soil microbes and mixed cultures of induced mutants have been found capable of oil degradation. Wild-type microorganisms in the sludge itself may also be contributing. Oxygen mass transfer, moisture concentration and nitrogen/phosphorus nutrient concentration must be controlled to optimize the rate of deoiling. Of course, temperature affects the rate of oil degradation, but temperature is not controlled in this open process. Contrary to municipal sludge practice, oxygen mass transfer through the pile is not a rate-limiting factor if the pile is not water-saturated. Oxygen mass transfer within the oil-containing agglomerates of sludge, however, is rate-limiting. Consequently, the objective of agitation in the industrial sludge process is to reduce the agglomerate size preferably to diameters on the order of 1 mm for which the chemical effectiveness factor approaches 1. Excessive aeration can be detrimental to the process if the material becomes too dry. A moisture content in the range of 10% to 15% is recommended. Ammonium sulfate and a phosphate fertilizer in the ratio of 10N:1P are being used for inorganic nutrients. Free ammonia concentrations on the order of 100 mg/L in the intraagglomerate pore water are desired. Excessive amounts of ammonia severely inhibit the microorganisms.

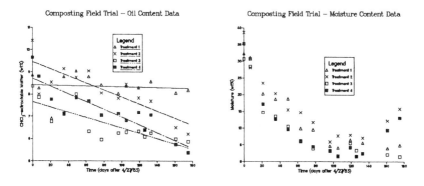

FIGURE 2. Oil and moisture transients in steel mill wastewater sludge composting experiment.

The results of the first six months of pilot testing are shown in Figure 2. Improved performance in future trials should come by more careful control of the moisture content and other engineering factors, and by improved microbiology to maintain an optimum population of microorganisms with rapid degradative capabilities.[15]

In another application of microbial destruction of industrial wastes, we are also investigating the possibility of treating cokemaking residuals, principally tars, to modify and detoxify them so they may be used for some valuable purpose or so that they will present no threat of adverse environmental impact.

The biological treatment of phenol and ammonia in coke plant waste waters is a well established process for the microbial destruction of industrial wastes. Kostenbader and Flecksteiner at Bethlehem Steel were among the leaders in the development of this technology.[16]

IMPACT OF GENETIC ENGINEERING

Considering the steel mill composting process in particular, the objective for process optimization is very similar to that of processes for making single cell protein from hydrocarbons or for cleaning up oil spills by microbial destruction. Coincidentally, one of the most well known triumphs of recombinant DNA technology was aimed at these last two processes. In 1975, Anandra Chakrabarty, then at General Electric, received the first U.S. patent on a life form for a recombinant mutant of *Pseudomonas putida* to which donor plasmids had been added to increase the organism's capability to metabolize hydrocarbons. Chakrabarty and others knew that the genes coding for a variety of hydrocarbon degradative enzyme sequences were located on plasmids commonly found in pseudomonads.

TABLE 1

Degradative Plasmids Found in Pseudomonads (partial list)

Plasmid	Apparent Primary Substrate	Number of Enzyme Steps
CAM	camphor	15-20
NAH	napthalene	>11
OCT	n-octane	3
SAL	salicylate	8
TOL	toluene/xylene	>11
XYL	xylene	>11

A list of these degradative plasmids and their apparent primary substrates is given in Table 1.[17] Accordingly, Chakrabarty produced a recombinant mutant by the following sequence of operations:[5]

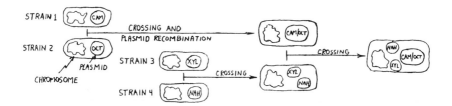

The final organism carried genetic programming to enable it to metabolize a wide spectrum of hydrocarbons. However, the patent holder, GE, has never allowed the commerical use of the organism because of concerns that use in an open environment might result in the spontaneous transfer of plasmids to a low activity pathogen which could create a high activity pathogen.[18]

The story of the barriers to commercialization of the first patented organism illustrates the promise and the concerns of the use of recombinant mutants in open environments. In principle, microbial technologies for treating industrial solid wastes could benefit through improvements to the enzyme producing capabilities of the microorganisms. However, it is not known whether recombinant organisms will ever be used in open environments. Test cases, such as the recent experiment in California with frost-protecting organisms, will set precedents.[19]

DISCUSSION AND CONCLUSIONS

The technical feasibility of microbial destruction of industrial organic wastes has been verified through our experiments and the work of others, most notably in the petroleum industry. Demonstrations of practicality in wide-ranging applications are still needed. However, there are several positive indicators. First, biological wastewater treatment has proven to be highly effective in a host of industries, e.g. cokemaking, petroleum refining and organic chemicals manufacturing. Second, compared to alternative technologies, microbial destruction processes offer the advantages of in-situ or remedial treatment of wastes, and of significantly lower capital cost requirements for plant and equipment. Third, the long experience of the petroleum industry with land treatment supports the economic attractiveness of microbial processes over alternative technologies, e.g. incineration. Extensions of microbial destruction technology to other wastes, such as chlorinated biphenyls (PCBs), pesticide residues and dioxin (TCDD), may be possible with additional research.[20]

Opportunities for process improvement exist both in the engineering and the microbiological aspects of the technology. The engineering improvements will come quickly; the microbiological improvements, e.g. identifying the best strains for a particular waste, are likely to require longer R&D periods. The impact of genetic engineering could be substantial. However, legal precedents for the use of recombinant mutant organisms in open environments are not in place and extensive political, emotional and technical questions regarding safety must be answered. The probability of using recombinant organisms in the near term for the applications discussed in this paper is low. Nevertheless, experts, such as Alan S. Michaels, consider the area of waste treatment in general as one of the fastest developing niches for the application of genetic engineering.[21]

ACKNOWLEDGEMENTS

The contributions of Jean T. Baschke, Deborah A. James-Yaney, Dr. Nicholas B. Franco and Thomas H. Weidner to the research on microbial treatment of solid waste at Bethlehem Steel are gratefully acknowledged.

REFERENCES

1. Grady, C. P. L. and Lim, H.C. 1980. *Biological Wastewater Treatment: Theory and Applications,* Marcel Dekker, New York.
2. DeRenzo, D. J. 1980. *Biodegradation Techniques for Industrial Organic Wastes*, Noyes Data, Park Ridge, NJ.
3. Gaden, E. L. and Kirwan, D. J. 1982. *Introduction to Biotechnology and Bioprocesses*, AIChE Today Series, American Institute of Chemical Engineers, New York, 2-24.
4. Pollock, M. R. 1962. Exoenzymes. in: *The Bacteria,* I. C. Gunsalus and R. Y. Stanier, Eds., Academic Press, New York, 4: 121-170.
5. Hopwood, D. A. 1981. The Genetic Programming of Industrial Microorganisms. *Scientific American*, 245(3): 91-102.
6. Doell, H. W. 1975. *Bacterial Metabolism*, Academic Press, New York, 490-518.
7. Dagley, S. 1975. A Biochemical Approach to Some Problems of Environmental Pollution. in: *Essays in Biochemistry*, P. N. Campbell and W. N. Aldridge, Eds., Academic Press, London, 11: 81-138.
8. Dagley, S. 1975. Microbial Degradation of Organic Compounds in the Biosphere. *American Scientist*, 63: 681.
9. Ribbons, D. W. and Eaton, R. W. 1982. Chemical Transformations of Aromatic Hydrocarbons that Support the Growth of Microorganisms. in: *Biodegradation and Detoxification of Environmental Pollutants*, A. M. Chakrabarty, Ed., CRC Press, Boca Raton, FL., 60-84.
10. Furukawa, K. 1982. Microbial Degradation of Polychlorinated Biphenyls (PCBs). *ibid.*, 33-58.

11. Chakrabarty, A. M. 1982. Genetic Mechanisms in the Dissimilation of Chlorinated Compounds. *ibid.*, 127-140.
12. Huddleston, R. L. 1979. Solid-Waste Disposal: Landfarming. *Chemical Engineering*, 86(4): 119-124.
13. 40 CFR 264, *Federal Register*, July 26, 1982, 47(143): 32361-32365.
14. 40 CFR 265, *Federal Register*, May 19, 1980, 45(98): 33247-33249.
15. Bartusiak, R. D. and Butz, J. A. 1983. Applications of Biotechnology for Treating Hydrocarbons in Steel Plant Solid Waste. in: *Proceedings of the EPA/AISI Symposium on Iron and Steel Pollution Abatement Technology for 1983*, J. S. Ruppersberger, Ed., U.S. EPA, Research Triangle Park, NC.
16. Kostenbader, P. D. and Flecksteiner, J. W. 1968. Biological Oxidation of Coke Plant Weak Ammonia Liquor. *Blast Furnace and Steel Plant*, June 1968.
17. Broda, P. 1979. *Plasmids*, W. H. Freeman, Oxford, U. K., 129-133.
18. Anon. 1976. Genetic Engineering: Two-edged Sword. *Chemical Week,* 118(19): 65-68.
19. Anon. 1983. Groups Seek to Halt Release of Bacteria with Altered Genes. *New York Times*, Sept. 15, 1983.
20. Chakrabarty, A. M. 1983. personal communication, Aug. 16, 1983.
21. Michaels, A. S. 1983. The Impact of Genetic Engineering on the Food, Drug, and Chemical Industry: Current Status and Future Prospects. Paper presented at the AIChE 1983 Summer National Meeting, Denver, CO. Aug 28-31, 1983.

Solid and Liquid Wastes: Management, Methods and Socioeconomic Considerations. Edited by
S. K. Majumdar and E. Willard Miller. © 1984, The Pennsylvania Academy of Science.

Chapter Ten

Solid And Liquid Waste Control Techniques

Hsai-Yang Fang[1], Jeffrey C. Evans[2] and Irwin J. Kugelman[3]

[1]Professor and Director, Geotechnical Engineering Division
Department of Civil Engineering, Lehigh University, PA 18015
[2]Senior Project Engineer, Woodward-Clyde Consultants
Plymouth Meeting, PA
[3]Professor of Civil Engineering, and
Director, Center for Marine and Environmental Studies
Lehigh University, PA 18015

Interdisciplinary knowledge from the geotechnical, hydrogeological and environmental fields is required to analyze, design and construct waste containment systems. Conventional passive groundwater and surface water barriers are frequently adapted as barriers for waste containment; however, special considerations are required in the design and construction of these barriers when used in waste containment systems. In this chapter, we will present a systematic engineering approach to the design and construction of passive containment alternatives which can be used to mitigate contaminant migration. Active waste control technologies, such as pyrolysis and thermal reduction are discussed elsewhere in this volume.

The three major passive waste control components are top seals (caps), barrier walls, and bottom seals (liners). Top and bottom seals include covers and liners of native clay, bentonite clay and polymeric membranes. Barrier walls include soil-bentonite slurry trench walls, cement-bentonite slurry trench walls, and vibrating beam cutoff walls. Design aspects and construction practices of waste containment are discussed below. The functions of each of these components are examined and the design process is then detailed with emphasis on the difference between conventional applications versus waste control applications.

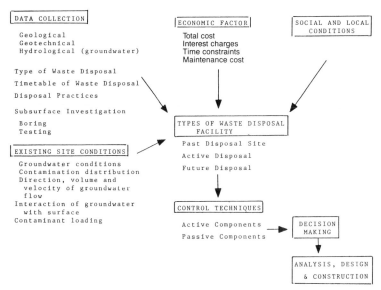

FIGURE 1. Systematic Approach for Planning and Design of Solid and Liquid Waste Control Systems.

Systematic Approach to Solid and Liquid Waste Control

Waste disposal systems can be grouped into three categories, as shown in Table 1. A systematic approach to the control of wastes in the categories given in Table 1 requires the engineer to fully assess both the site and subsurface conditions and evaluate the applicability of containment alternatives. The basic requirements for controlling the wastes are shown in Fig. 1. There are four steps involved, with each step closely related.

Step 1—Review existing information including historical site data, geological and groundwater data. For past disposal sites, it is necessary to obtain as much information as possible on the types of waste disposed, the timetable of waste disposal, and the previous disposal practices, i.e., drum, solid waste, lagoons.

Aerial photos can be extremely useful. Information regarding the subsurface conditions can be obtained from previous records and borings, from the site construction history, and from geological information. Details regarding subsurface investigations are discussed in Lowe and Zaccheo (1975).

Step 2—Assess in detail the existing site conditions including geological conditions, groundwater conditions and contamination distribution. Field investigation will probably be required at this step. The use of geophysical instruments prior to test boring or monitoring well installations can provide valuable insight into the subsurface conditions and will typically result in a more complete and cost effective boring and sampling program (Kolmer, 1981).

Step 3—Quantify site conditions including the direction, volume and velocity

TABLE 1

Waste Disposal Facility Classification

Facility Category	Facility Description	Generic Names	Control Wastes Disposed
I	Past Disposal Site	abandoned inactive retired midnight dump uncontrolled site orphaned	Little to none
II	Active Disposal	secure landfill sanitary landfill	Some
III	Future Disposal	waste treatment complex recycling facility	Well-controlled

of groundwater flow, the interaction of groundwater with surface water, the distribution of contamination in the groundwater system and the contaminant loading. The degree of sophistication of this quantification phase may vary from a simple conceptual model to a complex computer model.

Step 4—Develop the containment/treatment program: the portion of the program where the application of environmental geotechnology receives the major emphasis.

In many cases, subsurface investigations are primarily geohydrologic investigations. Geotechnical properties of soil may not normally be part of the routine investigation. It is desirable to have geotechnical engineering input during the site investigation phase (Evans, Kugelman and Fang, 1983) to avoid future data gaps. Thus, testing should include tests for both physical and engineering properties. The latter should include strength, compressibility and permeability (Olson and Daniel, 1979; Zimmie, Doynow and Wardell, 1981; Evans, Fang and Kugelman, 1983).

Waste Control System Components

Solid and Liquid waste control systems can consist of a wide range of components as shown in Fig. 2. These components can be classified in two general categories: (1) active and (2) passive (Evans and Fang, 1982). Active components of a containment system are those which require on-going energy input. Examples of active components include disposal wells, pumping wells, and treatment plants. Conversely, passive components of a containment system are those which do not require on-going energy input. Typical examples of passive components include drain tile collection systems, liners, covers and barrier walls. While an active system requires on-going energy input, the passive components typically require maintenance. For example, it is necessary to keep vegetation taproots from penetrating a clay cap.

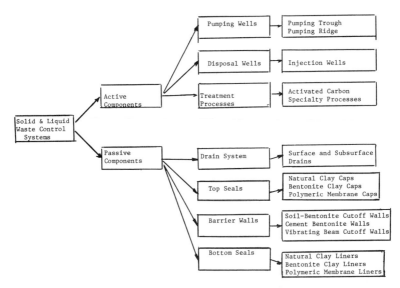

FIGURE 2. Solid and Liquid Waste Control Systems.

Passive Waste Control Systems

As indicated in Fig. 2, there are four major passive components and numerous possible subsystems. However, as indicated earlier, only three of these (top seals, barrier wells, and bottom seals) will be covered here. The physical structure of each system is illustrated in Figures 3, 4, and 5. A brief discussion of each system follows.

Surface Seal for Controlling Infiltration

The function of top seals or surface seals (caps) is to control surface water so as to minimize infiltration, thereby reducing subsequent leachate production and/or contaminant transport potential.

Surface sealing is typically accomplished utilizing caps or covers of native clays, processed clays or synthetic membranes. In general, clays will last longer than synthetic materials, and clay caps rather than synthetic caps are usually chosen. However, to avoid the bathtub effect (more water entering the facility than can drain out), the use of a synthetic membrane cap whenever the bottom liner is also a synthetic membrane is required (USEPA, 1982). The most cost effective surface seal usually is native clay material from locally available sources. If suitable native clays are not available, imported materials or synthetic membranes must be used.

Asphalt cements, concrete, Portland cement concrete, liquid and emulsified asphalts, tars and bituminous fabrics (Asphalt Institute, 1979) and other stabilized soil may also be considered for surface seals but are rarely used.

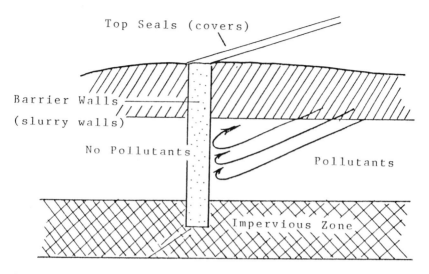

FIGURE 3. Components of Passive Remedial Systems for Controlling Solid and Liquid Wastes.

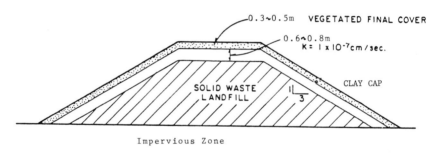

FIGURE 4. Typical Native Clay Top Seal.

Natural Clay Surface Seals: Selection of clay materials for clay covers is based on the hydraulic conductivity of available material. Tap water permeant is preferred to the conventional use of distilled water (Olson and Daniel, 1979). As shown in Fig. 4, the thickness of the clay cap is typically from 0.6 m to 0.8 m. Some factors to be considered in design are:

(a) The top several inches of clay cannot be as well compacted as the remainder of the thickness. Further, it may be difficult, in the long-term, to maintain the clay density in the top few inches due to potential desiccation cracking and frost action (Fang, et al., 1983).

(b) The bottom of the clay cap may become somewhat intermixed with subgrade material during installation. Therefore, the effective thickness of nominal 0.6 m (2′) cap may be on the order of only 0.3 m (1′). Native clay caps

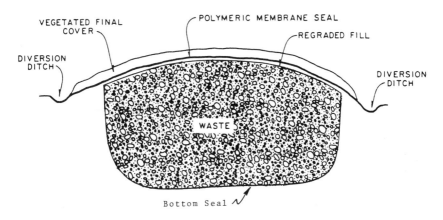

FIGURE 5. Typical Section: Polymeric Membrane Topseal.

must be protected from degradation due to erosion by rainwater, cracking due to drying, cracking due to differential subgrade movement, penetration by deep taproots of vegetation and rutting from moving vehicles. Hence, the native clay caps are virtually always used in conjunction with an additional cover. This cover can be utilized to enhance runoff and as a base of revegetation.

Bentonite Clay Surface Seals: Where local natural clay is not available in sufficient quantity, quality or price, processed clay is a commonly utilized alternative. Processed clay is typically sodium montmorillonite from Wyoming, South Dakota and other western states commonly called bentonite (Grim, 1953).

Bentonite is a hydrophilic colloidal clay which swells in water. Some bentonite can absorb as much as five times its own weight of water. The construction of a processed clay cap requires the application of the bentonite at a controlled rate (e.g., about two pounds per square foot), adequate mixing with the in-place soils to a predetermined loose thickness and compaction. A 10-15 cm (4″ to 6″) layer of clayey soil with low hydraulic conductivity (1 x 10^{-7} cm/sec) can thereby be formed. The main advantage of this method is low cost with relatively low hydraulic conductivity. The disadvantage is that it is difficult to obtain uniform application at the blending rate. Further, since these seals are typically thinner than native clay seals, there is little margin for error. Therefore, the effectiveness of a bentonite surface seal in reducing infiltration requires close control of subgrade materials, preparation, application rate, application uniformity, mixing uniformity, compaction, and subsequent grading cover. Deficiencies in any step of these aspects could result in a reduced surface sealer effectiveness.

Membrane Surface Seals: Synthetic membranes (see Fig. 5) can be utilized as top seals for waste containment (Emrich and Beck, 1980). Membranes are available in a wide range of materials from numerous manfuacturers and distributors. Factors to be considered in the design and construction of surface seals utilizing membranes include: types of subgrade soil, method of subgrade preparation, slope of grade, final cover of finished slope, membrane type, installation procedures, jointing of seams, and continuity of membrane covers. Subgrade soils must be free of materials such as sticks, large rocks and debris that can puncture the membrane top seals. The subgrade should be graded and compacted to provide rapid runoff and prevent liquid ponding. Soil cover is necessary to protect the membrane from traffic and ultraviolet degradation. Proper design is needed for side slopes (Fang, 1975) and to prevent movement of cover materials along the membrane top seal. As precipitation infiltrates the cover and flows along the membrane, a saturated weakened layer in the cover material may develop causing a slump of the cover material. A final cover of the finished layer is vegetated to minimize erosion (see Fig. 5).

Other Surface Water Controls: In addition to surface sealing, surface water diversion and collection systems can provide short and long term measures to isolate waste disposal sites from surface water inputs. Techniques used to control flooding and offsite erosion transport of cover and surface seal materials include dikes and berms, interceptor ditches, diversion dikes and berms, terraces and benches, sheets and downpipes, levees, seepage ditches and sedimentation basins and ponds (Cedergren, 1967).

Surface seals provide multiple functions in the overall liquid and solid waste containment control system. The main function has been discussed as the control of infiltation by minimizing water infiltration and/or maximizing surface runoff away from the site as well as the minimization of erosion. Surface seals also minimize vector breeding areas and animal attractions to the site by sealing off the putrescible materials from contact with insects and burrowing animals. Other advantages include the minimization of unsightly debris, aid in the control of noxious odors and the overall improvement of the appearance of the contained site (Lutton, et al. 1979, Lutton, 1980).

Barrier Walls for Control of Contaminant Migration
    The containment of contaminant migration from existing disposal sites or impoundments may necessitate a subsurface barrier to horizontal groundwater flow. Barriers are commonly constructed as soil-bentonite slurry trench cutoff walls or cement-bentonite slurry trench cutoff walls. Vibrating beam cutoff walls, grout curtains or sheet piling are also used. For barrier walls to be effective, they generally (although not always) key into an impermeable stratum of natural materials beneath the site.

Soil Bentonite Slurry Wall: The procedure for design and construction fo a soil-bentonite slurry trench cutoff wall is well documented (Xanthakos, 1979; Evans and Fang, 1982). As shown in Fig. 6, a trench is excavated below the ground surface and trench stability is maintained utilizing the slurry of bentonite and water. This slurry maintains trench stability in much the same way as drilling

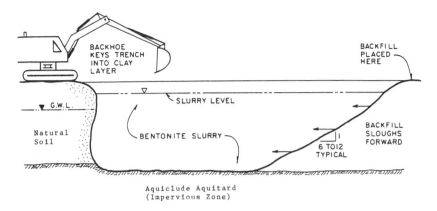

FIGURE 6. Schematic Section of Slurry Trench Excavation and Backfill.

FIGURE 7. Photo Shows the Excavation of Slurry Wall Trench.

fluid maintains borehole stability. Fig. 7 is a photograph of an actual installation. The bentonite water-slurry is designed by the geotechnical engineer to have certain density, viscosity and filtrate loss properties which allow for the formation of a filter cake along the walls of the trench resulting in a computed factor of safety for trench stability greater than one (D'Appolonia, 1980).

Trench depths are generally limited to about 10 m (33 ft) using conventional backhoes. In order to achieve greater depths, a modified dipper stick is required

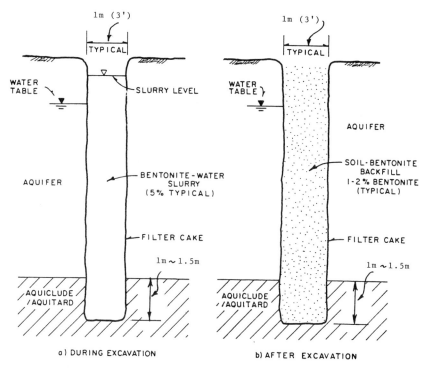

FIGURE 8. Typical Section: Soil-Bentonite Slurry Trench Cutoff Wall.

which can be provided by specialty slurry wall contractors. To go deeper than about 17 m (51 ft) usually requires the utilization of a clam shell, although an extended backhoe capable of excavating 22 m (73 ft) has been developed (Case, 1981). Clam shell digging is typically slower and can increase the cost of the barrier wall.

After excavation and addition of the slurry, the trench is backfilled with a matrix of soil materials. Shown in Fig. 8 is a typical cross-section of a slurry wall during excavation and after backfilling. The mixture of soil and bentonite-water slurry typically has a consistency similar to that of high slump concrete. During trench backfilling, care must be taken to achieve a uniform mixing of the backfill and avoid entrapment of pockets of pure bentonite water slurry in the trench.

Much has been written regarding the design of soil-bentonite cutoff walls for conventional groundwater control applications (Boyes, 1975, Ryan, 1977, Xanthakos, 1979, D'Appolonia, 1980, and Millet and Perez, 1981). In order to design a soil-bentonite slurry wall for waste containment, studies are required beyond those typically required for other applications. Chemical analysis of samples of onsite materials considered as potential backfill materials may be required. If onsite materials are contaminated, the effects of this contamination upon the properties of the soil-bentonite mix must be evaluated. Contaminated soils may inhibit or reverse the hydration of the bentonite. Several bentonites are available which are identified as being contaminant resistant. However, it should be recognized that treated bentonite is not totally contamination resistant to all contaminants at all concentrations. If the onsite soil is not satisfactory, the soil characteristics at offsite borrow areas must be surveyed to establish potential utility. Compatibility testing is required for all the potential sources of backfill material.

The question of compatibility between the waste and the cutoff wall is approached in two steps. First, the chemical characteristics of the waste are determined. A review of the published information on pore fluid effects upon clay behavior (D'Appolonia, 1980, Green, et a., 1980, Anderson and Brown, 1981 and Evans, et al., 1983) can be used to allow a preliminary assessment of the compatibility. After the preliminary compatibility assessment is made, a laboratory testing program can be designed and conducted to provide site-specific soil-waste compatibility data.

As part of the soil-waste compatibility program, samples of backfill need to be tested in the laboratory utilizing the bentonites under consideration. The most important of these laboratory tests is the triaxial permeability test (Zimmie, et al., 1981, Evans, Fang, and Kugelman, 1983). The test is typically conducted utilizing the site leachate, liquid waste, or groundwater as the final permeant. The samples can be initially set up in a triaxial cell, consolidated, and permeated with the water planned for use during construction; not distilled water or water from some other source. Offsite water may be necessary if available onsite water inhibits adequate hydration of the bentonite. The bentonite utilized in the testing program must also be that planned for project usage. After initial permeation with water, the waste is introduced and the test continued till two or four pore volumes of liquid permeate through the sample. Permeability and volume change versus time can then be calculated from the test data and these relationships studied to yield conclusions regarding the effect of the waste upon the soil-bentonite mixture.

Proper application of cutoff wall technology to waste containment requires analysis of the interaction between the bentonite, mixing water, backfill, and site groundwater/leachate. The potential long-term performance of a soil-bentonite cutoff wall for waste containment must therefore be fully investigated.

Should a soil-bentonite slurry trench wall prove to be practicable, detailed construction specifications must be written. These specifications must include the source of the mixing water, required hydration time or slurry properties, allowable methods of mixing the bentonite slurry and the backfill, the bentonite-water viscosity and density limits, approved sources of the backfill, the bentonite-water viscosity and density limits, approved sources of the backfill, and allowable methods of backfill placement. An excellent treatment of slurry wall specifications has been written by Millet and Perez, (1981).

Close control is required to ensure construction consistent with design assumptions and intent. A resident geotechnical engineer should document depth to key material, test the backfill and slurry, and provide onsite technical representation for the owner/engineer. Consideration must be given to the disposal of excavated soils should they be categorized as contaminated. Finally, proper planning for work safety is essential should a barrier wall be planned for containment of previously disposed wastes.

Cement-Bentonite Slurry Trench Cutoff Walls: As an alternative to soil-bentonite cutoff walls, cement-bentonite cutoff walls can be utilized. The trenches are excavated in a manner similar to soil-bentonite walls utilizing a slurry to maintain trench stability. However, cement rather than soil is used in the slurry. The slurry is left in the trench and allowed to harden. A strength equivalent to stiff clay can be obtained after a period of about a month. Design considerations include the cement and bentonite content and type, and their relationship to the strength and permeability of the backfill.

Leachate compatibility tests must be conducted utilizing the site contaminant as permeant. The overall permeability of a cement-bentonite cutoff wall is generally higher than for soil-bentonite walls.

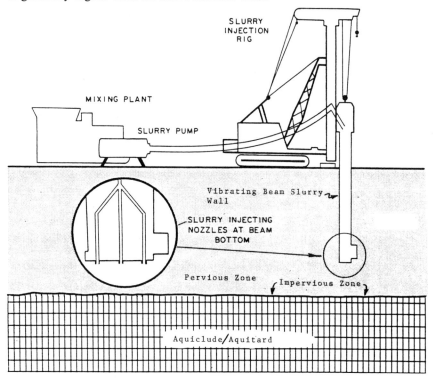

FIGURE 9. Schematic of Vibrating Beam Slurry Wall.

Vibrating Beam Walls: Barriers to horizontal groundwater flow have been designed and constructed using the vibrating beam injection method. This technique utilizes a vibratory-type pile driver (see Fig. 9) to cause the penetration of a beam of specified dimensions to the design depth. Slurry is added through injection nozzles as the beam penetrates the subsurface soils and as the beam is withdrawn.

The slurry utilized with the vibrating beam technique is generally either of two types, cement-bentonite, or bituminous grout. Mix design considerations for cement-bentonite were previously discussed in this paper. Bituminous grouts are prepared as a homogeneous blend of asphalt emulsions, sand, portland cement and water. Flyash may also be included. It is reported that this bituminous grout can resist strong acids and high saline content wastes.

The engineer must be aware of the detailed aspects of thin wall barriers installed by the vibrating beam technique in order to assure an adequate waste containment design and installation. The specification may include the slurry mix design, installation equipment requirements, batch mixing equipment requirements, vertical limits, injection pressure, overlap, depth, injection and extraction rates and procedures, and wall thickness.

Control of the beam tip location cannot be guaranteed, particularly with deep penetrations. For example, the presence of cobbles and boulders may cause a deflection of the beam tip. As in the case of conventional slurry walls, compatibility testing is necessary to investigate the slurry resistance to the contaminant being contained. A principal advantage of this technique is the elimination of the need to excavate potentially contaminated soils, an important safety consideration when dealing with a barrier wall around active or retired facilities.

In lieu of slurry trench or vibrating beam cutoff walls, grout curtains and sheet pile cutoffs may provide attractive alternatives. These techniques are usually more expensive and have not commonly been utilized.

Bottom Seal (Liners) for Controlling Waste Leakage

In new containment facilities, it is frequently necessary to provide a liner system. The major function of a liner is to prevent leachate or waste from entering the groundwater regime. Liners, as with covers, can consist of native clays, processed clays, or synthetic membrane liners. It is important to note that under the recent "Interim Final Regulations," the USEPA, 1982, considers a synthetic membrane liner best to "prevent" migration of wastes; whereas a clay liner will "minimize" migration of wastes. With all materials, compatibility testing is essential to determine the liner resistance to the waste or leachate to be contained.

Native Clay Liners: The compatibility between the natural clays and the waste is an important design consideration for the use of natural clays as liners. It is important to ascertain the volume change and permeability characteristics of the proposed clay liner material. The bulk transport of liquid waste through cracks in the liner must be precluded. Bulk transport of liquid through clay liners could occur due to differential settlement of the foundation materials. Tensile stresses within the liner could also result in cracking and subsequent bulk transport of liquid waste through the liner. Finally, and probably most importantly, physical-chemical stresses due to the pore-fluid-clay interaction could cause cracking (Fang, et al., 1983).

The determination of liner/waste compatibility requires site-specific studies. The compatibility is a function of both waste type and concentration. Studies of clay-waste compatibility conducted to date (Evans, et al., 1983, Green, et al., 1980, Anderson and Brown, 1981, Fong and Haxo, 1981) have shed considerable light upon the subject. Despite these recent advances, laboratory tests under triaxial stress and gradient conditions are required for site-specific data from which to evaluate the suitability of a natural clay liner.

Bentonite Clay Liners: Design and construction considerations for the use of processed clay for liners must include waste-liner compatibility as well as considerations previously discussed. The volume change characteristics of the processed clay are especially important. Generally, the processed clay is mixed with the subgrade material to form the impermeable liner. The impedance to groundwater flow is primarily due to the processed clays, especially when the matrix soil is relatively free of natural fines. Hence, if the processed clay shrinks upon exposure to the waste, large increases in permeability can occur. Even greater flow can occur if bulk transport of liquid waste occurs due to liner cracking. The hydration of a processed clay liner with uncontaminated water prior to waste disposal is recommended (Hughes, 1975). The triaxial permeability tests with the actual processed clay subgrade material, groundwater and leachate contaminants discussed previously should be conducted in this situation.

Synthetic Membrane Liners: As with other liner types, waste compatibility is a major design consideration. However, the permeability of a polymeric liner can also increase due to liner stretching. Thus, total and differential foundation settlement can impact the liner design. Close construction control is essential to the overall system performance. The "permeability" of an installed membrane liner system is generally a function of bulk transport through seam, joints, tears, holes and pinholes. Additional aspects of polymeric membrane liners can be found elsewhere (Gunkel, 1981).

Summary and Conclusions

Passive techniques can and have been used to mitigate contaminant migration. Consideration must be given to the identification of all contaminant pathways and selection of the most appropriate control technique. Each technique then must be evaluated as to its effectiveness and the design and construction must incorporate all site specific and technique specific considerations.

References:

1. Asphalt Institute (1979) A Basic Asphalt Emulsion Manual, *Manual Series No. 19* (MS-19) March, 260 p.
2. Anderson, D. and Brown, K. W. (1981) Organic Leachate Effects on the Permeability of Clay Liners, *Proceedings, 7th Annual Reserach Symposium Land Disposal: Hazardous Waste. USEPA Report* No. 600/9-81-002b, March, 119-130.
3. Case International Co. (1981) "Case Study No. 5," *Case Slurry Wall Notebook.*
4. Cedergren, H. R. (1967) *Seepage, Drainage, and Flow Nets*, John Wiley and Sons, Inc., NY, 489 p.
5. D'Appolonia, D. J. (1980) Soil-Bentonite Slurry Trench Cutoffs, *Journal of the Geotechnical Engineering Division, ASCE*, 106, GT4, 399-417.
6. Emrich, G. H. and Beck, W. (1981) Top Sealing to Minimize Leachate Generation—Status Report, *Proceedings, 7th Annual Research Symposium Land Disposal: Hazardous Waste. USEPA Report* No. 600/9-81-002b, pp. 291-297.
7. Evans, J. C. and Fang, H. Y. (1982) Geotechnical Aspects of the Design and Construction of Waste Containment Systems, *Proceedings, National Conf. on Management of Uncontrolled Hazardous Waste Sites,* NOV. Washington, DC p 175-182.
8. Evans, J. C. and Fang, H. Y. (1983) Passive Techniques to Control Containment Migration, *Proceedings, Conf. on the Disposal of Solid, Liquid and Hazardous Wastes,* ASCE, Lehigh University, p. 9-1 to 9-16.
9. Evans, J. C., Kugelman, I. J., and Fang, H. Y. (1983) Influence of Industrial Wastes on the Geotechnical Properties of Soils, *Industrial Waste Conference,* Bucknell University, Lewisburg, PA.
10. Evans, J. C., Fang, H. Y., and Kugelman, I. J. (1983) Influence of Hazardous and Toxic Wastes upon the Engineering Behavior of Soils, *Proceedings of the Third Ohio Environmental Engineering Conference,* Columbus, OH.
11. Fang, H. Y., Chaney, R. C., Failmezger, R. A. and Evans, J. C. (1983) Mechanism of Soil Cracking, *Proceedings of the 20th Meeting of the Soc. of Eng. Science,* Newark, DE.
12. Fang, H. Y. (1975) Stability of Earth Slopes, Chapter 10, *Foundation Engineering Handbook,* Van Nostrans Reinhold Co., NY, p. 354-372.
13. Fong, M. A. and Haxo, H. E., Jr. (1981) Assessment of Liner Materials for Municipal Solid Waste Landfills, *Proceedings 7th Annual Research Symposium, Land Disposal: Municipal Solid Waste. USEPA Report* No. 600/9-81-002a, 138-162.
14. Green, W. J. , Lee, G. F. and Jones, R. A. (1980) The Permeability of Clay Soils to Water and Organic Solvents: Implications for the Storage of Hazardous Wastes, *Report submitted to USEPA,* October.

15. Grim, R. E. (1953) *Clay Mineralogy,* McGraw-Hill, NY.
16. Gunkel, R. C. (1981) Membrane Liner Systems for Hazardous Waste Land-fills, *Proceedings, 7th Annual Research Symposium, Land Disposal: Hazardous Waste. USEPA Report* No. 60/9-81-002b, 131-139.
17. Hughes, J. (1975) Use of Bentonite as a Soil Sealant for Leachate Control in Sanitary Landfills, *Volclay Soil Engineering Report Data 280-E.*
18. Kolmer, J. R. (1981) Investigation of LiPari Landfill using Geophysical Techniques, *Proceedings, 7th Annual Research Symposium, Land Disposal: Hazardous Waste, EPA Report* No. 600/9-81-002b, March, 298-311.
19. Lowe, J. III and Zaccheo, P. F. (1975) Surface Explorations and Sampling, Chapter 1, *Foundation Engineering Handbook*, Van Nostrand Reinhold Co., p. 1-66.
20. Lutton, R. J. (1980) Evaluating Cover System for Solid and Hazardous Waste, *USEPA Report* No. SW-867, Sept., 57 p.
21. Lutton, R. J., Regan, G. L. and L. W. Jones (1979) Design and Construction of Covers for Solid Waste Landfills, *USEPA Report* No. 600/12-79-165. Aug. 249 p.
22. Millet, R. A. and Perez, J. Y. (1981) Current USA Practice, Slurry Wall Specifications, *Journal of the Geotechnical Engineering Division, ASCE,* 107, GT8, 1041-1056.
23. Olson, R. C. and Daniel, D. E. (1979) Field and Laboratory Measurement of the Permeability of Saturated and Partially Saturated Fine-grained Soils, *Geotechnical Engineering Report 80-5,* University of Texas, Austin, 78 pp.
24. USEPA (1982) 40 CFR Part 260, July 26.
25. Xanthakos, P. (1979) *Slurry Walls,* McGraw-Hill, NY.
26. Zimmie, T. F., Doynow, J. S. and Wardell, J. T. (1981) Permeability Testing of Clay Liners, *Proceedings, 7th Annual Research Symposium, Land Disposal: Hazardous Waste. USEPA Report* No. 600/9-81-002b, March, 119-130.

Solid and Liquid Wastes: Management, Methods and Socioeconomic Considerations. Edited by
S. K. Majumdar and E. Willard Miller. © 1984, The Pennsylvania Academy of Science.

Chapter Eleven

Solid Waste Land Treatment Systems

Michael R. Overcash, Ph.D.

Professor, Chemical Engineering Department
Professor, Biological and Agricultural Engineering Department
North Carolina State University
Raleigh, NC 27695-7905

The use of land treatment technology for solid wastes is primarily centered on sludges and liquid residues and not on refuse and related municipal sanitation materials. In this context, land application has served a steadily increasing role in the management of sludges, primarily as an alternative to placing such materials in landfills. The advantage of land treatment as a means of managing sludges is that the waste is treated or assimilated thus obviating the long-term implications of storage associated with a landfill. It should be noted that land treatment, if not designed and operated correctly, can be disadvantageous and uneconomical and therefore can not be a panacea for all sludges. Equally noteworthy however is that an increasing number of diversity of solid wastes are being managed with land treatment as a method of technical and economic choice.

LAND TREATMENT CONCEPT

It will be appropriate to start by defining the concept of land treatment to be used in the following discussions. Land treatment is the use of a plant-soil system to do two tasks: (1) to treat or to renovate wastes, and (2) to serve as an ultimate receiver of wastes. This combination of objectives is what really differentiates land treatment from almost all of the other solid waste practices, hazardous and non hazardous. There are very few technologies that serve as treatment and as the ultimate disposal site for waste materials. A landfill is only an ultimate repository; and not a treament system. Incineration or biological treatment have an effluent or residue that must be accommodated. Land treatment, therefore, really offers substantive advantages from an operative point of view.

The treatment or assimilation calculations done relative to industrial land treatment are summarized in three types. The first are those which account for chemical and biological reactions to break down a portion of the waste. Here, the objective is to land apply the material at roughly the rate at which it is going to decompose to prevent long-term accumulation in the system. The second class of criteria will account for adsorption and fixation of other fractions of the waste. These are the conservative constituents such as metals for which there is a long-term objective. Finally, one must calculate and account for the controlled migration of certain inorganic fractions, anionic species, most typically nitrates, chlorides, etc., to assure that this part of the overall system is adequately accommodated.

A basic objective by which to measure how close a land treatment system comes to success or failure is stated as follows: For those areas that are going to be used for land treatment, and as the ultimate receiver of the waste, the rate of application (lb/acre/year) or the use period of (50 years, 100 years, or whatever period) be such that when the application is discontinued, these areas still can be used for other societal uses. This is basically a nondegradation constraint. It says that at any point in time you should be able to stop the industrial waste land treatment system, and within a short buffer period, convert the site over to growing major agricultural crops; silviculture,; or build a plant on it; any of these logical uses or societal uses of the land.

With the previous basic definition and constraints it is important to view the overall design procedure for a land treatment system for solid wastes. In fact this procedure has substantial similarity for municipal and industrial solid wastes as well as more liquid wastes and effluents. Similar design procedures have been proposed by Brown (1980), Loehr (1979) and Overcash (1977, 1979). The purpose of the design procedure is to provide a method that for any type of waste, and for any type of site, provides a logical procedure for putting the main factors together to come up with the design for a land treatment system. Briefly, it is divided into four stages, Figure 1. Stage one involves evaluation of (1) site information and the basic nondegradation constraints as well as (2) a variety of waste characteristics. These two are put together in a manner to arrive at the amount of land needed and specific chemical constituents which control the required land area.

The second stage is basically engineering or detailed specification of the system. That is, with the amount of land known, what are the components that are necessary to get the waste from the plant out to the land area and do it satisfac-

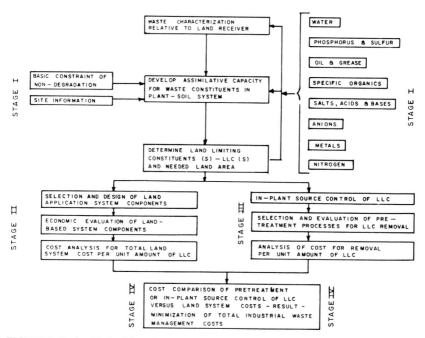

FIGURE 1. Design Methodology for Industrial Waste Pretreatment—Land Application System

torily; and what are the costs? In parallel, stage three begins with the critical constituents, goes back into the plant or evaluates pretreatment options which will reduce those specific parameters and arrive at the costs for such reduction. Stage four is a very simple-minded economic comparison between in-plant or pretreatment processes versus land treatment.

Many times the balance between land treatment and pretreatment or in-plant source control is dictated only by economics. There are no arbitrary standards that say one must have tertiary treated material before you can apply it to the land. Instead, sometimes 20 percent of the job is in-plant and 80 percent on the land treatment area, or other times the reverse. The exact percentage is dependent on the specific conditions. This design procedure allows one to apply any type of waste from any industry to any type of site conditions, and operate the system in a rationale mode. This procedure is being converted by several states into their land treatment guidelines for industrial waste. It is a procedural approach and not a detailed quantitative approach in which one must evaluate all the particular parts of the system.

The design procedure, until the engineering part (stage 2), is virtually insensitive to whether the waste is an effluent, sludge, solid waste, or a raw waste. It just does not make any difference. These are physical descriptors of the waste. In fact, all that matters are the chemical descriptors of the waste and site. Then

TABLE 1

*Existing Wastes for Which Land Treatment Design and/or Systems Exist
(Earth Systems Associates, Athens, GA)*

Type of Waste	Location
Metal Plating	Georgia, Virginia
Metal Shop	Georgia
Electroplating	Georgia, Puerto Rico
Chemical Processing	Georgia
Chemical Manufacturing	South Carolina
Specially Organic Chemical Syntheses	North Carolina
Textile Sludge	South Carolina, North Carolina
Poultry Processing	Georgia, North Carolina
Food Wastes	Georgia, North Carolina
Printing Plant	North Carolina
Commercial Land Treatment	Illinois, Virginia
Leather and Tanning	New Hampshire
Dye Sludge	North Carolina
Paper Mill Wastewater	Georgia
Drilling Muds	Louisiana
Integrated Chemical Complex	Louisiana
Metal Foundry	Alabama
Chemical Manufacturing	Alabama
Pesticide Intermediate Waste	Tennessee
Pharmaceutical	North Carolina, Puerto Rico
Tobacco Processing	Puerto Rico
Textile Finishing	Georgia, North Carolina
Herbicide Residue	Georgia
Chemical Manufacturing	Virginia

the design procedure allows one to handle any type of wastes as these are generated as solid wastes from municipal or industrial sources.

Land treatment has already been used, or the design criteria developed, for virtually every industrial category, an exception being radioactive-containing waste. This includes metal plating wastes, microelectronic assembly wastes, petroleum complex wastes, municipal sludge, organic chemical complex wastes, and even difficult to treat wastes like "swine manure."

It should be noted that there are certainly many unknowns relative to some of the specific criteria. But, overall, a large diversity of industrial and municipal solid wastes can be managed with this technology. Table 1 illustrates, from just one organization (Earth Systems Associates), the diverse systems that they have designed for land treatment. This further reinforces the fact that land treatment is growing from the early developmental stages in the petroleum and food industries.

In examining a number of municipal and industrial solid waste land treatment systems, there evolve a number of observations that assist in evaluating the potential use of land treatment. Manufacturing facilities that do not have land immediately adjacent to the plant are already using land treatment. Pumping liquids one to two miles is a routine part of a number of systems and hauling wastes of two percent solids and higher goes routinely up to 12 to 30 miles. This often opens up the options for land treatment at a large number of facilities that do not have land immediately adjacent to the facility. The transportation portion of the overall costs is often not so prohibitive that municipalities or industries can not haul substantial distances to find amenable land areas.

Industrial plants often have spills, process upsets, and other malfunctions, which generate large pulse loads of wastes. Land treatment has been found to be quite tolerant of these excursions, particularly because of the relatively low rates of application and hence, wide variations in tolerable waste concentrations. From a plant operational point of view and the generation of solid wastes, this is advantageous because these waste pulses always happen.

ASSIMILATIVE CAPACITY DETERMINATIONS

The three calculational procedures referred to in the definition of land treatment constitute the pathways and procedures for the assimilation of individual constituents when applied to land. The first is organic species for which data are needed on the rate of decomposition, phytotoxic effects, and leaching/volatilization. There are two principal sources for such information. The first is the direct published literature. Since this source is quite large we have developed a search system for terrestrial systems. It is called the Terrestrial Behavior and Organic Chemicals catalogue and is developed by going through article by article about 300 journals for a given year and writing down every organic chemical in the terrestrial system that are listed. This then allows us to go back and search that article. From the literature information for two years, we have found about 2,000 compounds. Extending this search would lead to a very good base which we could use to start looking at a large number of organic chemicals in industrial and municipal solid wastes as a means of assuring that the nondegradation approach is matched in the system.

The information that shows up in this search include data on volatilization, decomposition, adsorption, plant uptake, and phytotoxicity. These are the typical broad information groupings. Classes of compounds for which we have found substantial information includes phenols and phenolics, polycyclic aromatics, nitrosamines, chlorinated biphenyls, substituted aromatics, cellulosics, and a variety of developed polymers, solvents, alcohols, antibiotics, organic acids and phthalic acid esters.

The second approach is to conduct actual measurements of decomposition or other effects. In these treatability studies, one looks at small test plots where leaching can occur. The waste is mixed with the soil; sampled over a period of time, and the phytotoxicity level established. These studies are routine procedures. While the tests are routine, there are also not that many groups which are involved in land treatment to conduct them.

In the migrational category, one determines the amount of land area required such that with rainfall and the water which comes with the wastes, the anionic species, e.g., chlorides, nitrates, fluorides, etc., that move through the root zone of the soil, emerge at drinking water standards. Thus assimilation implies meeting drinking water standards.

Metals in the soil system fit in the accumulation, precipitation, and adsorption categories, the third assimilative mechanism. The main environmental concern is to not reach soil concentrations or levels that are above what are now termed acceptable. The criteria relative to plants are those levels which do not affect either the concentration in the plant or the ability to grow. For specific quantitative determinations of these three assimilative pathways the reader should consult several detailed references which cover in detail the assimilative capacity for land treatment (Overcash 1979 and 1981, Loehr 1979, Brown 1980).

ENGINEERING SPECIFICATIONS

While the equipment and physical systems for land treatment of solid wastes are not as diverse as the complete alternatives for all types of wastes applied to land, there is still a range of options to be specified. These selections are generally referred to as the engineering phase and are typically the third stage in the overall process of implementation for land treatment systems, Table 2.

The engineering effort translates information developed in site testing, waste characterization, and concept design into: (a) specification of size, dimensions, operating characteristics, locations, and type of equipment, and (b) drawings which convey this information in an accepted form for regulatory review and implementation. This effort is done for each of the proposed land treatment system components:
- application system and equipment (liquid or sludge/solid waste),
- conveyance from waste source to application site,
- storage, as needed,
- vegetative system and associated agricultural equipment,
- site modification (diversions, terraces, runoff control, etc.),
- buffer zones, community acceptance, and beautification,
- control systems,
- monitoring.

In addition, as indicated in Table 1, there are numerous engineering tasks performed in the design phase. These are sufficiently site-specific that the reader should consult Overcash (1979) for detailed information and references to other sources of information.

TABLE 2

Development of Land Treatment Systems

Stage	Activity
1	Feasibility Assessment
2	Concept Design
3	Engineering Specifications
4	Acquisition of Permits
5	Bid Document Preparation and Specifications
6	Construction
7	Operator Training
8	Long Term Monitoring

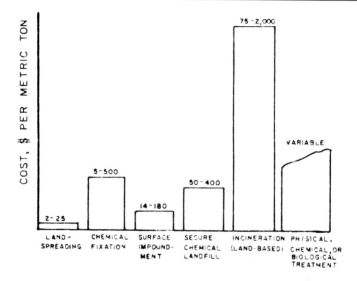

TECHNOLOGY

FIGURE 2. Environmentally sound technologies are available for treatment and disposal of hazardous waste. Costs vary widely, according to type and volume of waste handled, and are substantially in excess of unsound practices.

Source: EPA, Draft Economic Impact Analysis, 1979.

ECONOMICS

As described earlier, the trend toward utilization of land treatment is a primary indication that this technology should be considered in detail for most cases since there is a reasonable probability of being the least cost alternative. Specific cost comparisons to alternatives are difficult to apply generally, so the information in Figure 2 and Table 3 should be used as rationale for considering land treatment and not as guaranteed cost-effectiveness. The land treatment alternative was 10%-70% of more conventional alternative again illustrating the impetus for assessing land treatment as an option.

TABLE 3

Comparative Industrial Waste Management Alternatives

System	Land Treatment Alternative		Comparative Treatment Alternative	Ratio
	Size (ha)	Investment Cost ($)	Investment Cost ($)	
Pharmaceutical	49	490,000	745,000	0.66
Poultry Processing		220,000	720,000	0.31
Potato Processing	67	350,000	780,000	0.45
Nylon and Polyester	12	140,000	1,300,000	0.11
Refinery	10	6.50/wet ton	11.50/wet ton (Incineration)	0.56

SUMMARY

The use of land treatment for municipal and industrial solid wastes represents a viable option.

The design of land treatment systems is based primarily upon the limiting constituent or limiting parameter concept. This concept has important advantages:

• It provides a standard methodology for considering important site-specific soil data, climatic conditions, waste characteristics, and environmental concerns,

• It is based opon the nondegradation concept and embodies information from many disciplines to avoid adverse environmental impacts,

• It provides a direct relationship between waste characteristics and land assimilative capacity,

• The methodology is general so that any waste or site can be investigated.

It is apparent that the cost-effectiveness of land treatment is a factor that has increased the attention given this solid waste management technique.

REFERENCES CITED

1. Brown, K. W. and Associates, Inc., Hazardous Waste Land Treatment, U.S.E.P.A., SW-874, Municipal Environmental Research Laboratory, Cincinnati, OH, 974 p., 1980.
2. Loehr, R. C., W. J. Jewell, J. D. Novak, W. W. Clarkson, and G. S. Friedman, Land Application of Wastes, Van Nostrand Reinhold Co., New York, 308 p., 1979.
3. Overcash, M. R. and D. Pal. Industrial Waste Land Application, Today Series, Amer. Inst. of Chem. Engr., New York, NY, 512 p. 1977.

4. Overcash, M. R. and D. Pal. Design of Land Treatment for Industrial Wastes, Theory and Practice. Ann Arbor Science Publishers, Ann Arbor, MI, 684 p. 1979.

5. Overcash, M. R. Decomposition of Toxic and Nontoxic Organic Compounds in Soils, Ann Arbor Publishers, Ann Arbor MI, 485 p. 1981.

6. Sommers, L. E., Chemical Composition of Sewage Sludges and Analysis of Their Potential Use as Fertilizers, J. Environ. Qual. 6(2):225-232 (1977).

7. U. S. Environmental Protection Agency, Process design manual, land treatment of municipal wastewater, EPA 625/1-81-013, Cincinnati, OH, 458 p. 1981.

8. U. S. Environmental Protection Agency, Draft Economic Impact Analysis (Regulatory Analysis Supplement) for Subtitle C, Resource Conservation and Recovery Act of 1976, Office of Solid Waste, U.S.E.P.A., Washington, DC, January 1979.

Part 3
Environmental and Health Impacts

In recent years it has been recognized that liquid and solid wastes have many environmental and health implications. Many of these wastes have been proven to adversely affect health. The exposure of humans to waste substances can occur in varied ways, such as in contaminated water.

The initial paper of this section presents a discussion and analysis of the behavior of toxic and nontoxic organic compounds when applied to the land. This is followed by a study on cadmium and other trace elements in the environment. At present, the disposal of sewage sludge on agricultural land represents a primary means through which concentrations of trace elements may be increased in soils. The detrimental effects of elevated levels of cadmium and other trace elements found in sludge are evaluated.

Because oil is increasingly polluting the oceans, the marine environment is threatened. The sources of oil pollution, control of oil spills, ecological damage and the development of environmental laws of the ocean are important aspects to be considered.

As water pollution has increased there have been many questions raised as to its effect or different aspects of recreation. As an example the impact on seasonal variations in fish species at the Brunner Island Steam Electric Station at York Haven, Pennsylvania is studied. This investigation revealed that the facility had no appreciable effect on the population of the species, but that the plume from the facility has both negative and positive effects on the recreational fish catch due to heat.

The practice of disposing domestic wastewater into streams carries with it an element of health risk. The use of epidemiological methods to establish a relationship between swimming—associated illness and water quality was first attempted in the early 1950's. Continued studies have revealed a direct linear relationship between water quality and swimming—associated gastrointestinal illness.

As the use of flourides has increased in water supplies to decrease tooth decay, questions have arisen on the effects of fluorides on other health and environmental issues. As a consequence, this chapter points out the continued need to determine the sources of fluorides in the environment and the effects they have on humans and ecosystems.

It is now recognized that acid rain has reached a critical stage in most of eastern North America. It has great potential of harming aquatic life as well as the natural vegetation. This study recognizes the need for future research, not only to monitor the effects of acid precipitation, but to develop technologies that will lessen the environmental damage.

Solid and Liquid Wastes: Management, Methods and Socioeconomic Considerations. Edited by
S. K. Majumdar and E. Willard Miller. © 1984, The Pennsylvania Academy of Science.

Chapter Twelve

Organic Compounds in the Terrestrial Environment*

Michael R. Overcash, Ph.D.

Professor, Chemical Engineering Department
Professor, Biological and Agricultural Engineering Department
North Carolina State University
Raleigh, NC 27695

The study of organic chemicals (industrial and agricultural) is a broad field that encompasses research interests in pesticides and residues, other agricultural chemicals, soil sterilization, response of chemical spills, organic fertilizers, and fundamental investigations of soil-plant behavior. With this broader perspective, it is not surprising that there is a very large data base and substantial delineation of mechanisms operative for specific organic compounds in soil-plant systems. This information began accumulating in early 1900 and the article by Buddin (1914) represents (even at that early date) a sound understanding of germicidal effects on soils related to organic chemicals (some of which are now considered priority pollutants). Thus the research into the behavior of toxic and nontoxic organic compounds when applied to land has been ongoing for 70 to 80 years, is fairly well developed, and has continued as a significant research area today.

The magnitude and diversity of the organic species that have been evaluated in the context of a terrestrial system was estimated recently and put in catalogue form, called TERRETOX (Overcash 1981). From an article-by-article search of 240 appropriate journals for the period 1976-1981, over 4,000 organic compounds were found (many non-pesticides). These citations contained one or more evaluations of the following classes of terrestrial behavior of specific organics: volatilization, photolysis, decomposition, adsorption, plant uptake, phytotoxicity. Thus there exists a substantial (and to many surprisingly large) data base from which to begin evaluation of specific organic compounds as these might impact land application of municipal effluents and sludges.

---

*This paper is based on a similar presentation at EPA Conference on Utilization of Municipal Wastewater and Sludge on Land, Denver, CO, February 1983.

ORGANIC CHEMICAL BEHAVIOR IN TERRESTRIAL SYSTEMS

A specific organic species undergoes a variety of chemical and biological processes when applied to a soil or soil-vegetation system. Although a lot of research has not been conducted to verify preliminary observations, it is felt that the presence of the overall organic and inorganic fractions of a waste merely attenuates the various pathways by which an organic species is treated in the terrestrial system. Thus at this stage of development it remains important to understand and quantify the assimilative pathways (from whatever system was used to generate such data) and to project the probable acceptable rates for each organic constituent of interest in the terrestrial system.

Volatilization

An organic compound irrigated in a wastewater, surface spread with a sludge, present in a spill or contaminated site, or injected into the soil will partition between the gas and liquid phases to exert a vapor pressure. This vapor may be rapidly lost as might occur with irrigation, may be re-adsorbed if present beneath the soil surface, or may exhibit any behavior in between these two extremes. Both the factors, the manner of occurrence in a soil system as well as the inherent organic compound volatility, are important in quantifying how such material is lost through volatilization. It must also be remembered that volatile losses occur when municipal effluents are discharged to streams and when sludges are put in landfills as well as when these wastes are managed in soil systems.

The level of compound vapor pressure at which volatile losses are known to be significant (from pesticide research) is usually taken as 5×10^{-6} mmHg at $25°C$. If one tries to identify vapor pressure characteristics as a measure of whether volatile losses should be evaluated, this threshold is probably a good first estimate. However the adsorption to organic matter and soils could greatly reduce such initial concerns. For example, the substantially volatile toluene is the second most prevalent organic priority pollutant in municipal sludge even after opportunities for volatile loss during aerobic treatment, hence illustrating the strong adsorption effects.

Decomposition

This is a rather broad category which basically reflects the conversion of specific organic compounds (the parent species) into metabolites, chemical intermediates, biological material, and small molecules (primarily CO_2). Microbiological as well as chemical reactions are usually acting simultaneously and when measured as loss of parent compound are usually undifferentiated. The decompositon pathway is probably the most important facet of the organic constituent behavior in terrestrial systems and differentiates this group of chemical constituents most clearly from that of metals or other inorganics.

Chemical reactions (abiotic routes) are also a part of the overall measure of compound decomposition. Two typical reactions are hydrolysis and neutralization of the parent organic species. While these have been measured, it is not typical that these be differentiated from the overall decomposition process, probably since these chemical reactions leave the bulk of the parent structure still intact. Thus the separate quantification of abiotic reaction rates are uncommon.

Decomposition of specific organics is measured by four common methods:

1. CO_2 evolution,
2. O_2 consumption,
3. dynamics of intermediates or final products,
4. loss of parent compound.

Such experiments are not measuring the same phenomena in precisely the same manner and so results do not always agree. The intricacies of alternative techniques are not discussed here, but the reader must note that differences exist (Overcash 1979, 1981).

From the course of extensive literature searches concerning the decomposition of specific organics in the terrestrial environment, one concludes that very few organic compounds can be said to be non-degradable. This large decomposition potential of soil systems is not generally understood, but is significant in the national program to treat organic priority pollutants. Conventional aerobic treatment reviews list numerous organics as non-degradable. However this categorization most often should be interpreted as applicable for the relatively short residence times found in conventional processes (6 hours to 30 days). Given very long time periods typical in soil systems, it appears that two classes of compounds could be regarded as non-degradable.

1. synthetic polymers manufactured for stability,
2. very insoluble large molecules e.g. 5-10 chlorinated biphenyls.

In both of these cases, the soil microorganisms appear to have the potential to decompose the compounds (deamination, dehalogenation, etc.) but simply can not achieve the molecular level conditions necessary for decomposition.

It does appear that decomposition rate (kg/ha/yr or ppm/day) is dependent on the concentration of the organic species applied to the soil. This is illustrated in Figure 1 for polynuclear and heterocyclic aromatics. Such relationships are first order and the measurement of the decomposition rate or half life then become the essential data for design. However some compounds exhibit zero order decomposition rates (anthraquinone dyes, Tilchin 1983) under the conditions studied. Thus it is sufficient to state that there exists several reaction models and a large number of specific organic compound decomposition reaction rate constants from which to assess loading criteria and behavior in municipal waste land treatment systems. Table 1 contains a range organic decomposition half life values to illustrate the assimilative capacity data for specific organics.

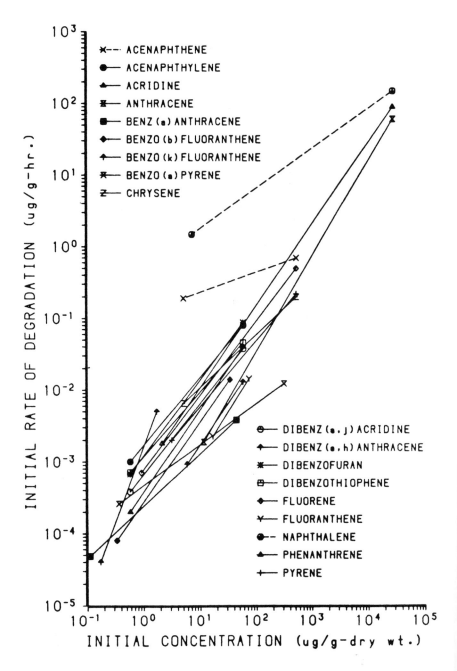

FIGURE 1. Decomposition of Polynuclear and Heterocyclic Aromatics (Sims and Overcash, 1983).

TABLE 1

Illustrative Range of Decomposition Half Life for Organic Compounds.

Compound	Approximate Half-Life
Aminoanthroquinone dyes	100-2,200d
Anthracene	110-180d
Benzo(a)pyrene	60-420d
Di n-butylphthalate ester	80'180d
Nonionic surfactants	300-600d
2,4 methyaniline	1.5d
n-Nitrosodiethylamine	40d
Phenol	1.3d
Pyrocatechin	12 hrs.
Cellulose	35d
Acetic Acid	5-8d
Hydroquinone	12 hr

Adsorption, Leaching, and Runoff

Among the physical processes governing the behavior of specific organics in soil-plant systems these three physical phenomena represent a balanced relationship. The attraction charge behavior between the oganic species and either the mineral or the organic soil phases results in a partitioning into adsorbed and solution states. This partitioning may occur immediately upon application to the soil or may be delayed until separation from the waste medium occurs. Competing decomposition reactions can vary the actual amount of a given organic compound that resides on the soil/waste phase or in the soil-water solution.

The adsorption or partitioning processes occur within very short distances from the waste application location (molecular or soil pore-size distances). In contrast, the occurrence of a soil water, matrix potential (for example from a rain event), migration of soil-water along appropriate gradients can cause transport of organics in the soil solution phase. This leaching can be viewed as theoretically capable of moving a dissolved organic (i.e. all organics) to any location receiving that soil water. That is, at least one molecule could be leached. As a practical matter, leaching of organics is insignificant if (1) the municipal effluent or sludge land treatment occurs at normal application rates, (2) a reasonable drainage and cyclic establishment of sustained aerobic soil conditions occur, and (3) groundwater remains deeper than 1-2 feet from the soil surface. That is, for most usual municipal land application situations or low level contamination situations, leaching or soil migration of specific organics does not appear to be significant or detectable. Rapid infiltration systems and overland flow systems are possible exceptions to this limited leaching behavior as well as landfills containing substantial amounts of organic constituents.

Organic compounds present at or near the soil suface (whether adsorbed or

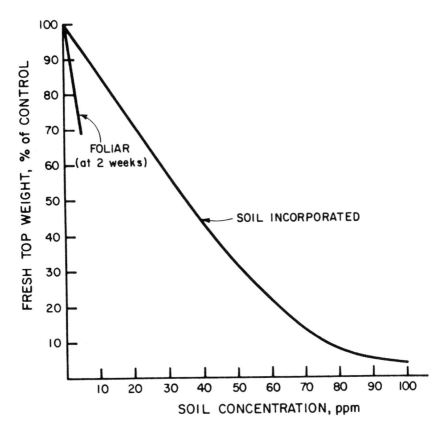

FIGURE 2. Foliar versus soil impact of 2,4 dinitro phenol on corn in Lakeland sand (Overcash et al. 1981).

in the soil-water solution) can also be transported in rainfall runoff. The mechanisms and the algorithms to estimate the magnitude in runoff of organic compounds are available in existing non-point source models and research related to pesticides (Donigian 1976, Frere 1975, Overcash 1983). Some modification to account for the organic sludge phase would be necessary to establish the runoff impact of organic priority pollutants from land treatment sites. The reader may utilize this extensive non-point source information to assess the runoff pathway.

Plant Response
   The final pathway describing the terrestrial behavior of a specific organic compound centers on the interaction with vegetation. Vegetation may be continuously present (e.g. forest, grasses) or may be planted and harvested in cycles (e.g. corn, soybeans). An organic species impacts vegetation in two broad ways.

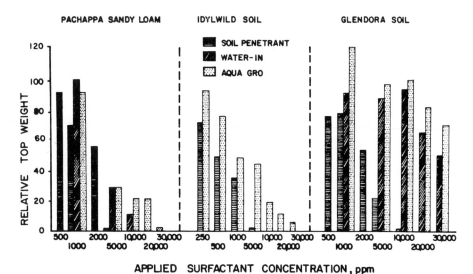

APPLIED SURFACTANT CONCENTRATION, ppm

FIGURE 3. Relative top weight of barley grown in soils receiving various applications of nonionic surfactants (Valoras et al 1976).

TABLE 2

Summary of critical phytotoxic levels for surfactants (Overcash and Pal, 1979)

Surfactant	Critical Soil Level (kg/ha) Yield < 80% of Control		
Three cationics	>		
Water-In (nonionic)	50	< > 250	Pachappa sl
	500	< >	
		1,000	Glendora
Aqua Gro (nonionic)	50	< > 250	Pachappa sl
	15	< > 25	Idylwild
	1,000	< >	
		3,000	Glendora
Soil Penetrant (nonionic)	25	< > 50	Pachappa sl
	~15		Idylwild
	~50		Glendora
Three anionics	< •ππ		

1. At low concentrations, a compound partitioned into the soil water solution is available for uptake by the vegetation root system.
2. At high concentrations a phytotoxic response occurs.

The magnitude of these plant responses is dependent upon both the organic chemical and the vegetative species. As with decomposition rates, there exists substantial direct and indirect information to allow evaluation of plant response to numerous organic compounds.

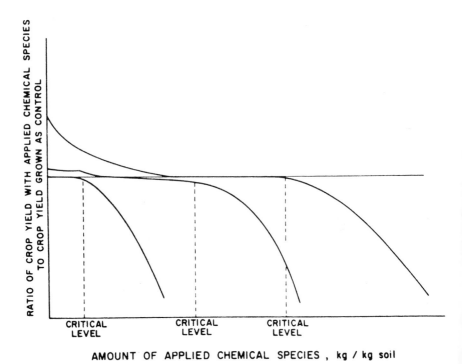

FIGURE 4. Illustrative stimulation and phytotoxic response of organic chemicals applied to soils.

Phytotoxic response can occur when the foliar portion of vegetation is exposed to an organic compound or when the root system contacts an organic species. The applied concentrations at which phytotoxicity occurs is typically different for foliar versus root contact. In the latter pathway the competing adsorption and decomposition phenomena allow greater tolerance to elevated concentrations of organic compounds. As an example of this plant phytotoxicity, Figure 2 illustrates the critical levels of the organic priority pollutant 2,4 dinitro phenol.

As a converse to the phytotoxicity response to organic compounds, many of these same compounds actually stimulate plant growth when present at low concentrations in the soil. That is, the chemical acts like a herbicide or insecticide when applied in small amounts. An example organic compound that illustrates this stimulatory effect at low soil concentrations is a nonionic surfactant, Figure 3. At higher levels the phytotoxic response is seen, Table 2. Figure 4 depicts the generalized plant growth behavior to applied organic species. Depending on the compound, there are different degrees of low level stimulation and varying critical soil concentration at which phytotoxic responses are evidenced.

TABLE 3

Benzo(a)pyrene Levels in Soils and Plants (Sims and Overcash, 1983)

Type	Plant Concentration	Soil Concentration	Plant:Soil
	-------mg/kg-dry weight-------		
Corn (above ground)	5.1	39.9	0.13
Oats (grains)	0.2	39.9	0.005
Oats (above ground)	4.6	39.9	0.12
Tomato fruit	0.1	39.9	0.003
Potato tubers	0.2	39.9	0.005
Wheat (grains)	0.1	39.9	0.003
Wheat (above ground)	4.6	39.9	0.12
Cotton seed	1.6	6.0	0.28
Potato(total)	0.08	5.0	0.016
Potato(total)	0.12	10.0	0.012
Potato (peeling)	0.36	5.0	0.072
Potato (peeling)	3.72	25.0	0.15

The second broad impact on vegetation from specific organic compounds is the uptake from the soil-water solution. This is a complex process that can involve transfer into the roots, translocation to upper portion of the vegetation, decomposition intra-plant, and even migration back to the roots and soil (as growth conditions in the vegetation changes). Most frequently, data are available on the measured concentrations of a specific organic in the vegetation (sometimes differentiated by parts) as related to the concentrations found in the soil. Examples of these data are given in Table 3. In a review of a number of organic compounds and uptake by various vegetative species, there does not appear to be any precise predictive mechanism for determining plant levels. It does appear that with very low detection limits that most organics can be taken into plant roots or tissues, however with simultaneous decomposition and adsorption the issue of uptake remains quantitatively unclear. If the absence of other soil processes (e.g. in hydroponic cultures) there does appear to be an uptake preference for cationic > nonionic > anionic. Since the soil adsorption operate with a similar preference (cationic > nonionic > anionic) the actual response of vegetation concentration is quite soil-and chemical-specific.

SUMMARY

There are a number of pathways in the plant-soil system which serve to completely assimilate and thus prevent adverse environmental impact of organic chemicals in terrestrial areas. These observations are based on the relatively low concentrations of these chemicals in land treatment systems, in residues after spill or contaminated site clean-up, or from atmospheric deposition on soils. Such minimal impact can not be assumed for these same organic chemicals when present in highly concentrated forms, such as landfills or massive spill areas. The reader should consult the large volume of literature available for landfill, spill area, and groundwater literature to assess the behavior of organics under those conditions. It is also important to continue reasearch into defining organic chemicals which are least well assimilated even under the low application rate conditions described above.

REFERENCES CITED

1. Buddin, W. "Partial Sterilization of Soil by Volatile and Nonvolatile Antiseptics," J. Agric. Sci. 6:417-451 (1914).
2. Casanova, M. and J. Dubroca. "Residues of Pentachloronitrobenzene and its Impurity Hexachlorobenzene in Soils and Lettuce," Academie d'Agric. de France, 12: 990-998 (1972).*
3. Dejonckheere, W., W. Steurbaut and R. H. Kips. "Residues of Quintozene, Hexachlorobenzene, Dichloran, and Pentachloraniline in Soil and Lettuce," Bull, Enviorn. Contam. Toxicol. 13:720-729 (1975).*
4. Donigan, A. S., Jr. and N. H. Crawford. Modeling Pesticides and Nutrients on Agricultural Lands. EPA-600/2-76-043, 332 pp. (1976).
5. Frere, M. H., C. a. Onstad and H. N. Holtan. "An Agricultural Chemical Transport Model." USDA ARS-H-3. U. S. Govt. Printing Office. 54 pp. (1975).
6. Hafner, V. M. "Hexachlorbevzolruckstande in Gemuse - bedingt durch Aufnahme des hexachlorokenzols aus dem Boden," Gesunde Pflanzen 27(3):37-48 (1975).*
7. Overcash, M. R., R. Khaleel, K. R. Reddy and P. W. Westerman. "Nonpoint Source Model: Watershed Inputs From Land Areas Receiving Animal Wastes," U.S. E.P.A. report, in review, (1983).
8. Overcash, M. R., J. B. Weber and M. L. Miles. "Behavior of Organic Priority Pollutants in the Terrstrial System: Di-n-butyl Phthalate Ester, Toluene, and 2,4 Dinitrophenol," Water Resource Research Institute of North Carolina, Report 171, 94 pp. (1982).
9. Overcash, M. R. and R. Sims, 1981, Terrestrial behavior of organic chemicals catalogue, Chemical Engineering Department, North Carolina State University.
10. Overcash, M. R. "Decomposition of Toxic and Nontoxic Organic Comounds in Soils, Ann Arbor Publishers, Ann Arbor, MI, 485 pp. (1981a).
11. Overcash, M. R. and D. Pal. "Design of Land Treatment for Industrial Wastes, Theory and Practice," Ann Arbor Science Publishers, Ann Arbor, IL, 684 pp. (1979).

12. Sims, R. C. and M. R. Overcash. "Fate of Polynuclear Aromatic Compounds (PNAs) in Soil-Plant Systems," Residue Review, in press (1983).
13. Smelt, J. H. "Behavior of Quintozene and Hexachlorbenzene in the Soil and their Absorption in Crops," Gewasbeschermung 7(3):49-58 (1976).
14. Tilchin, M. Land Treatment of Textile Dyeing Wastewaters, M.S. thesis, Biological and Agricultural Engineering Department, North Carolina State University, 1982.
15. Valoras, N., J. Letey and J. Osborn. "Nonionic Surfactant-Soil Interaction Effects on Barley Growth," Argon, J. 68(4):591-595 (1976).
16. Wallnofer, P., M. Roniger and G. Englehardt. "Fate of Xenobiotic Chlorinated Hydrocarbons (HCB and PCBs) in Plants and Soils," Zeit. Pflanzen. Planzenschutz 82(2):11-100 (1975).*

*For translated version of this article, see Overcash, M. R., "Decomposition of Toxic and Nontoxic Organic Compounds in Soil," Ann Arbor Science Publishers, Ann Arbor, MI, 485 p. (1981).

Solid and Liquid Wastes: Management, Methods and Socioeconomic Considerations. Edited by
S. K. Majumdar and E. Willard Miller. © 1984, The Pennsylvania Academy of Science.

Chapter Thirteen

Cadmium and Other Trace Elements in the Environment[1]

Ann M. Wolf[2] and Dale E. Baker[3]

[2]Research Assistant in Soil Chemistry and
[3]Professor of Soil Chemistry
Department of Agronomy
221 Tyson Building
The Pennsylvania State University
University Park, PA 16802

The goal of this paper is to present an overview of the most important principles involved in understanding the movement and potential toxicity of trace elements in the food chain. At present, the disposal of sewage sludge on agricultural land represents a primary means through which concentrations of trace elements may be increased in soils. The discussion which follows will be limited to the potentially detrimental effects that elevated levels of Cd and other trace elements found in sludge may have on plant growth and on animal and human health. From an environmental and ecological standpoint, the effects of excessive levels of trace elements on other life forms, such as the soil biota and the wildlife species, are also important. However, because of the limited data available and the limitations of space, a discussion of these effects will not be included.

------

[1]Contribution of the Pennsylvania Agri. Exp. Sta., University Park, PA. Authorized for publication on December 8, 1983 as paper number 6838 in the Journal Series.

MOVEMENT OF TRACE ELEMENTS THROUGH THE
SOIL-PLANT-ANIMAL SYSTEM: OVERVIEW

The most important pathways by which trace elements applied to the soil move through the soil-plant-animal system are shown in Figure 1. Examples are included of the pathways taken by several of the potentially harmful trace elements found in sludge. These elements include Cd, Cu, Ni, Mo, Pb, and Zn (CAST, 1976; Sommers, 1980; Baker et al., 1984b). As shown in Figure 1, a trace element applied to the soil may (1) be held in a form that is available to plants, (2) be leached through the soil profile to the ground water, or (3) react with the soil components in such a way that it is effectively immobilized. If taken up by plants, the element may produce toxicity symptoms when present in large enough quantities or it may have little or no effect on plant growth. If plant growth is not adversely affected, the element may continue its path through the food chain to be ingested by domestic animals or human beings. Alternative routes which trace elements applied to soils may enter the food chain are through ground water contamination, through the direct ingestion of soils by grazing livestock animals and through the ingestion of soil by children during play activities (see Figure 1).

To evaluate the impact of increasing the level of a trace element in the soil, one must have an understanding of the manner in which the element interacts with each phase of the soil-plant-animal system. In this section, a brief overview will be presented of the factors most important in influencing the behavior of trace elements applied to soils and their subsequent movement through the food chain. In section III, the specific factors influencing the movement and toxicity of the most potentially harmful trace elements found in sewage sludge will be presented.

A. Trace element reactions with soils

A summary of the reactions occurring between trace elements in the soil solution and the solid phase is shown in Figure 2. In general, the most important reactions involving retension of trace elements in soils are those of electrostatic and specific adsorption (Ellis and Knezek, 1972; Hodgson, 1963). Electrostatic adsorption involves the attraction of a charged species of one sign for a charged soil component of the opposite sign. Specific adsorption is a general term which refers to the ability of some species to be preferentially adsorbed by the soil components over others. It is attributed to additional forces of attraction between the trace element species and the soil surface and is generally distinguished from electrostatic adsorption by its higher adsorption energy (Bolt and Bruggenwert, 1978). In some instances precipitation of trace elements may also be important (Lindsay, 1972; Hodgson, 1963). However, in soils of neutral or lower pH, the concentration of most trace elements is too low for precipitation reactions to occur (Lindsay, 1972).

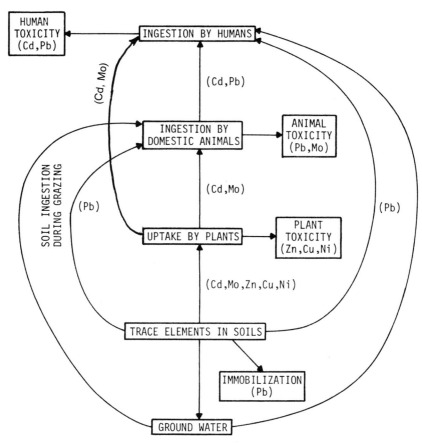

FIGURE 1. Pathways through which trace elements applied to soils may impact plants, domestic animals and human beings. (Pathways which may be taken by several of the potentially harmful trace elements are shown in parentheses.)

The specific nature and extent of the reactions involving trace element in soils will determine the element's distribution between the solid and solution phases. The activity of a trace element maintained in the soil solution will be important in determining its availability to plants as well as its ability to be leached from the soil profile. Factors which influence the retention of trace elements in soils include (1) the nature of the soil components, (2) the forms and concentration of the trace elements present, (3) the concentrations of organic and inorganic species in the soil solution, and (4) the soil pH (Ellis and Knezek, 1972; Hodgson, 1963; Stevenson and Ardakani, 1972; Page et al., 1982; Jenne, 1968). In addition the properties of the waste with which the trace element behavior in soils (CAST, 1976).

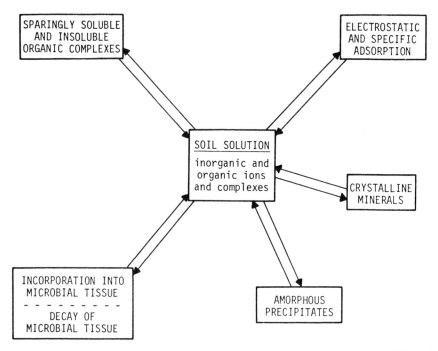

FIGURE 2. Reactions occurring between a trace element in solution and components of the soil solid phase (adapted from Page et al., 1981).

The soil components of primary importance in trace element adsorption are the clay minerals, organic matter and the amorphous oxides of Fe, Al, and Mn (Ellis and Knezek, 1972; Jenne, 1968; Page et al., 1981). While many of the clay minerals carry a permanent charge, the charge characteristics of the organic matter fraction, the oxides and other clay minerals will be determined by the pH of the soil solution.

The forms and quantities of trace elements applied to soils are of obvious importance in influencing their adsorption. The form of a trace element will determine its charge characteristics as well as its ability to undergo specific adsorption reactions. Factors influencing the form of the trace element found in the soil are the presence of organic and inorganic complexing reagents (Page, et al., 1981; Stevenson and Ardakani, 1972), the redox condition of the soil (Jenne, 1968; Hodgson, 1963) and, for those elements which undergo hydrolysis reactions, the pH of the soil solution (Ellis and Knezek, 1972). The quantity of a trace element applied is important in relation to the soil's adsorption or "loading" capacity for the element (Baker et al., 1984a) as is the presence of other trace elements or inorganic constituents which are competitive for the same adsorption sites (Page, et al., 1981).

As a consequence of the pH effects mentioned above, the overall effect of increasing the pH on trace element retention by soils is, in general, to increase the adsorption of the positively charged species and to decrease the adsorption of species which are negatively charged (Page, et al., 1981).

Because of the complexity of the factors listed above, it is difficult to predict the specific reactions which will occur when a given trace element is applied to a soil. However, based on our understanding of these factors, generalizations on the behavior of individual trace elements applied to soils may be made.

B. Movement and Impact of Trace Elements in the Food Chain

While the solution activity of the trace element maintained by the soil will be critical in determining its availability to plants, the actual quantity adsorbed will be determined by specific plant properties. In monitoring the movement of a trace element through the soil-plant-animal system, it is necessary to understand the potential for the trace element to be adsorbed by plants in relation to the quantities considered potentially harmful to plant growth and to animal and human health.

Many of the trace elements applied to the soil with wastes are essential to plants, animals, and man but are toxic if taken up or ingested in excessive quantities. Not infrequently, there is only a small margin between the sufficiency and toxicity of an element (Bowen, 1979). Other elements which are not essential to plants or animals may be tolerated at low levels but beyond a "critical" concentration will have a detrimental effect on the health of the organism (Bowen, 1979).

These concepts are illustrated graphically in Figures 3a and 3b which show idealized response curves of an organism to increasing supplies of essential and non-essential elements. In general, when the margin between the sufficency (or inactivity) and toxicity of an element is narrow, increasing the supply of the element to the organism presents a greater hazard than increasing the supply of an element for which the margin is wide. Thus, the relative toxicity of an element may be judged on the basis of the margin that exists between the levels which are sufficient (or inactive) and those which are toxic.

In evaluating the movement of trace elements through the soil-plant-animal system, it is helpful to have information on the "critical" levels at which the element becomes toxic to plants, domestic animals and man. A comparison of these levels allows one to determine where in the food chain the impact of the elevated trace element concentrations will be greatest. However, because of the diversity and complexity of a biological system, it is difficult to establish single toxicity levels. Plant species and varieties vary in their abilities to absorb trace elements from soils as well as in their sensitivities to excessive trace element concentrations (Loneragan, 1975; Brown et al., 1972). The tolerance of

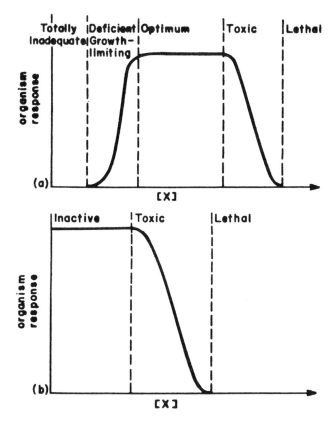

FIGURE 3. Response curves of an organism to increasing concentrations of an element X: (a) X is essential; (b) X is non-essential (from Bowen, 1979).

a plant to concentrations of one trace element may also be influenced by the presence or absence of other elements in the plant tissues (Loneragan, 1975). Similarly, animal species differ with respect to the levels at which trace elements become toxic, and like plants, this level will vary with the nutrient status of the animal (National Research Council, 1980; Underwood, 1977).

Despite these difficulties, attempts have been made to establish general levels at which many of the trace elements become toxic to plants, animals and man (Allaway, 1968; Bowen, 1979; National Research Council, 1980; US Bureau of Foods, 1975). A list of these values for several of the potentially harmful trace elements present in sewage sludge are given in Table 1. Reference to this table will be made when the impact of applying each of these elements to soils is evaluated in the section that follows.

TABLE 1

Essentiality and toxicity of several trace elements to plants, domestic animals and man.

Element	PLANTS			DOMESTIC ANIMALS			MAN			
	Essentiality	Normal concentration[1] (ppm)	Toxic concentrations[2] (relative or ppm)	Essentiality	Relative toxicity[4]	Maximum tolerable dietary limits for cattle, sheep, and swine[5] (ppm)	Essentiality	Current dietary intake[6] (mg/day)	Toxicity[7] (relative or mg/day)	WHO/ FAO tolerable dietary limits[8] (mg/day)
Cd	non-essential	0.2-0.8	7-160[3]	non-essential	mod-high	0.5	non-essential	0.039	high	0.055-0.071
Cu	essential	5-20	>20	essential	low	25-250	essential	0.5-6.0	low	—
Mo	essential	1-100	low toxicity	essential	high	10-20	essential	0.05-0.35	low	—
Ni	non-essential	1-5	>50	essential	low-mod	50-100	essential	0.30-0.50	low	—
Pb	non-essential	0.1-1.0	low toxicity	non-essential	mod	30	non-essential	0.254-0.30	high (1)	0.429
Zn	essential	25-150	>200	essential	low	300-1000	essential	18	low (150-600)	—

[1]From Allaway, 1968; Chapman, 1966; and Jones, 1972.
[2]Values for all elements except Cd from Allaway, 1968.
[3]Concentration of Cd in diagnostic leaf tissues of different platn species associated with a 25% decreasein yield. From Bingham et al., 1975, and Bingham et al., 1976.
[4]From Bowen, 1979.
[5]From Natinoal Research Council, 1980. Values for cadmium and lead based on human food residue considerations.
[6]From Bowen, 1979, and U. S. Bureau of Foods, 1975.
[7]From Bowen, 1979; Gough et al., 1979; and Underwood, 1977.
[8]From U. S. Bureau of Foods, 1975.

MOVEMENT AND IMPACT OF Cd, Cu, Mo, Ni, Pb AND Zn IN THE SOIL-PLANT-ANIMAL SYSTEM

In the previous section, general factors influencing the movement and toxicity of trace elements in the soil-plant-animal system were discussed. In this section, these factors will be considered in relation to the impact of applying excessive quantities of Cd, Cu, Mo, Ni, Pb, and Zn to soils on plant growth and on animal and human health.

A. Cadmium

Of the trace elements commonly applied to soils with sludge, Cd is recognized as the one presenting the greatest hazard to human health (CAST, 1976; Chaney, 1980; Baker et al., 1984b). The long-term accumulation of Cd in the human body has been associated with kidney failure (Bousquet, 1979; Goyer et al., 1978), hypertension (Schroeder, 1965), and cardiovascular disease (Bousquet, 1979). In addition, evidence suggests that Cd may also be an oncogen (Malcolm, 1979) and a teratogen (Rohrer et al., 1979).

While Cd may be toxic to plants when taken up in large quantities, the levels at which toxicity will occur are too high to prevent unacceptable increases of Cd from occurring in the diets of domestic animals and human beings. As shown on Table 1, Cd is normally present in the vegetative tissues of plants in the range of 0.2 to 0.8 ppm. However, depending on the plant species, greater than 100 fold increases of Cd in the plant tissues may be tolerated before toxicity is observed. In contrast, the World Health Organization's recommended maximun dietary intake of .055 to .071 mg Cd/person/day is on the order of only two times greater than the estimated current dietary intake of .039 mg/person/day (Table 1). Because the margin between the current and the acceptable levels of Cd in the diet of humans is so small in comparison to the quantities that may be accumulated by plants, it is important to understand the soil and plant factors which influence the movement of Cd through the food chain.

In uncontaminated soils, Cd is present in quantities ranging from 0.01 to .070 ppm with an average of 0.06 ppm (Lindsay, 1979). When added to soils, the activity of Cd maintained in the soil solution will be influenced by the soils clay content, the percent organic matter, the presence of hydrous oxides and the soil pH (Jenne, 1968; Korte et al., 1976, Gardiner, 1973; Andersson and Nilsson, 1974). In general, as pH increases, availability of Cd to plants decreases (Andersson and Nilsson, 1974). This effect has been attributed to a decrease in competition for exchange sites by H^+ ions and to an increase in the number of pH dependent cation exchange sites on the clays, organic matter and hydrous oxides. In addition, others (Andersson and Nilsson, 1974; Iwai et al., 1975) have suggested that competition of Ca and Cd at the root surface may be partially responsible for decreased uptake when lime is added to a soil. While adsorption of Cd is known to occur on specific soil surfaces, the chemistry of Cd in soils is still not well understood and it is not possible to evaluate the adsorption capacity of a soil for Cd on the basis of its known physical and chemical properties. Currently, U.S.E.P.A. guidelines for determining the maximum loading limits for Cd in soils are based on the cation exchange capacity (CEC) of the soil and the soil pH (U.S.E.P.A., 1979). In the Northeast, where the soils are predominantly acid, loading limits for Cd are based on soil texture (Baker et al., 1984b). Both CEC and soil texture reflect the relative abundance of some soil components responsible for absorbing Cd. Alternatively, Baker et al. (1984a) have proposed the use of a laboratory procedure to determine the quantity of Cd that may be added to a soil while still maintaining safe levels of Cd in the soil solution. The loading limits for Cd proposed by these groups are based on the potential for Cd applied to different soils to be taken up by plants and to move through the food chain in quantities which may be harmful to human health.

For a given set of soil conditions, it is generally agreed that additions of Cd to the soil will result in increased concentrations of Cd in the plant in quantities proportional to the amounts added (Ryan, 1982). However, there exist large differences among both plant species and varieties as to the amounts of Cd accumulated in the plant tissues (Bingham et al.,1975; Hinesly et al., 1978). In addition, the accumulation of Cd in various plant parts has been found to vary considerably. Generally, leafy vegetables such as lettuce and spinach have been found to accumulate the greatest amount of Cd as have the vegetative parts of plants in comparison to the fruit and seed tissues (Bingham et al., 1975). Thus in addition to the importance of soil properties, the selection of crop species and/or variety to be grown on a soil enriched with Cd may also be important in controlling the quantities of Cd which enter the food chain.

From the above discussion, it is apparent that increases of Cd in the soil will result in increases of Cd in the food chain. In evaluating the impact of applying wastes containing Cd to soils it is necessary to consider the soil and plant factors discussed above in relation to the increases of Cd in the diet that will not adversely affect human health. While the many factors affecting the movement of Cd through the soil-plant-animal system makes this task difficult, an understanding of these factors is essential in establishing loading limits for Cd in soils that will prevent excessive quantities of this element from entering the food chain.

B. Copper

Copper is found naturally in soils in the range of 2 to 100 ppm with an average value of 30 ppm (Lindsay, 1979). The most important reactions involving Cu in soils are those with organic matter and with the hydrous oxides of Fe, Al and Mn (Ellis and Knezek, 1972; Jenne, 1968). While Cu is strongly adsorbed by organic matter, it will also form soluble organic complexes (Ellis and Knezek, 1972; Stevenson and Ardakani, 1972) which in many soils is the predominant form of Cu found in the soil solution (Hodgson, 1963). In general, increasing the soil pH will decrease the activity of Cu in the soil solution (Ellis and Knezek, 1972; Lindsay, 1972).

Copper is an essential element to plants but only a narrow margin exists between the concentrations normally present in the plant tissues and those at which toxicity is observed (Table 1). Part of this narrow range may be attributed to the fact that excessive quantities of Cu will commonly accumulate in the plant root before being translocated to the aerial portions of the plant (CAST, 1976). The level at which toxicity occurs will be influenced by the levels of P, Fe, Mo and Zn that are available to the plant (Olsen, 1972).

Copper is also an essential element to animals but the margin between essentiality and toxicity varies with the individual species as well as with the levels of other trace elements present in the diet (National Research Council, 1980). Copper does not appear to be highly toxic to man (Gough, 1979) and it is unlikely that increases in dietary Cu that could occur through plant uptake would have a detrimental effect on human health. Of the domestic animals, ruminants, especially sheep, are the most sensitive to Cu toxicity and the maximum tolerable level of dietary Cu recommended in the diet of sheep is 25 ppm (National Research Council, 1980). In comparison, the tolerable levels recommended for cattle and swine are 200 and 250 ppm, respectively. A comparison of these levels with those resulting in toxicity to plants (Table 1) indicates that applications of excessive quantities of Cu to soils will in most instances be more detrimental to plant growth than to animal health.

C. Lead

The concentration of Pb in uncontaminated soils averages 10 ppm although it may be found in concentrations ranging from 2 to 100 ppm (Lindsay, 1972). Present in the soil solution primarily as the divalent cation,, Pb^+,Pb is strongly sorbed to the soil surfaces and may also undergo precipitation reactions forming relatively insoluble hydroxide and phosphate compounds (CAST, 1976; Lagerwerff, 1972). The availability of Pb to plants will thus be influenced by the pH of the soil as well as the phosphorus content.

Lead is a non-essential element to plants and tends to accumulate in the plant roots rather than be transported to the above-ground portions (Chapman, 1966). The toxicity of this element to plants is considered low and when soils are limed to pH levels of 5.5 or higher, the impact of Pb on plant growth and its movement into the aerial portions of the plant should be minimized (CAST, 1976). However, as shown in Figure 1, Pb may also enter the food chain through routes other than plant uptake. One of these is through the direct ingestion of soil particles by grazing animals, and another is the ingestion of soil by children during play activities (Chaney, 1980). In general, lead is considered moderately toxic to livestock animals (Table 1) through the quantities required for toxicity will be influenced by the concentrations of Ca, Fe, Zn and P present in the diet

(Lagerwerff, 1972). The maximum concentration of Pb recommended in the diets of domestic animals is 30 ppm (Table 1).

Food represents a major source of Pb ingested by humans, and of the food sources approximately 30% is estimated to come from meat and chicken (Lagerwerff, 1972). Thus, an increase in the accumulation of Pb by livestock could likely result in an increase of Pb in the diet of man. The World Health Organization (US Bureau of Foods, 1975) has determined that the maximum safe dietary intake of Pb by humans is .429 mg/day. In a study performed by the US Bureau of Foods (1975), the current dietary intake of Pb by teenagers was estimated to be .254 mg/person/day indicating that only a small margin exists betwen the current and the maximum safe levels of Pb in the diet. While it is difficult to evaluate the movement of Pb from the soil through the food chain because of the routes by which it may enter, its potential toxicity to animals and man makes it an element with which we must be concerned when Pb-containing wastes are applied to soils.

Molybdenum

Molybdenum is commonly found in soils in the range of 0.2 to 5 ppm (Lindsay, 1979) and is present in the soil solution as the oxyanion, MoO_4^{-2} (Lindsay, 1972). In acid soils, Mo is associated predominantly with the hydrous oxides of iron, and to a lesser extent, aluminum (Ellis and Knezek, 1972; Russell, 1973). As a consequence of the increasing negative charge on the oxide surface with increasing pH, Mo becomes more available to plants as the pH of the soil rises (Ellis and Knezek, 1972).

Molybdenum is an essential element to plants and may accumulate in quantities far in excess of those required before toxicity symptoms are observed (see Table 1). Molybdenum is also essential to livestock, but the margin between sufficiency and toxicity to livestock animals is much smaller than for plants (see Table 1). The concentration at which Mo becomes toxic to animals will largely be dependent upon the concentrations of Cu and P in the diet as well as upon the animal species (National Research Council, 1980; Underwood, 1977). Cattle are among the most susceptible of the livestock species to Mo toxicity and will tolerate a level of up to 10 ppm in their diet (National Research Council, 1980). As shown in Table 1, plants may accumulate up to 10 times this quantity before toxicity is observed. As a consequence, the applications of excessive quantities of Mo to soils will have its greatest impact on animal health. While there are no established values for the tolerable level of Mo in the diet of humans, this element does not appear to be highly toxic to man (Gough et al., 1979; Underwood, 1977).

E. Nickel

Background levels of Ni in soils may range from 5 to 500 ppm (Lindsay, 1979). Nickel is normally present in the soil solution as divalent cation Ni^{2+}, and its retention in the soil will be influenced primarily by its reactions with the hydrous oxides and organic matter (CAST, 1976; Jenne, 1968; Stevenson and Ardakani, 1972). As a divalent cation the solubility of Ni and hence its availability to plants will decrease with an increase in pH (Chapman, 1966).

Nickel is non-essential to plants but is an essential element to animals and man. A comparison of the levels at which Ni becomes toxic to plants with the concentration of Ni tolerated in the diets of domestic animals (Table 1) indicates that the impact of adding excessive quantities of Ni to soils will be greater on plant growth than on animal health. Similarly, Ni is a relatively non-toxic element to man (Underwood, 1977), and the movement of this element through the soil-plant-animal system is unlikely to occur in quantities that would prove detrimental to human health.

F. Zinc

The average content of Zn in soils is 50 ppm though it may range in content from 10 to 300 ppm (Lindsay, 1979). The most predominant form of this element in the soil solution is Zn^{2+} but it may also be present in soluble organic complexes (Lindsay, 1972; Ellis and Knezek, 1972). The activity of Zn maintained in the soil solution will be controlled by its reaction with organic matter (Ellis and Knezek, 1972), the hydrous oxides of Fe and Mn (Jenne, 1968), and with the clay minerals where it may be incorporated in the clay structure as well as adsorbed onto the clay surface (Ellis and Knezek, 1972). In general, the activity of Zn in the soil solution and its availability to plants decreases with an increase in pH (Lindsay, 1972).

Zinc is an essential element to plants, animals and man. A comparison of the margins between sufficiency and toxicity of this element indicates that Zn will reach toxic concentrations in plants before it reaches concentrations in the diets of animals or humans that would be detrimental to their health. In general, Zn will become toxic to plants when present in concentrations greater than 200 ppm in the plant tissue although the actual concentration will vary with the individual plant species as well as with the concentrations of Fe, Cu and P available to the plant (Olson, 1972). In contrast, sheep, which are among the more sensitive animals to Zn toxicity, can tolerate levels up to 300 ppm in their diets while levels as high as 1000 ppm can be tolerated by less sensitive species (National Research Council, 1980). The relative toxicity of Zn to man is low (Underwood, 1977) and large increases in dietary Zn would have to occur before this element becomes hazardous to human health.

SUMMARY AND CONCLUSIONS

The impact of applying wastes containing trace elements to soils will be determined by the ability of the element to move through the soil-plant-animal system in relation to the quantities of the element considered toxic to plants animals and man. An overview of the most important factors influencing the movement of trace elements through the food chain has been considered and the potential impact of increasing the soil content of Cd, Cu, Mo, Ni, Pb and Zn has been discussed. Although these are only a few of the many trace elements that could be applied to soils with wastes, the approach taken in this paper may be used to evaluate the potential impact of increasing the content of other trace elements in soils as well.

While generalizations concerning the behavior and impact of applying large quantities of trace elements to soils may be made, much of the knowledge required to critically evaluate the long-term impact of trace element additions to soils is still not available. Consequently, the application to soils of wastes containing trace elements known to be hazardous to plants, animals or man should be done with caution until the full implications of increasing the trace element content of soils can be more critically evaluated.

REFERENCES

Allaway, W. H. 1968 Agronomic controls over the environmental cycling of trace elements. Adv. Agron. 20:235-274.

Andersson, A. and K. O. Nilsson. 1974. Influence of lime and soil pH on Cd availability to plants. Ambio. 3:198-200.

Baker, D. E., D. S. Rasmussen and J. Kotuby. 1984a. Trace metal interactions affecting soil loading capacities for cadmium. IN: Industrial and Hazardous Waste. Proc. Int. Symposium, ASTM Committee D-34, Philadelphia, PA (in press).

Baker, D. E., et al. 1984b. Criteria and recommendations for land application of sludges in the Northeast. Pa. Agri. Exp. Sta. Bul. (in press).

Bingham, F. T. and A. L. Page, R. J. Mahler, and T. J. Ganje. 1975. Growth and Cd accumulation of plants grown on a soil treated with Cd enriched sewage sludge. J. Environ. Qual. 4:207-211.

Bingham, F. T. and A. L. Page, R. J. Mahler, and T. J. Ganje. 1976. Yield and cadmium accumulation of forage species in relation to cadmium content of sludge-amended soil. J. Environ. Qual. 5:57-59.

Bolt, G. H. and M. G. M. Bruggenwert. 1978. Soil Chemistry. Elsevier Scientific Publishing Co. NY. 281pp.

Bousquet, W. F. 1979. Cardiovascular and renal effects of cadmium. IN: H. H. Menner (ed.), Cadmium Toxicity. Marcel Dekker, Inc., NY. pp. 133-157.

Bowen, H. J. M. 1979. Environmental Chemistry of the Elements. Academic Press, NY. 333 pp.

Brown, J. C., J. E. Ambler, R. L. Chaney and C. D. Foy. 1972. Differential responses of plant genotypes to micronutriets. IN: J. J. Mortvedt, P. M. Giordano, and W. L. Lindsay (eds.) Micronutrients in Agriculture. Soil Sci. Soc. Amer., Inc., Madison, WI. pp. 389-418.

Chaney, R. L. 1980. Health Risks associated with toxic metals in municipal sludge. IN: G. Bitton, B. L. Damron, G. T. Edds, and J. M. Davidson (eds.), Sludge—Health Risks of Land Application. Ann Arbor Science Pub., Ann Arbor, MI. 367 pp.

Chapman, H. D. (ed.). 1966. Diagnostic Criteria for Plants and Soils. Quality Printing Co., Inc., Abilene, TX

Council for Agricultural Science and Technology (CAST). 1976. Application of sewage sludge to cropland: Appraisal of potential hazards of the heavy metals to the plants and animals. U. S. Environmental Protection Agency (EPA-430/9-76-013,) Washington, DC. 63 pp.

Ellis, B. G. and B. D. Knezek. 1972. Adsorption reactions of micronutrients in soils. IN: J. J. Mortvedt, P. M. Giordano, and W. L. Lindsay (eds.), Micronutrients in Agriculture. Soil. Sci. Soc. of Amer., Inc., Madison, WI. pp. 59-78.

Gardiner, J. 1973. The chemistry of cadmium in natural water—II. The adsorption of cadmium on river muds and naturally occurring solids. Water Research 8:157-164.

Gough, L. P., H. T. Shacklette , and A. A. Case. 1979. Element concentrations toxic to plants, animal, and man. Geological Survey Bulletin 1466. US Government Printing Office, Washington, DC.

Goyer, R. A., M. G. Cherian and L. D. Richardson. 1978. Renal effects of cadmium. IN: Proc. First Int. Cadmium Conf. Metals Bulletin, Limited, London. pp. 183-185.

Hinesly, T. D., D. E. Alexander, E. L. Zeigler, and G. L. Barrett. 1978. Zinc and cadmium accumulation by corn inbreds grown on sludge-amended soil. Argon. J. 70:425-428

Iwai, I., T. Haro, and Y. Sonoda. 1975. Factors affecting Cd uptake by the corn plant. Soil Sci. Plant Nutr. 21:37-46

Jenne, E. A. 1968. Controls on Mn, Fe, Co, Ni, Cu, and Zn concentrations in soils and waters. The significant role of hydrous Mn and Fe oxides. IN: R. J. F. Gould (ed.), Trace Inorganics in Water. Amer. Chem. Soc., Washington, DC.

Jones, J. B. 1972. Plant tissue analysis for micronutrients. IN: J. J. Mortvedt, P. M. Giordano and W. L. Lindsay (eds.), Micronutrients in Agriculture. Soil Sci. Soc. Amer., Madison, WI. pp. 319-346.

Korte, N. E., J. Skopp, W. H. Fuller, E. E. Niebla, and B. A. Alesii. 1976. Trace element movement in soils: Influence of soil physical and chemical properties. Soil Sci. 122:350-359.

Lagerwerff, J. V. 1972. Lead, mercury and cadmium as environmental contaminants. IN: J. J. Mortvedt, P. M. Giordano, and W. L. Lindsay (eds.), Micronutrients in Agriculture. Soil Sci. Soc Amer., Madison, WI. pp. 593-636.

Lindsay, W. L. 1979. Chemical Equilibria in Soils. Wiley-Interscience, Somerset, NJ. 416 pp.

Lindsay, W. L. 1972. Inorganic phase equilibria of micronutrients in soils. IN: J. J. Mortvedt, P. M. Giordano, and W. L. Lindsay (eds.), Micronutrients in Agriculture. Soil Sci. Soc. Amer., Madison, WI. pp. 41-57.

Loneragan, J. F. 1975. The availability and adsorption of trace elements in soil-plant systems and their relation to movement and concentration of trace elements in plants. IN: D. J. D. Nicholas and A. R. Egan (eds.), Trace elements in Soil-Plant Systems. Academic Press, Inc., NY. pp. 417.

National Research Council. 1980. Mineral Tolerance of Domestic Animals. National Academy of Sciences, Washington, DC 577 pp.

Malcolm, D. 1979. Cadmium as a carcinogen. IN J. H. Mennear (ed.), Cadmium Toxicity. Marcel Dekker, Inc., NY.

Olsen, S, R, 1972. Micronutrient interactions. IN: J. J. Mortvedt, P. M. Giordano, and W. L. Lindsay (eds.), Micronutrients in Agriculture. Soil Sci. Soc. Amer. Inc., Madison, WI. pp. 243-264.

Page, A. L., A. C. Chang, G. Sposito, and S. Mattigod. 1981. Trace elements in wastewater: their effects on plant growth and composition and their behavior in soils. IN: I. K. Iskander (ed), Modeling Wastewater Renovation by Land Application. US Army Cold Regions Res. and Engr. Labs. Monograph, John Wiley and Sons., NY. pp. 182-222.

Rohrer, S. R., S. M. Shaw and D. C. VanSickle. 1979. Cadmium teratogenesis and placental transfer. IN: J. H. Mennear (ed.), Cadmium Toxicity. Marcel Dekker, Inc., NY.

Russell, E. W. 1973. Soil Conditions and Plant Growth. 10th edition. William Clowes and Sons, Ltd., London.

Ryan, J. A., H. R. Pahren, and J. B. Lucus. 1982. Controlling cadmium in the human food chain: A review and rationale based on health effects. Environ. Res. 28:251-302.

Schroeder, H. A. 1965. Cd as a factor in hypertension. J. Chron. Dis. 18:647-656.

Sommers, L. E. 1980. Toxic metals in agricultural crops. IN: G. Bitton, B. L. Damron, G. T. Edds, and J. M. Davidson (eds.), Sludge—Health Risks of Land Application. Ann Arbor Science Pub., Ann Arbor, MI. 367 pp.

Stevenson, F. J. and M. S. Ardakani. 1972. Organic matter reactions involving micronutrients in soils. IN: J. J. Mortvedt, P. M. Giordano, and W. L. Lindsay (eds.), Micronutrients in Agriculture. Soil Sci. Soc. Amer., Inc., Madison, WI. pp. 79-114.

Underwood, E. J. 1977. Trace Elements in Human and Animal Nutrition. 4th Edition. Academic Press, NY. 545 pp.

U.S. Bureau of Foods. 1975. Compliance program evaluation. FY 1974 Heavy metals in foods survey. (7320, 13C) Food and Drug Administration, H. E. W., Washington, DC. 21 pp.

U.S. Environmental Protection Agency. 1979. Criteria for classification of solid waste disposal facilities and practices. Fed. Reg. 44(179):53462.

Solid and Liquid Wastes: Management, Methods and Socioeconomic Considerations. Edited by S. K. Majumdar and E. Willard Miller. © 1984, The Pennsylvania Academy of Science.

Chapter Fourteen

Oil Pollution in the Ocean: Sources, Problems and Control

E. Willard Miller, Ph.D.

Prof. of Geography and Assoc. Dean
for Resident Instruction (Emeritus)
College of Earth and Mineral Sciences
318 Walker Building
THE PENNSYLVANIA STATE UNIVERSITY
University Park, Pennsylvania 16802

Oil has always polluted the oceans but has not occurred in significant amounts affecting the environment until recent decades. Only when oil was produced in large quantities has the environment been threatened. However, because oceanic pollution affected few people directly, the dangers were long ignored. Even as late as the early 1950's no international law protected the oceans. Only when pollution became a "social cost" did it become a political issue. The first international controls were initiated in 1954 and since then, a vast body of regulations has evolved. Most of these regulations have been initiated by the Intergovernmental Maritime Consultative Organization (IMCO). There is now a body of rules and regulations on accidental pollution, rules for preventing intentional pollution and a detailed intergovernmental scheme for assigning liability when prevention fails. While progress has been made, enforcement of the rules and regulations has been slow, and remains inadequate.

OCEANIC SOURCES OF OIL POLLUTION

There are four major sources of oceanic oil pollution—natural seepages from underground deposits, pollution from land sources, spills from oil exploitation, and oil pollution from tankers due to operational activities and accidents(Table 1).

TABLE 1

Sources of Oil in the Marine Environment (Annual)

Marine operational losses		
Tanker		1,000,000
Bilge discharge		300,000
Marine accidental discharge		
from all sources		350,000
Offshore operations		
Production		150,000
Natural seepages		200,000-600,000
Land based discharges		
Refineries, petrochemical plants and waste oils		1,300,000-3,000,000
	Total	3,300,000-5,400,000

Source: Sources of Pollution, J. Wardley-Smith in Wardley-Smith, J. ed., 1979. *The Prevention of Oil Pollution.* New York: John Wiley, p. 11.

Natural Seepages from Underground Deposits

In many places in the world oil is able to penetrate the surface through fissures from oil sands. Because many oil sands lie beneath the oceanic waters, this oil escapes into the water. This is a natural phenomenon and will continue indefinitely. No reliable figures are available for the amount of oil escaping into the oceans each year from fissures, but estimates place the amount between 200,000 and 600,000 tons.

Pollution from Land Sources

The major source of oil pollution in the oceans is from the land. Much of the oil spilled on land is ultimately carried to the seas by streams. Many factories discharge their oily wastes directly into rivers. Principal sources of such wastes are refineries and petrochemical plants. These oily waters from plants may be discharged accidentally or deliberately into the river systems.

Large quantities of sump oil are poured into drains or onto the ground during the maintenance of motor vehicles. In the operation of vehicles some oil is expelled through the exhaust system which drips into the roadbed. Much of this oil is ultimately deposited into the oceans. The amount of oil from these sources that enters the oceans annually can only be speculated, but estimates vary from as low as 1,300,000 tons to several million tons.

Spills from Oil Exploitation

In the early days of offshore drilling it was not uncommon for the oil pressure in the producing formation to be greater than expected and for spills to occur as the oil-well casing fractured. Little attention was given to these occurrences until the oil spillage occurred in the waters of Santa Barbara, California, beginning on January 28, 1969. In 1977 another major oil spill occurred at Ekofisk in a drilling operation. In each of these spills less than 100,000 tons of oil were spilled, but the public awareness was great and the demands to protect the environment were spurred.

Recent data indicate that in the offshore waters of the United States, one well in 1,000 will spill small amounts of oil, but only one well in 10,000 will spill sufficient quantities of oil to create an environmental hazard. Under the best control conditions some oil will thus be spilled. It is now estimated that the spills from the operations of undersea oil fields average about 150,000 tons annually.

Operational Pollution From Tankers

Most of the oil fields of the world are located far from the major consuming countries. The oil must thus be transported as either crude or refined products. For most of these oil fields, such as those in the Middle East, north Africa, Nigeria, and Venezuela, the only means of shipment is by tanker. As a result of the growing demand for petroleum, the world movement of oil increased from a little less then 500 million metric tons in 1960 to about 1,750 million tons in 1973. Since then the amount of oil transported by tankers has increased slowly due to the great increase in oil prices and, since about 1979, the world industrial depression.

The oil discharged into the oceans from tankers comes from operation of the vessels and accidental spills. There are four major sources of oil pollution from operational activities—ballasting and deballasting, tank cleaning, discharging bilge water, and transferring oil from one bunker to another. The operational spills are rarely, if ever, spectacular in nature but estimates indicate these sources may contribute 1,300,000 tons of oil to the oceanic waters each year. In contrast to the spills due to operational activities, a tanker accident receives worldwide attention for it is a rare event that may discharge tens of thousands of tons of oil quickly into the sea with devastating effects on nearby shores.

Bilge Water Discharge. Small amounts of water, fuel and lubricating oils collect in the bottom of all vessels. After a given time these oily liquids must be pumped out of the vessel. Because these waste liquids are easily observable in the water, their discharge has been forbidden in most ports and harbors. If the vessel must remove these oily waters while in a port, the vessels are usually provided with a discharge line from the bilge pump to a cofferdam where the water is stored for later processing.

In order to provide international standards the IMCO 1973 Pollution Convention established strict guidelines for the disposal of bilge water. While these guidelines have not been ratified by some governments, they are slowly being implemented. The International Chamber of Shipping (ICS) has encouraged their acceptance. Under these guidelines bilge water discharge is permitted only when a vessel is enroute and is more than 12 miles from land and only when the oil and water is separated so that the oil content does not exceed 100 ppm. Oil-water separators are a standard feature of most modern-day vessels. Nevertheless, these voluntary regulations are frequently ignored in the open ocean. It is estimated that 300,000 tons of oil are deposited from bilge waters annually in the oceans.

Tank cleaning. Tank cleaning is a major source of oceanic pollution. When oils, particularly those with heavy waxy or asphaltic components, are contained in a vessel a small portion of these substances will collect on the tank walls. These layers may be several centimeters thick and commonly constist of 0.3 to 0.5 percent of the oil originally found in the tank. These residues are refinable oils but were long regarded as waste because of their difficult recovery. In a large tanker they can amount to 1,000 tons or more in the oil tanks and perhaps as much as 300 tons in pipelines and pumps. In the 1970's the oil tankers transported about 1.8 thousand million tons of oil. This provided the potential for 5.4 to 9.0 million tons of oil to be discharged into the sea annually. This is obviously an unacceptable quantity and has been greatly reduced by the controls developed by several international conventions.

The cleaning of the cargo tanks in each voyage is required in order to prepare them for carrying clean ballast and to remove the accumulation of oily residues. The ballast water is pumped into selected tanks as soon as the vessel is emptied of oil in order that it can go to sea immediately. It is too costly in time lost to clean the tanks in port. The cleaning of the tanks is thus done while the vessel is at sea.

The attempts to control tank cleaning pollution began in the 1920's, but as late as the 1950's it was customary to discharge tank washings and bilge water directly into the ocean. Significant advances were made by the 1954 Oil Pollution Convention where large areas adjacent to coasts were defined where discharge of these oily waters was prohibited. Since then with increased movement of oil on the oceans the regulations have been strengthened. The 1969 Intergovernmental Maritime Consultative Organization's regulations have been approved so that no oily substances may be discharged except under the following conditions:

1. The tanker is proceeding enroute;
2. The instantaneous rate of discharge of oil content of any mixture does not exceed 60 liters per mile;
3. The total quantity of oil discharged on a ballast voyage must not exceed 1/15,000 of the total carrying capacity; and
4. The tanker must be more than 50 miles from the nearest land whenever any oily mixture is being discharged.

When these regulations are observed the small amount of oil that enters the oceans is rapidly dispersed and degraded.

Ballasting. An oil tanker is a single product vehicle carrying either crude or refined oils. As a consequence the vessel carries oil in one direction only and then returns unloaded to its source area. When the oil is unloaded the tanker floats very high in the water for the weight of the vessel is not sufficient to immerse it deeply. In many unloaded vessels the draught of the bow is only a few feet, and half of the propeller may be exposed. In order to remedy this situation a number of the ships tanks are filled with water. Since the tanks and pumps originally contained oil, the ballast water is contaminated.

Because the ballast water must be removed before a new shipment of oil is received, it must be cleaned at sea for contaminated ballast water cannot be discharged when the vessel is in port. This requires days of activity in cleaning the tanks and pumps and the excess oily residue that cannot be dumped into the sea is stored in the vessel. All vessels are required by international regulations to keep a record of this activity.

Oil Transfer Operations. There are a number of operations associated with tankers where spills may occur. At times oil must be transferred within the vessel. This transfer is performed by using bunker transfer pumps. In this process, the most common cause of spilling is overfilling the tanks for this operation may require several hours and the pumping rate may be misjudged.

Oil may also be spilled when it is transferred from ship to ship or from ship to a barge. This practice has greatly increased in recent years as tankers have grown in size and cannot enter shallow harbors. The oil is thus unloaded from the supertanker in the open sea into smaller vessels that can enter the shallow ports. These operations are normally carried out at times of good weather and in areas where vessels can easily maneuver. Although minimum equipment standards have been established, there is always the potential for oil spills when hoses and pipes connect two vessels. Berthing often occurs with both tankers underway and the supertanker maintains course and speed while the smaller tanker maneuvers alongside using its "fenders" to absorb the berthing impact. During this process, oil can be spilled.

Spills from Tanker Accidents. Because of the large volume of oil carried and the number of tankers engaged in this activity, occasional accidents occur (Table 2). The most common accident creating an oil spill is due to structural failure of the vessel. This single cause accounts for over one-third of the accidents. To illustrate, on March 16, 1978 the Amoco Cadiz, carrying a cargo of 220,000 tons of light Arabian crude oil, was approaching the English Channel. About 10 miles offshore there was rudder failure and, before a tug could aid the vessel, the weather deteriorated and the tanker grounded on the Brittany coast. Although the crew was rescued, on March 24 the vessel split in two spilling its entire cargo. A large proportion of the oil was carried onto the coast of Brittany.

The second most important cause of accidents is grounding the vessel. The first supertanker to ground was the Torrey Canyon on the Seven Stones Shoal about 21 miles off Cornwall's Land End on March 18,1967. The tanker spilled 119,000 tons of Kuwait oil into the sea. This was the first major supertanker accident in the world. Since then there has been the grounding of many large tankers. For example, the Ocean Eagle grounded at the entrance to the San Juan, Puerto Rico harbor on March 3, 1968, and in 1976 the Urquiola struck an under-

TABLE 2

Tanker Accidents Resulting in Oil Spills 1969-1973

Accident type	Number	Percent	Oil Spill (long tons)	Percent
Structural Factor	94	20.8	339,181	35.6
Grounding	123	27.3	230,806	24.3
Collision	126	27.8	185,088	19.4
Fire and Explosion	48	10.6	97,738	10.3
Breakdown	11	2.4	29,940	3.1
Ramming	46	10.2	13,645	1.5
Other	4	0.9	54,911	5.8

Source: Sources of Pollution, J. Wardley-Smith in Wardley-Smith, J. ed., 1979. *The Prevention of Oil Pollution.* New York: John Wiley, p. 10.

water obstruction while entering the harbor of La Coruna, Spain. Many of the groundings are due to human error. In the case of the Ocean Eagle the Captain and officers, although highly experienced, ignored basic precautions for navigating unfamiliar waters.

Collision of tankers is the third major cause of accidents causing oil spills. A spectacular collision between two tankers, the Ven Oil and the Ven Pat, occurred December 16, 1977 off Port Elizabeth, South Africa. Each vessel was of 325,000 registered gross tons. One vessel was loaded, but the other was in ballast. As a result of the collision some 30,000 tons of oil were spilled, polluting a considerable area of the South African coast. Collisions are usually due to negligence. If the sophisticated navigation equipment carried by tankers, particularly the supertankers, is used properly, these collisions could be avoided.

Explosion and fire constitute the fourth major hazard causing tanker spills. These hazards frequently occur after another accident has damaged the vessel. For example, the Jacob Maersh struck an underwater obstruction while trying to birth at Oporto, Portugal. This accident triggered an explosion and fire and the entire cargo of about 80,000 tons of crude oil was either burnt or spilled in the sea.

At times the fire remains undetermined. On August 6, 1983 a Spanish supertanker laden with about 175,000 tons of Persian Gulf oil burst into flame about 80 miles northeast of Capetown, South Africa. Some eight hours after the fire started the ship broke apart in choppy seas. The Benguela Current swept the oil toward South Africa's Langebaan Lagoon, a 15-mile long coastal sanctuary for marine and bird life. Although a large percentage of the oil burned, oil streamed from the bow and stern spreading for about a mile until it joined to form a 3-mile wide slick. Further, it was feared that if the bow and stern sank close to shore, oil would ooze from the wreckage for weeks. At the height of the fire, black smoke rose 8,000 feet into the sky and could be seen 180 miles away. Farmers along the Atlantic coast, where wind delivered a choking day-long smoke, reported an oily film that covered grainfields and vineyards. Newly shorn sheep became oil slicked, and black goo accumulated on houses and cars. Other causes of accidents include breakdowns, ramming and reasons that cannot be determined.

CONTROL OF OIL SPILLS

Whenever an oil spill occurs near a coast, the oil endangers the shoreline. The Torrey Canyon disaster off the Cornwall coast and the Santa Barbara spill illustrate these types of spills. It was in these areas that the initial methods to combat the effects of the oil spills were developed.

The attempts to disperse the oil at sea before it reaches the coastal areas have not been successful. As oil spewed from the wreck at the Torrey Canyon the Royal Navy used about 700,000 gallons of detergents to break up the oil while at sea. At Santa Barbara chemical dispersants were also used to disperse the oil. The conclusion reached was that these types of chemicals were ineffective in controlling large amounts of spillage in the open sea.

Mechanical devices to control the spread of oil in the open sea were also ineffective. These included booms, skimming devices, and suction pumps. Booms proved wholly ineffective in open water for they broke up in even modest seas and they were difficult to maneuver. In harbors booms were more successful in reducing the impact of the oil on shore by repelling the oil before it could enter the protected areas.

When oil from the sea reaches the coastal areas, a number of measures have been used to limit the effects of the spill. In the early coastal deposits from the Santa Barbara and the Torrey Canyon spills, detergents were utilized. In the Torrey Canyon disaster over 1,000,000 gallons of detergent were sprayed on Cornwall beaches and cliffs. It was hoped that the detergent would break up not only the crude oil but also the mousse, a water and oil emulsion. It is thought that mousse is formed by wave and wind action and is bound together by the asphaltic oil residues. When detergent is poured on the beach, the ensuing tides cause the detergent, oil and mousse to move downward in the sand to depths as much as 5 feet. As a consequence, there is a structural breakdown of the compounds creating a quicksand. Further, the interaction of the oil and detergent causes a caustic effect on the skin as well as a stench. As a result of these early endeavors it was found that detergents are only effective in cleaning oil from rock surfaces.

There have also been attempts to find materials that will absorb the oil as it reaches the shore. Talc and perlite have been somewhat successful, but straw and sawdust have proved to be most effective. On rocky areas pressurized hot and cold water and steam cleaning have been somewhat successful. However, sandblasting has proved to be the only really satisfactory method of cleaning oil-stained rocks.

In the beach clean-up the process used depends considerably on whether tourism or the saving of the fishing industry is most important. When there are oil spills along the coast of Brittany in France, they have the potential of destroying an important fishing industry. The French, to save the fishing waters, have attempted to control the oil and mousse in two ways. The collection of the oil and mousse by pumping it into scavenging vessels has been somewhat successful. Even more successful has been the use of stearated chalk to sink the mousse to the ocean floor. The mousse, once deposited on the bottom, does not surface for it has a density close to seawater and remains in place. In a short time sediments solidify the entire mass of material.

There is strong evidence from the scores of coastal spills that if a cleanup does not occur, there will be evidence of oil on the shore for many years. In time, the oil will be dissipated into the sea by wave turbulence, blowing new sand, and gradual penetration into the beach sand. In estuaries and protected harbors the rate of change is much slower and limited to the aging and hardening of the oil by air exposure and deeper penetration into the sand. Rapid removal of oil occurs only when there is great water energy, such as a high velocity current that continually scours the beach.

ECOLOGICAL DAMAGE

The ecological damage to fauna and flora after an oil spill varies greatly. Although a large number of studies have been made on the effects of oil spills, conflicting results do not provide a sound basis for the development of controls. Most of the studies have not been by independent researchers and most have had a limited life span. A continuing effort to study the effects of oil pollution in the oceans and coastal areas over an extended time span is needed.

Birds
As oil reaches a shoreline birds suffer the most evident immediate loss. In the Santa Barbara oil spill it was estimated that a minimum of 6,000, and possibly as many as 15,000 birds, died as a result of oil contamination. In the Torrey Canyon spill it was thought that at least 30,000 birds were killed on the Cornwall and Brittany coasts.

In all coastal disasters there have been major attempts to save the birds. However, the survival rate is amazingly low, usually between one to 11 percent. The death of the seabirds is due to shock, improper cleaning and handling, and ingestion of oil and toxic chemicals by preening. Many birds die of starvation and exhaustion, being unable to hunt for food.

Birds also die from pneumonia because the insulating ability of the feathers is destroyed by oil accumulation. In the Torrey Canyon spill the Plymouth laboratory reports indicated that man-applied pollutants, such as detergent, were far more damaging to the ecology than the oil from the spill.

Although the initial impact on the bird population was devastating, there is no indication that permanent damage has occurred to the present. Nevertheless ecological disaster is possible. The coasts of southern England and Brittany are the breeding ground of such seabirds as the guillemots, rajorbills, shags, preffins and the great northern divers. The Santa Barbara area is the breeding ground of the brant. If these birds had migrated to these areas at the height of the oil spills, entire species could have been exposed to extermination.

Marine Life

Of the oil spills affecting coastal United States, the Santa Barbara oil spill is best documented. Except for birds, the effect on marine life varied greatly. Barnacles and surf grass suffered significantly. Some colonies of acorn and goose-necked barnacles were virtually destroyed. In contrast anemones, mussels, limpets and starfish were little affected. Common amphipods and bloodworms dug into the sand and escaped injury. Bottom studies revealed no damage to sea- floor plant or animal life attributable to the oil spill.

In the early period after the spill the plankton count declined greatly. However, this situation did not exist over an extended time. The initial low plankton count may have been due to several causes. The loss of plankton could have been caused by foraging of fish and crustaceans. The locally observed decrease of dissolved oxygen may have been caused by the concentration of bacteria feeding on the oil.

The damage to fish and shellfish was not conclusive. Fishermen claimed that the fish and shellfish production declined. For example, in 1970, Leon Durden, a fisherman of long experience, indicated that in the first six months of 1970 his catch was three times greater than in the first six months after the spill, but was still only half of the pre-spill years. Nevertheless, statistical surveys conducted by the state, federal and academic investigators showed no significant damage to the fish population due to the spill. The major problem in determining the effect of the Santa Barbara oil spill on the fish population was that there was little information on the movements of migratory fish. Large numbers of anchovy, mackerel, and bonito entered and then left the channel during the spill's early months. But, no data are available to compare pre- and post fish movements and numbers. Dr. Herbert W. Frey, a senior marine biologist of the California Fish and Game Commission stated, "We have detected no impairment of numbers, reproduction or feeding activities attributable to oil pollution. That's not saying that five years from now(1970) we won't find something different."

A number of studies were conducted on the effect of the oil spill on the ecology of the coastal area. The $240,000 study by Dr. Dale Straughan of the University of Southern California, a 12-month field study completed in the fall of 1970, found that the spill had a negligible effect on the marine life of the area. The study was criticized for it was primarily visual and qualitative. It was indicated by a number of scientists that oil is a chemical that creates a severe biological reaction and that meaningful oil pollution research must combine chemical and biological studies. In order to determine the effects of oil in sediments on organisms, gas chromatography and mass spectrometry evaluations are necessary techniques.

Mammals

The effect of the Santa Barbara oil spill on the mammals of the area was not conclusive. Of the elephant seals, at least 74 were contaminated with oil, but there was no conclusive evidence that pollution caused any deaths. However, Drs. Burney Le Boeuf and Richard Peterson, of the California State Fish and Game Department tempered their findings by indicating that they could not prove that oil pollution did not affect the elephant seal mortality.

The Santa Barbara area is noted for its California sea lions. Evidence indicated that the oil spill did affect these mammals. Of the 881 living and dead sea lion pups observed on San Miguel Island by the scientists from the State Fish and Game Commission, slightly less than half were oily, but about two-thirds of all dead pups were oily. Thus many more pups that were oily died than would be expected.

Only three dead dolphins were found in the Santa Barbara channel during the early days of the oil spill. While an autopsy of one did not reveal it died from oil contamination, eyewitnesses indicated its breathing hole was filled with oil and its lungs were ruptured. The dolphins, however, escaped before great damage was inflicted on their poplulation.

During the oil spill five dead whales were washed ashore on the California coast. Of these, three were grey whales, one was a sperm and one was a pilot whale. As with the other mammals the effect of the oil contamination was not conclusive. It appeared that an autopsy on one of the whales linked its death with oil contamination. Whales and other sea mammals are highly susceptible to respiratory ailments and any interference with their respiratory system could cause ailments, such as pneumonia, to occur.

Ecological Evaluation

The minor influence on the marine life after the Santa Barbara spill may have been due to several conditions. Of possibly greatest importance, the amount of oil reaching the coastal areas varied greatly during the early stages of the spill. The greatest quantities of oil, about 12 pounds per square yard, washed onto the Santa Barbara beach; only a short distance to the west at Ellwood and El Capitan, the beaches had less than one-fiftieth of that amount. About 390,000 gallons of oil came ashore in a seven-mile stretch at Santa Barbara, and in the total 55 miles of coast affected, the amount totaled about 1,300,000 gallons. Thus, only limited areas were heavily polluted.

Another factor that mitigated the spill's effect on marine life was the extended period the oil remained at sea. During this time much of the toxic qualities of the oil was dissipated. The Santa Barbara oil was less soluble in sea water than many other crude oils so that the water was less contaminated. There was also the use of chemicals which had a modest ameliorating affect on the oil.

The Santa Barbara spill occurring in winter came during the dormant stage of most plant and marine life species. For example, the damage could have been far greater to the intertidal zone life if it had occurred during the breeding season. One exception may have been the California sea lions that were caught during their May/June pupping season on San Miguel and San Nicholas Islands.

A number of observors also felt that the exposure of marine life to the oil spill was lessened due to the natural seepage of oil in the seawater. Where natural seepage occurred, marine life was normally reduced. When seepages occur there is a gradual and permanent reduction of intertidal organisms.

Associated with the Santa Barbara oil spill were two unique factors that made its effect on flora and fauna atypical. First, it occurred during an abnormally cold year so that the coastal waters were colder than normal, and second, the winter of 1969 experienced heavy rainfall so that the maximum runoff carried large amounts of pesticides into the coastal waters. This activity may have contaminated the seawater temporarily, reducing marine life.

DEVELOPMENT OF ENVIRONMENTAL LAWS OF THE OCEAN

In order to develop oceanic pollution control rules and regulations there must be an understanding of the pollution problems, the economics and politics involved by each of the world's nations, and finally the effectiveness of the organization that implements the rules and regulations.

Defining the Problems

There are three aspects that must be considered in the development of international oil pollution. First, the importance of oil pollution in the oceanic environment must be determined through sound scientific analysis. Second, the place of oil pollution in the total global economic system is of major significance, and thirdly, the control of oil pollution depends on the technology available at any one time.

Most studies on the impact of oil on the oceanic environment do not provide conclusive evidence of the harm done to the environment. In localized areas the immediate damage of the oil spill is evident, but the incremental effect of gradual pollution has gone unnoticed. As a consequence, there has been little urgency for politicians to develop stringent controls that are frequently costly.

When conflicting scientific studies appear, legislators have had a tendency to support the least costly approach. To the present, there has been little independent research, such as by an international agency as the Intergovernmental Maritime Consultative Organization (IMCO). Because of differences in scientific opinion, the basic questions as to what the pollution problems are has created difficulties of legislating adequate control laws. For example, legislators are reluctant to accept costly "solutions" to problems that may not exist.

The second problem in developing regulations to control oil pollution is that it is sometimes considered a rather insignificant aspect of the total global energy system. The great growth in oil consumption due to economic development is the ultimate cause of the oil pollution problem. As a consequence, the means of controlling pollution are limited to what the larger economic system will permit. Because the system is international in scope there is not only great competition but also an interdependence within the system. The system is thus highly competitive and interdependent. Consequently, each nation seeks to ensure that the regulations will not favor a competitor. No nation will permit a regulation which gives it a comparative disadvantage in the total system.

Intertwined with the economic problem is that of technological development. Because industries possess much of the technology necessary to control pollution, they exert a major influence on the development of regulations. Since 1954, when the first pollution control conference occurred, progress in technical standards reflect the constraints imposed by the dependence on technologies provided by the shipping and oil companies. For example, the crude oil washing system to recover oil from ballast water was developed as early as 1967, but was long rejected as uneconomical. It came into use only after the OPEC price rise made it profitable and was then recognized for its environmental advantages. Even then, the tanker owners accepted it as a mandatory requirement only because of the more expensive proposal for the retrofitting of segregated ballast tanks. Even today there is no monitoring system for light oils that has been revealed by the industry. In general, industry is unwilling to act in developing new control systems, except under great political pressure.

DEVELOPMENT OF OIL POLLUTION REGULATIONS

The first international controls on shipping began in the World War II period when the United Maritime Authority was established. In 1948, the UMA was succeeded by the United Maritime Consultative Council (UMCC). In that year governments were invited to attend the United Nations Maritime Conference in Geneva to consider the establishment of an international body within the United Nations. Within 17 days the Convention created the Intergovernmental Maritime Consultative Organization. However, it was not until 1958 that the required 21 nations ratified the Convention and IMCO began to function.

IMCO does not have the power to enact laws. As the word "consultative" in its title implies it can pass recommendations, convene conferences, draw up conventions, and facilitate consultations among member states. The resistance to IMCO as a law-making body was based on the grounds that this action would introduce political distractions into an area that was fundamentally technical in nature. Although the initial Convention of IMCO did not mention a pollution control function, IMCO extended its activities in this direction at an early date. In 1965, the pollution control aspects had become so important that IMCO established a Subcommittee on Oil Pollution under its Maritime Safety Committee.

With the Torrey Canyon oil spill in 1967 IMCO increased its activity to control oil pollution in the oceans. For example, it organized the Conference on Marine Pollution in 1973 and the Tanker Safety and Pollution Prevention Conference in 1978. IMCO now possesses a permanent committee, the Marine Environment Protection Committee, to examine the legal aspects of accidents at sea that create pollution problems. Within IMCO the Maritime Safety and Pollution Control Committees are now of equal importance. In the middle 1960's it was questioned whether IMCO could survive as a specializd agency within the United Nations' system if its involvement was limited to maritime safety alone. The increased involvement in pollution control has given the organization vitality.

In the control of oil pollution of the oceans, IMCO has recommended a vast number of regulations. The individual nations have made many of these regulations laws and have even extended them to protect their own waters.

CONCLUSIONS

Since the 1950's considerable progress has been made in controlling oil pollution in the oceans, The major disasters, such as the Santa Barbara oil spill and the Torrey Canyon accident, made the public and the politicians aware that oil pollution was a major problem. The Intergovernmental Maritime Consultative Organization has come to be the single international agency that focuses on the development of environmental regulations. A vast body of laws now exists. However, international control of oil pollution has not yet been achieved. Many of the regulations proposed by IMCO lack the necessary ratifications to bring them into operation. Unfortunately, because of long delays in ratification, many of the regulations provide dated solutions when they become effective. While oceanic oil pollution will not be eliminated for a considerable time in the future, the problem of oil pollution in the cnvironment is recognized and progress is occurring toward this goal.

BIBLIOGRAPHY

1. Baldwin, Malcom F., May, 1970, "The Santa Barbara Oil Spill," *University of California Law Review,* 42, 33-76.
2. Boesch, Donald F., Carl H.Hershner and Jerome H. Milgram, 1974, *Oil Spills and Marine Environment,* Cambridge, MA: Ballinger, 114 pp.
3. Brown, Joseph E., 1978, *Oil Spills: Danger in the Sea,* New York: Dodd, Mead, 123 pp.
4. Card J. C., P. V. Ponce and W. D. Snider, 1975, *Tankership Accidents and Resulting Oil Outflows 1969-1973,* San Francisco, CA: Conference on Prevention and Control of Oil Pollution
5. Cheek, Lestie, November/December 1981, "Pollution: the Peril Around Us," *Journel of Insurance,* 42: 2-7
6. Conrad, Jon M., Spring 1980, *"Oil Spills: Policies for Prevention, Recovery, and Compensation," Public Policy,* 28: 143-170
7. Cormack, D., et al., 1982, *Oil Mop Device for Oil Recovery on the Open Sea, 1979,* New York: State Mutual Bank
8. Cox, Geraldine V., ed., 1979, *Oil Spill Studies: Strategies and Techniques,* Park Forest South, IL: Chem-Orbital, 148 pp.
9. Cycon, Dean E., October 1981. "Calming Troubled Waters: The Developing International Regime to Control Operational Pollution," *Journal of Maritime Law and Commerce,* 13: 35-51
10. Degler, Stanley, ed., 1969, *Oil Pollution: Problems and Policies,* Washington, DC: BNA Books, 142 pp.
11. Dempsey, Paul Stephen and Lisa L. Helling, Fall 1980, "Oil Pollution by Ocean Vessels—An Environmental Tragedy: The Legal Regime of Flags of Convenience, Multilateral Conventions, and Coastal States," *Denver Journal of International Law and Policy,* 10: 37-87
12. Easton, Robert, 1972, *Black Tide: The Santa Barbara Oil Spill and Its Consequences,* New York: Delacorte Press, 336 pp.
13. Fairhall, David and Philip Jordan, 1980, *The Wreck of the Amoco Cadiz,* Briarcliff Manor, NY: Stein and Day, 256 pp.
14. Grigalunas, Thomas A. and John Meade, 1983, *The Economic Cost of an Oil Spill: A Case Study of the Supertanker Amoco Cadiz,* South Hadley, MA: J. F. Bergin, 352 pp.
15. Harris, L. M. and W. T. Ilfrey, January 1970, "Drilling in 1,300 Feet of Water—Santa Barbara Channel, California," *Journal of Petroleum Technology,* 22: 27-37
16. Hoult, D. P., 1969, *Oil on the Sea* New York: Plenum, 112 pp.
17. Jordon, Randolph E. and James R. Payne, 1980, *Fate and Weathering of Petroleum Spills in the Marine Environment: A Literature Review and Synopsis,* Ann Arbor, MI: Ann Arbor Science, 170 pp.

18. Lynch, B. and J. Smith, 1981, *Dispersants for Oil Spill Clean-Up Operations at Sea, on Coastal Waters and Beaches, 1979,* New York: State Mutual Bank.
19. McCaull, Julian, November, 1969, "The Black Tide," *Environment,* 11: 2-16.
20. McGonigle, R. Michael and Mark Zacher, 1979, *Pollution, Politics and International Law, Tankers at Sea,* Berkeley, CA: University of California Press, 394 pp.
21. Moghissi, A. A., ed., 1980, *Oil Spills,* Elmsford, NY: Pergamon, 80 pp.
22. Ohlendorf, Harry M., Robert W. Risebraugh and Kees Vermeer, 1978, *Exposure of Marine Birds to Environmental Pollution,* Washington, DC: US Department of Interior, Fish and Wildlife Service, 40 pp.
23. O'Neill, Trevor, 1980, "Oil Spill Contingency Plans and Policies in Norway and the United Kingdom," *Coastal Zone Management Journal* 8: 289-317.
24. Parker, H. D. and D. Cormack, 1981, *Remote Sensing of Oil on the Sea,* New York: State Mutual Bank.
25. Petrow, Richard, 1968, *In the Wake of Torrey Canyon,* New York: D. McKay, 256 pp.
26 Ross, William M., 1973, *Oil Pollution As An International Problem,* Seattle, WA: University of Washington Press, 279 pp.
27. Sittig, Marshall, 1978, *Petroleum Transportation and Production: Oil Spill and Pollution Control,* Park Ridge, NJ: Noyes Data Corp., 360 pp.
28. Smets, Henri, January 1983, "The Oil Spill Risk: Economic Assessment and Compensation Limit," *Journal of Maritime Law and Commerce,* 14: 23-43
29. Smith, James, ed., 1968, *Torrey Canyon — Pollution and Marine Life,* Cambridge, England: Cambridge University Press, 196 pp.
30. Waters, W. G., et al., 1980, *Oil Pollution from Tanker Operations: Causes, Costs and Controls,* Vancouver, British Columbia: Centre for Transportation Studies, University of British Columbia, 216 pp.
31. Wardley-Smith, J., ed., 1982, *The Control of Oil Pollution on the Sea and Inland Waters,* New York: State Mutual Bank, 264 pp.
32. _____, 1979, *The Prevention of Oil Pollution,* New York: Halsted Press, 309 pp.
33. Wolfe, Douglas A., J. W. Anderson, et al., eds., 1977, *Fate and Effects of Petroleum Hydrocarbons in Marine Ecosystems and Organisms,* Proceedings of a Symposium, November 10-12, 1976, Seattle, Washington, New York: Pergamon Press, 478 pp.
34. Organization for Economic Cooperation and Development, 1982, *Combatting Oil Spills: Some Economic Aspects,* Paris, France: 140 pp.
35. _____, 1982, *The Cost of Oil Spills,* Paris, France: 252 pp.

Solid and Liquid Wastes: Management, Methods and Socioeconomic Considerations. Edited by
S. K. Majumdar and E. Willard Miller. © 1984, The Pennsylvania Academy of Science.

Chapter Fifteen

Recreational/Sport Fishery Benefits Associated with a Fossil Fuel Generating Station

Robert F. Denoncourt, Ph.D.

Biological Sciences Department
York College of Pennsylvania
York, PA 17405

A total of 48, 12 hour-fishing pressure and creel census surveys were conducted 1977-1978 to estimate the recreational/sport fishery associated with Brunner Island Stream Electric Station, York Haven, Pennsylvania. A second year of data was gathered in 1980 emphasizing comparison of the first mile below BISES and the same reach on the other side of the Susquahanna River.

An estimated 9,587 angler fished 27,267 hours and caught 19,860 fishes within a one-mile (1.6 km) reach of "affected" shore zone below the BISES discharge in 1977-1978. Some 70% fished February through June, 93% gave a residence within 30 or less straight-line miles (48 km), and recorded a predominant catch of channel catfish, smallmouth bass, shorthead redhorse, carp and walleys. Nearby Metropolitan Edison Hydro Electric Station (ambient temperatures) host 9,928 anglers for 22,357 hours. They caught 14,130 fishes during the same time period but with 59% of the pressure March through June. A general reverse trend of fishing pressure and catch was evident on a seasonal basis and catch per unit effort was generally better in the warmer waters. Determination of relative fishing pressure within a 6.5 mile (10.4 km) reach of the Susquahanna River from MEHES to Codorus Creek indicated 59% of the angling along the west shore occurred in waters directly affected by BISES. Data from the 1980 surveys allowed extrapolation for both sides of the river and to 1977-1978. This gave mean actual estimates for a 4.8 mile (7.7 km) reach of shore zone below BISES, over and above the opposite shore zone; 13,907 fishermen, 44, 984 hours and 31,897 fishes.

It was obvious from interviews of the 2253 fishermen over the two years of study that most had no strong preference and were primarily enjoying a recreational activity. The best estimate of this recreational/sport fishery resulting from the BISES cooling water discharge on the basis of the annual "dollar value" from the literature gave $658,635. this was significant and gave another aspect that deserves consideration during an impact assessment of similar facilities.

INTRODUCTION

Brunner Island Steam Electric Station (BISES) in York County, Pennsylvania (Figure 1) has been the focus of considerable environmental research since the mid 1970's. Licensing procedures directed by new environmental regulations and an interest in possible impacts caused Pennsylvania Power and Light Company to initiate a series of assessment studies in the aquatic ecosystems. Background to these studies and a description to the facility were given in Denoncourt (1983).

This paper reports results and implications of creel census and fishing pressure surveys conducted 1977-1978 and 1980. Specific purposes include determination of relative fishing pressure in the Susquehanna River directly influenced by the cooling water discharge and estimation of the recreational benefit related to BISES.

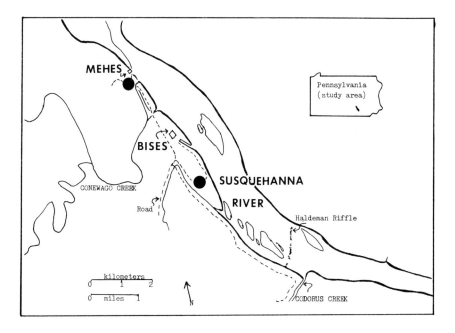

FIGURE 1. Location of creel-census check-stations (dots) used in the 1977-1978 and 1980 surveys.

METHODS

Angler use and catch data were obtained from fishermen on 48 12-hour days, two weekend days and two weekdays randomly selected from February 1977 to January 1978. Essentially 95-100% of all fishermen at two locations (BISES and MEHES) were stopped and interviewed as they exited on restricted roads. Angler use at the other areas within the research area were estimated by three or four "runs" via car throughout a 4.8 mile (7.7 km) section from York Haven Dam (MEHES) to Codorus Creek. Interview data included hours fished, species taken and their approximate sizes, and residence of angler. This allowed determination of total fishermen, total hours fished, number and kinds taken, catch per unit effort, straight line distance driven by anglers, and relative fishing pressure within segments of the study area (MEHES with special access area, opposite York Haven with special access area, first mile below BISES, and next 3.8 miles or 6.08 km below BISES).

A valid criticism of the 1977-78 study raised by the Pennsylvania Fish Commission was the lack of information from the east shore of the Susquehanna River opposite the one-mile section below BISES. Empirical knowledge suggested minimal use, but a study in 1980 obtained comparable data.

A random sampling period based upon probability obtained from the previous data, conservation officers, personal knowledge of investigators, or a combination of these was used in Pfeiffer (1966) and Euston and Mathur (1979). We had a full year of data from (1) the BISES plume, (2) west shore below MEHES, (3) east shore opposite MEHES, (4) west shore from Black Gut to Codorus Creek; as well as (5) combined personal experience and (6) communication with research scientists who have conducted fishing-pressure surveys from above Three-Mile Island Nuclear Station to below Conowingo Dam. This was used to give a sampling system that emphasized week-end days over week-days and PM hours over AM hours. A random method was used to pick specific days and time periods. Data on numbers of anglers seen and interviewed during these time periods was used to estimate total fishing pressure and catch as in Pfeiffer (1966), Malvestuto et al (1978) and Euston and Mathur (1979).

Malvestuto et al (1979) also used incomplete trip interviews obtained from non-uniform probability surveys to provide estimates of catch. Johnson and Wroblewski (1962) demonstrated that over- and under-estimation from fishermen balance each other. Our interviews involved both "incomplete trip" (east shore zone) and "complete trip" (west shore zone) data. These were utilized to estimate total pressure and catch. The 1977-78 data was re-examined and ratios from 1980 were used to extrapolate (1) and estimate annual use within the first mile of shore zone influenced by the BISES cooling-water discharge and (2) east versus west for both 1977-78 and 1980 data. Thus, 1977-78, personal knowledge and discussion from Malvestuto et. al. (1978) indicated a sample time-period of five week-end days and two week days within each "season"

(February-March, April-May, June-July, August-September-October) would be adequate for unbiased data.

Specifically: 1 - an instantaneous count of shore zone anglers was made for east and west at the beginning of each experimental time period, 2 - an investigator was stationed at the exit road from the west shore zone and obtained "complete trip" data, and 3 - an investigator used a boat to reach anglers on the east shore zone to obtain "incomplete trip" data. Anglers were asked about hours fished, fishes caught and preference of catch.

Raw data from the four sample seasons were organized to give date and time period; probabilities for the time period; number of fishermen interviewed; extrapolated estimates for the number of fishermen, hours of fishing and fishes caught; instantaneous counts of anglers within the west and east shore zone sections; catch by species; preference of catch; and extrapolation to 1978.

Finally, information was found upon which to base a financial estimate of the recreational value derived from the sport fishery at BISES.

RESULTS AND DISCUSSION

Creel Census at Brunner Island Steam Electric Station, 1977-78 —

There were 1681 anglers interviewed on the 48 days of the survey from February 1977 through January 1978 (Table 1). These data allowed the following estimates for the year: 9,587 anglers; 27,267 hours; and 19,860 fishes taken. Monthly estimates were used to obtain catch per effort and hours fished. Grand means from these indicated 0.75 fish per hour (or 1.9 hours per fish) and an average of 2.7 hours per angler for the year. Some 70% or 1099 of the anglers fished from February through June. Estimates of monthly variations in number of anglers, number of hours, and number of fishes taken are graphically depicted in Figures 2. The anglers (Fig. 2A) and hours fished (Fig. 2B) were high in winter and early spring, dropped in summer, and began to increase in fall (Post-survey observations revealed a large increase in anglers in March and April 1978). Catch (Fig. 2C) had similar trends with an increase occurring in early fall and a suggestion of decrease in the winter of 1977-78. (This latter was influenced by "muskie fishermen" who spent long hours, but caught few fishes; and by unusually high waters and cold weather.) Catch per unit effort (Fig. 3A) and effort per fisherman (Fig. 3B) were more variable and were often correlated with milder weather conditions.

These results indicate an extensive fishery which had not previously been evaluated. However, data from other publications (Lingerfelter and Summerfelt 1972, TVA 1977, Moore and Frisbee 1972) suggested it would be substantial. Pfeiffer (1982) recommended "power plant" fishing at 28 different locations in the East, including BISES at York Haven. That local fishermen take advantage of this is well depicted in Figure 4 which shows some 93% of the fishermen

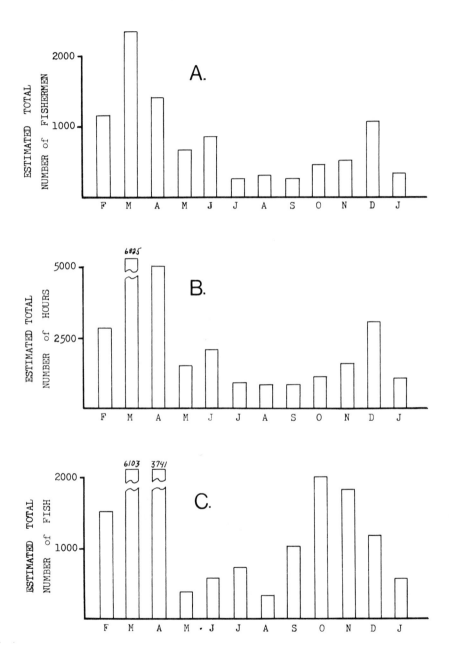

FIGURE 2. Total estimated number of fishermen (A), total estimated number of hours (B) and total estimated number of fishes caught (C) for each month from February 1977 through January 1978 at Brunner Island Steam Electric Station for the first mile below the cooling-water discharge.

TABLE 1

Summary of interviewed and estimated[1] number of fishermen, hours fished, fish captured, number of fish per hour and number of hours per fisherman in and below the cooling-water discharge canal at Brunner Island Steam Electric Station, York County, Pennsylvania, for the year February 1977 through January 1978.

	Feb.	Mar.	Apr.	May	June	July	Aug.	Sep.	Oct.	Nov.	Dec.	Jan.	Totals
Number of fishermen													
Interviewed	228	356	289	134	160	57	65	46	68	87	141	50	1681
Estimated total	1146	2354	1406	668	857	253	298	240	456	516	1064	329	9587
Number of hours fished													
From interviews	571	1035	708	297	364	175	186	138	$85	239	429	166	4493
Estimated total	2770	6825	4998	1441	2002	860	826	797	1117	1523	3140	968	27267
Number of fish caught													
From interviews	322	935	498	65	107	93	61	152	328	272	175	71	3079
Estimated	1510	6103	3741	371	568	695	304	993	2008	1823	1184	560	19860
Number of hours per fish[2]	2.5	1.1	1.3	3.9	3.5	1.2	2.7	0.8	0.6	0.8	2.7	1.7	x = 1.37
Number of fish per hour[2]	0.40	0.91	0.77	0.26	0.29	0.83	0.37	1.25	1.70	1.25	0.37	0.59	x = 0.73
Number of hours per fisherman[2]	2.2	2.9	2.5	2.2	2.3	3.4	2.8	3.3	2.4	3.0	3.0	2.9	x = 2.84

[1]Estimated based upon weighted data = mean for week days times number of week days times number of week days in the month + mean for week-end days times number of week-end days in the month.

[2]These values based upon the estimated data.

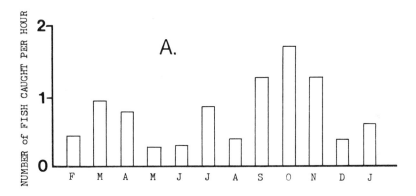

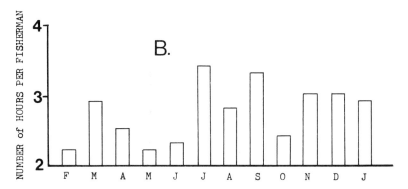

FIGURE 3. Number of fish caught per hour of angling (A) and number of angling-hours per fisherman (B) for each month from February 1977 through January 1978 at Brunner Island Steam Electric Station for the first mile below the cooling-water discharge.

interviewed gave a residence located 30 or less straight-line miles (48 kilometers) from BISES, 68% at 10 or less miles (16 kilometers).

The five most abundant of 15 fish species reported by anglers at BISES were: channel catfish, smallmouth bass, shorthead redhorse, carp, and walleye (Table 2). Monthly variation in percent abundance of six species-groups (Figure 5) showed suckers abundant in February and March, carp abundant in summer, catfish usually prevalent except in August, sunfishes more abundant in summer, and walleye common in winter. Although only six muskellunge were reported, it is included because of the importance to anglers who fished long hours hoping to catch this species.

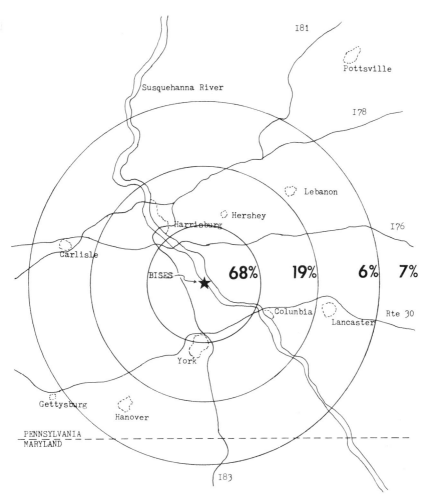

FIGURE 4. Percentages of fishermen interviewed from February 1977 through January 1978, arranged by residence within ten-mile (16 kilometers) straight-line distances of Brunner Island Steam Electric Station.

Creel Census at Metropolitan Edison Hydro Electric Station —

Some 1627 fishermen were interviewed from March 1977 through December 1978 (Table 3). Ice prevented fishing in February of 1977 and January of 1978. The fishermen reported 3,595 hours of angling and 2.123 fishes taken. These data gave an estimated total fishing effort and catch at MEHES for the year: 9,928 fishermen, 22,357 hours, and 14,130 fishes. Thus the average angler spent 2.3 hours fishing and took 1.58 hours to catch a fish, a somewhat lower catch per unit effort than in the warmer waters below BISES. The number of anglers and time spent fishing increased through April (almost double March); May

TABLE 2

Summary of the number of each fish taxa reported caught each month in and below the cooling water discharge canal at Brunner Island Steam Electric Station, York County, Pennsylvania, for the year February 1977 through January 1978.

	Feb.	Mar.	Apr.	May	June	July	Aug.	Sep.	Oct.	Nov.	Dec.	Jan.	Year's Total N	%
Muskellunge	—	1	2	—	—	—	—	—	1	1	1	—	6	+[1]
Carp	6	99	56	5	5	40	12	42	7	21	3	—	296	10
Suckers mostly shorthead redhorse some white suckers	153	323	94	3	2	2	—	—	9	8	10	1	605	20
quillback	—	—	4	—	—	—	—	—	10	—	—	—	14	+
Catfishes mostly channel catfish some bullhead	65	480	2661	34	59	28	4	69	126	92	44	30	1297	42
Rockbass	1	1	11	—	—	—	—	—	—	—	1	—	14	=
Sunfishes	1	—	6	12	5	13	12	3	11	7	1	—	71	+
Basses mostly smallmouth some largemouth	29	26	54	11	32	10	33	37	127	123	48	8	538	17
Crappie black and white	—	—	4	—	2	—	—	—	36	4	6	2	54	+
Yellow perch	—	2	—	—	—	—	—	—	—	—	—	1	3	+
Walleye	67	3	1	—	2	—	—	1	1	16	61	29	181	6
Total number per month	322	935	498	65	107	93	61	152	328	272	175	71	3079	—
% number for year	10	30	16	2	3	3	2	5	11	9	6	2	—	—

[1]The plus (+) means less than one percent.

TABLE 3

Summary of interviewed and estimated[1] number of fishermen, hours fished, fish captured, number of fish per hour and number of hours per fisherman at Met-Ed Hydro Electric Station, York County, Pennsylvania, for the year February 1977 through January 1978

	Feb.[2]	Mar.	Apr.	May	June	July	Aug.	Sep.	Oct.	Nov.	Dec.	Jan.	Totals
Number of fishermen													
Interviewed	—	173	197	267	331	143	197	141	80	65	33	—	1627
Estimated total	—	776	1324	1542	2052	952	1279	907	532	364	200	—	9928
Number of hours fished													
From interviews	—	298	407	634	378	434	296	236	151	78	—	3595	
Estimated total	—	2234	2618	3262	4475	2352	2801	1828	1495	856	436	—	22357
Number of fish caught													
From interviews	—	99	258	421	437	263	266	116	95	129	39	—	2123
Estimated	—	958	1635	2850	3085	1760	1604	835	530	691	182	—	14130
Number of hours per fish[3]	—	2.3	1.6	1.1	1.5	1.3	1.7	2.2	2.8	1.2	2.4	—	$x = 1.58$
Number of fish per hour[3]	—	0.43	0.63	0.91	0.67	0.77	0.59	0.45	0.36	0.83	0.42	—	$x = 0.63$
Number of hours per fisherman[3]	—	1.9	2.0	2.1	2.2	2.5	2.2	2.0	2.8	2.4	2.2	—	$x = 2.30$

[1]Estimated based upon weighted data = mean for week days times number of week days in the month + mean for week-end days times number of week-end days in the month.

[2]Ice

[3]These values based upon the estimated data.

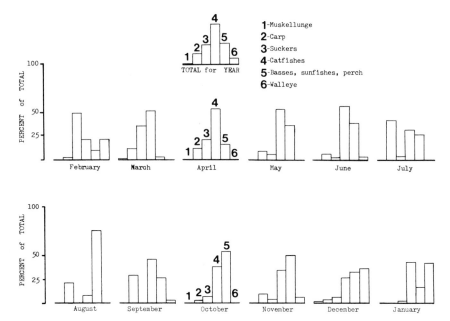

FIGURE 5. Percent of total catch for selected species or groups of fishes taken from waters in the first mile immediately affected by cooling-water discharge from Brunner Island Steam Electric Station from February 1977 through January 1978.

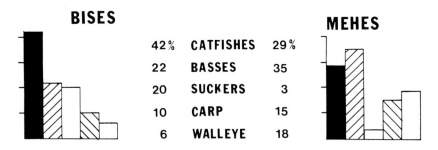

BISES			MEHES
42%	CATFISHES	29%	
22	BASSES	35	
20	SUCKERS	3	
10	CARP	15	
6	WALLEYE	18	

FIGURE 6. Comparison of the percent of catch for the dominant species taken by anglers at Brunner Island Steam Electric Station (BISES) and at Metropolitan Edison Hydroelectic Station (MEHES) for the year February 1977 through January 1978.

and June (amost triple March); decreased somewhat in July, August, and September (down approximately one half of June data); and then decreased steadily through October, November, and December. Number of anglers, effort, and catch followed these same trends which also correlated well with increasing ambient temperatures through July. The decrease in summer occurred at times of maximum water temperatures (25.6 to 28.9 in July and August).

TABLE 4

Summary of the number of each fish taxa reported caught each month at Met-Ed Hydro Electric Station, York County, Pennsylvania, for the year February 1977 through January 1978.

	Feb.	Mar.	Apr.	May	June	July	Aug.	Sep.	Oct.	Nov.	Dec.	Jan.	Year's Total N	%
Northern pike	—	1	1	—	—	—	—	—	—	—	—	—	2	+[1]
Muskellunge	—	—	2	—	—	—	2	—	—	—	—	—	4	+
Carp	—	3	36	129	18	40	72	10	3	4	1	—	316	15
Golden shiner	—	—	—	1	—	—	—	—	—	—	—	—	1	+
Creek chub	—	—	—	—	—	—	—	1	—	—	—	—	1	+
Suckers														
mostly shorthead redhorse some white suckers	—	—	1	—	—	—	—	—	—	—	—	—	1	—
quillback	—	—	4	—	—	—	—	—	10	—	—	—	14	+
Catfishes														
mostly channel catfish some bullhead	—	1	11	58	207	187	93	33	12	3	12	—	617	29
Rockbass	—	18	26	36	8	—	1	1	3	7	—	—	100	5
Sunfishes	—	—	7	10	15	6	9	1	3	2	—	—	53	2
Basses														
mostly smallmouth some largemouth	—	6	77	74	120	27	64	15	60	63	—	—	506	24
Crappie														
black and white	—	23	19	15	2	2	1	3	3	1	—	—	506	24
Yellow perch	—	5	—	1	—	—	1	—	—	—	—	—	7	+
Walleye	—	42	44	73	61	1	21	52	11	49	26	—	380	18
Total number per month	—	99	258	421	437	263	267	116	95	129	39	—	2123	—
% number for year	—	5	12	20	21	12	13	5	4	6	2	—	—	—

[1]The plus (+) means less than one percent.

Not only were water temperatures high, but low water levels combined with high air temperatures to discourage anglers in August and September. Some 59% or 968 of the anglers fished March through June 1978, and a total of 86% of the fishermen lived a straight-line distance of 20 or less miles (32 or less kilometers) from MEHES, 75% 10 or less miles (16 or less kilometers). As water temperatures dropped in the fall, catch per unit effort decreased. Fishermen began to turn their attentions to fishes known to move into the cooling water discharge below Brunner Island and an obvious shift of the fishermen occurred.

Fishermen at MEHES reported 18 species of fishes taken (Table 4). The most abundant of these were catfishes (mostly channel catfish), basses (small- and largemouth), walleye, carp, rockbass, and crappies. Definite trends of abundance in catches were observed: carp, channel catfish, and basses were more common in spring and summer; rockbass and particularly walleye were common in early spring and again in fall; suckers, crappie and sunfishes in spring.

A comparison of predominate fishes in the respective catches, MEHES vs. BISES, indicated more bass, walleye and carp at MEHES; more catfish and suckers at BISES (Fig. 6). A possible behavior mechanism may have resulted in the fact that more walleye at MEHES were larger than minimal legal length (14 inches), while smaller walleye were more common at BISES. Intra-specific competition may keep smaller individuals from the spawning areas until large enough to aggressively compete. The increase in bass catch at MEHES were likely related to more extensive and varied habitat (rocks, backwater, debris); while lower numbers of shorthead redhorse (sucker) were related to the fact that they migrated into nearby Conewago Creek to spawn and then dispensed throughout the Susquehanna River.

Creel Census and Survey 1980—

A total of 30 time-periods of investigation and 572 interviews (354 on west and 218 on east) gave data to estimate values for the one mile of shore zone immediately below the BISES cooling-water discharge and for a comparable section on the east shore (Table 5). Thus, nine months of investigation gave estimates of 7,383 anglers 21,703 hours of fishing; and 19,884 fishes caught for the one mile of west shore zone—over and above the east shore zone estimates of 1,500 anglers; 3,444 hours of fishing; and 1,340 fishes caught. Extrapolation to one full year gave 9,229 fishermen; 27,176 hours; and 23,284 fishes caught for the west shore zone. The estimates for 1977-78 allowed by this 1980 study indicated 8,715 fishermen; 23,802 hours of fishing; and 17,874 fishes caught—over and above the east zone (Table 6).

These annual estimates are conservative because (1) the 1977-78 data for November-December-January were based on a season of very limited access, and (2) no allowance was made for the next mile or more of west shore zone where fishermen were observed while none were observed for the east shore zone. Previous surveys (1977-1980) showed fishermen utilizing the warmer waters

TABLE 5

Summary of fishing pressure data for February-March, April-May, June-July, and August-September-October 1980; with an extrapolation to the 1977 survey.[1]

	February-March		April-May		June-July		Aug-Sept-Oct	
	West	East	West	East	West	East	West	East
Number of Fishermen								
Week-end Days								
Interviews	242	—	184	13	39	14	73	7
Estimated Total	1976	20	1567	226	415	296	840	210
Week Days								
Interviews	27	—	8	—	12	4	11	—
Estimated Total	1953	—	704	—	748	748	680	—
Combined:								
1980	3929	20	2271	226	1163	1044	1520	210
1977	2500	17	2074	196	1110	905	994	157
Number of Hours Fished								
Week-end Days	4799	48	4369	631	904	646	2228	557
Week Days	6447	—	3432	—	1562	1562	1406	—
Combined:								
1980	11246	48	7801	631	2466	2208	3634	557
1977	9595	47	6439	446	2862	2347	2740	466
Number of Fishes Caught								
Week-end Days	2884	29	2547	368	340	243	1919	480
Week Days	8694	—	3432	—	220	220	1188	—
Combined:								
1980	11578	29	5979	368	560	463	3107	480
1977	7613	36	4112	196	1263	1112	3305	474
Number of Hours/Fisherman								
1980	2.86	—	3.44	2.69	2.12	1.61	2.39	0.60
1977	2.74	—	3.10	—	2.58	—	2.76	—
Number of Hours/Fish								
1980	0.97	—	1.30	2.06	4.40	1.17	1.17	0.67
1977	1.26	—	1.57	—	2.27	—	0.83	—

[1]Estimates for the East shore zone were determined for number of fishermen, number of hours and number of fishes caught by utilizing the instantaneous count ratios. The determination of hours/fisherman and hours/fish were obtained from interview data of the complete-trip method on the West and the incomplete-trip method on East.

of the west shore zone (even in summer) to one or two miles below the railroad bridge at Codorus Creek.

All "runs" (I drove the extent of the study area, MEHES to Codorus Creek, and stopped wherever fishermen were observed) had the objective of temperature measurement and counts of "fishermen cars" (could be determined reliably). The counts were used to compare relative fishing pressure within (1) the one mile (1.6 km) immediately below the cooling-water discharge, (2) in the immediate vicinity of MEHES upstream, (3) an access area on the east shore and opposite MEHES, and (4) the next 2.5 miles (4.0 km) of shore zone along the west shore that is also influenced by the BISES effluent (Fig. 7). Relative fishing

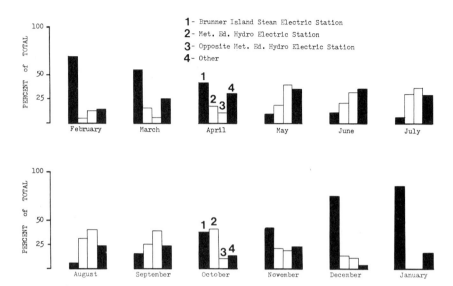

FIGURE 7. Percent of fishing pressure (based upon fisherman-car counts) from selected reaches of the Susquehanna River in the vicinity of Brunner Island Steam Electric Station from February 1977 through January 1978.

pressure for the reach under extensive study showed a high for the section closest to BISES in winter with a decrease into spring through summer (an examination of data from the fish distribution survey could have predicted this. Anglers fish when and where they anticipate a catch.) The reverse trend was evident for MEHES and the area opposite MEHES. If (1) and (4), both under direct "affect" from BISES, were combined, the contribution to recreational fishing for BISES would be considerably increased. For the year of the study, this amounted to 45% of the entire study area (both sides of the river considered) or over 59% of the recreational fishing along the west shore zone from MEHES to Codorus Creek. In addition, other studies (Denoncourt 1981) have shown an influence that attracts fishermen to at least another 2 km below Codorus Creek. A narrow and shallow plume of warmer water meanders through a 3-4 km long pool. Fishermen reported that they sought this "plume" and had success in and adjacent, catching channel catfish and smallmouth bass.

An examination of data on a sample-period basis (Tables 5 and 6) indicated a decrease in anglers in the west shore zone in June-July (and August, although not evident, due to combining) with a corresponding increase on the east shore zone. Discussion with some anglers on the east indicated they drove or boated to the west shore zone or "discharge waters" in other seasons. (Note: There is easier boat access on the east shore zone.) Data (Table 5) also indicated anglers spend more time per trip in the west shore zone. Empirical knowledge suggested this was true, but the values given in the table must be recognized as possibly

TABLE 6

Estimated fishing pressure and catch along the west shorezone Susquehanna River within the first mile influenced by Brunner Island S. E. S. cooling-water discharge and a comparable section of the east shorezone for February through October 1980, a similar period in 1977, and extrapolation to include one year. NOTE: numbers for the WEST have been corrected by subtraction of data for the EAST.[1]

	February-March		April-May		June-July		Aug-Sep-Oct		Total Feb-Oct		Nov-Dec-Jan		Yearly Total	
	West	East	West	East	West	East	West	East	West	East	West	East	West	East
Number of Fishermen														
1980	3909	20	2045	226	119	1044	1310	210	7383	1500	1846	63	9229	1563
1977-78	3483	17	2444	196	205	905	737	157	6869	1275	1846	63	8715	1338
Number of Hours														
1980	11198	48	7170	631	258	2208	3077	557	21703	3444	5473	158	27176	3602
1977	9548	47	5992	446	515	2347	2274	466	18329	3306	5473	158	23802	3464
Number of Fishes Caught														
1980	11549	29	5611	368	97	463	2627	480	19884	1340	3400	167	23284	1507
1977-78	7577	36	3915	196	151	1112	2831	474	14474	1818	3400	167	17874	1985

[1]Extrapolation for November-December-January was based upon observed fact in 1977-78 and 1980-81 that the Susquehanna River was frozen on the east side (thus no fishermen had ready access to fish the shorezone) in December and January, and the Aug-Sep-Oct ratios of 1980 for November.

TABLE 7

Summary of the percent of total catch for fishes in each time period (February-March, April-May, June-July, and August-September-October) in 1980, and the percent of species preference indicated by fishermen interviewed.

	February-March		April-May		June-July		Aug-Sept-Oct		Total	
	West	East	West	East	West	East	West	East	West	East
Fish Species Caught										
Catfish	43	—	48	82	23	19	23	67	40	50
Bass	10	—	8	12	44	67	72	17	19	39
Suckers	34	—	1	—	—	5	—	—	15	2
Carp	14	—	12	6	21	5	3	—	11	2
Sunfishes/rockbass	—	—	26	—	9	—	1	—	9	—
Walleye	7	—	3	—	3	5	3	17	5	5
Crappie	—	—	2	—	—	—	—	—	1	—
Preference										
Anything	49	—	84	100	59	67	55	88	62	82
Bass	9	—	2	—	33	33	31	—	12	15
Muskie/walleye	20	—	7	—	2	—	13	12	13	3
Catfish	13	—	5	—	—	—	1	—	8	—
Carp	6	—	2	—	6	—	—	—	4	—
Suckers	3	—	—	—	—	—	—	—	1	—
Sunfishes	—	—	+	—	—	—	—	—	+	—

distorted due to "incomplete-trip" data. The number of hours per fish, perhaps a more realistic comparison of catch per effort, indicated a better catch on the west in February-March and April-May, better on the east in June-July and August-September-October. Empirical knowledge indicated a better catch on the west in November-December-January.

Overall, 59 to 89% of the fishes caught were catfish (40% on the west, 50% on east) or bass (19% on west, 39% on east) (Table 7). Walleye which ranked sixth on the west and third on the east (5% for both) were beginning to migrate when the investigation concluded and would undoubtedly have increased in rank. Suckers (mostly shorthead redhorse) were common in February-March on the west (during their spawning "run" to Conewago Creek). Smallmouth bass were significant catches, particularly in June through October, in both shore zones. Anglers from boats (data not included because this was not shore zone and because of their mobility) successfully utilized the effluent-edge to catch significant numbers of smallmouth bass in July and August, this species apparently attracted perhaps by a food source. (This would add to our estimate of total recreational use given above.)

It was obvious from interviews that most anglers (49 to 100%) were fishing for recreational reasons and did not have a species preference (Table 9). As expected, there were seasonally related preferences: bass (31 to 33%) in summer, muskie and/or walleye (13 to 20%) and catfish (13%) in spring or fall.

A mean annual estimate based upon extrapolation can be obtained for 1980 and 1977-78: 8,972 fishermen, 25,489 hours of fishing, and 20,579 fishes caught. However, "best estimate" of "relative fishing pressure" from the 1977-78 data summarized in Figure 7 indicated a multiplication factor of 1.55 justified for the reach of shore zone BISES to Codorus Creek. Thus, the final estimates of annual recreational/sport fishing use for waters influenced by the BISES cooling water discharge were: 13,907 fishermen, 44,984 hours and 31,897 fishes caught (Table 8).

The benefits provided to fishermen by BISES are important in any evaluation of impact upon fishes. 316a demonstration involves evidence to support "no appreciable harm to population of resident important species." There may be some variations of opinion about definitions, but the RIS fishes always include game/pan species whose populations and availability to fishermen may be effected by the present operation of BISES. While a life stage of some species may be excluded from cooling-water plumes during certain seasons, other life stages find refuge in the same plume and are available to increased angling effort and success (Denoncourt 1978, 1980, 1981). Therefore, the plume has both positive and negative effects as a result of heat, both of which should logically receive due consideration. The recreational fishery assessed and described herein is thermally related and thus an important factor in total evaluation of impact.

Another way to examine the recreational/sport fishery would be to consider its value in dollars as a loss should the facility stop producing a thermal affect.

TABLE 8

Estimated number of fishermen, number of hours and number of fishes for the first mile (1.6 km) below Brunner Island Steam Electric Station and the total reach of "affected" shore zone for 1980 and 1977-78, with estimates of "mean annual dollar value".

| | ------------ First Mile ------------ | | | Total Reach |
	West	East	X W-E	X x 1.55[1]
Number of Fishermen				
1980	10792	1563	8972	13907
1977-78	10054	1339		
Number Hours				
1980	30778	3602	25489	44984
1977-78	27266	3464		
Number Fishes				
1980	24791	1507	20579	31897
1977-78	19859	1985		

| | | Estimated "Mean Annual Dollar Value" | |
Literature	Value	First Mile	Entire Reach
Poff 1972[2]			
Average trip expense	7.16	64,239	99,574
Spiragelli and Thommes 1976[2]			
Per pound, warm water	5.30	27,267	42,263
If half kept and weigh ½ lb.			
Ellefson 1973[2]			
Average trip	8.77	78,684	121,964
S. F. I. 1982[2]			
Average trip	11.00	98,692	152,977
Amer. Fish Soc. 1982[2]			
Unweighted x cash value per fish for fish kills	2.56	52,682	81,656
Johnson 1981			
Average trip (PA)	40.00	358,880	556,280
Stroud 1981			
Overall per trip	54.72	490,948	760,991
Warm water per trip	59.91	537,513	833,168
Lost pay value per trip	80.82	725,117	1,123,963

[1]Factor of 1.55 derived from data behind Figure 8.
[2]Not including accessory gear; car, food, etc.

Researchers have attempted to quantify the economic value of a sport/recreational fishery (Table 8). This was difficult at best and does not have a satisfactory method of "cost-effective evaluation." American Fisheries Society (1982—Monetary Values of Freshwater Fish...) mentions several phrases to emphasize economic importance: "fish are resources, and have tangible value to the public...," "when fish are destroyed...compensation...," "...used and upheld in three court cases.," and "...if financial costs of lost angler-days...can be estimated, these should be included in the total assessment." Johnson (1981) reported the serious concern of recreational loss due to acid rain, giving an angler-day value for Pennsylvania of $40. Estimates by others (Poff 1972, Ellefson 1973, Spirogelli and Thommes 1976) were attempted, but did not give values that include fishing gear, bait, food, etc. Probably the best estimates to date were those of Stroud (1981). Using these (Table 8) as a basis to estimate the economic value of BISES to the local communities as a sport/recreational fishery gives an annual range for the entire "affected reach" below BISES of $42,263 to $1,123,963.

Now, if one uses a mean of Johnson's (1981) estimate for Pennsylvania (probably trout oriented) and Strouds (1981) "overall estimate," the cost per day gives an annual economic value for the sport/recreational fishery provided by BISES of $658,635!

ACKNOWLEDGEMENTS

Pennsylvania Power and Light Company, Allentown, Pennsylvania initiated and financed these studies and allowed the release of data. The Environmental Management Division and especially W. F. Skinner of the Ecological Studies Laboratory encouraged all aspects, assisted in field and laboratory, and reviewed reports. E. G. & G. Environmental Consultants, Waltham, Massachusetts (Jerry Cura and Charles Menzie) assisted some field studies and released data. Kevin Rapp, Jack Fisher, Blaine D. Snyder, Dr. Richard B. Clark, Karin Norwig, E. C. Bollinger, Carol L. Denoncourt and Robert E. Denoncourt assisted in the field and/or laboratory. The permission and cooperation of the Pennsylvania Fish Commission was appreciated in this and many research studies by the author and his students.

LITERATURE CITED

American Fisheries Society. 1982. *Monetary values of freshwater fish and fish kill counting guidelines.* Amer. Fish. Soc. Publ. No. 13. 40pp.

Denoncourt, R. F. 1978. Ichthyofaunal and creel census/fishing pressure surveys in the vicinity of Brunner Island Steam Electric Station. Pennsylvania Power and Light Company, Allentown, PA. Typewritten report.

Denoncourt, R. F. 1980. An ichthyofaunal survey and creel census in the vicinity of Brunner Island Steam Electric Station. Paper Presentation. Sixth Annual EEI Biologists Workshop, Indianapolis, IN.

Denoncourt, R. F. 1981. Creel census. In: *An evaluation of the effects of thermal discharge from the Brunner Island Steam Electric Station on representative important fish species.* Prepared by E. G. & G. Environmental Consultants, Waltham, MA. App. 2:1-19.

Denoncourt, R. F. 1983. Fish distribution in the vicinity of Brunner Island Steam Electric Station. *Proc. Pa. Acad. Sci.* (in press).

Ellefson, P. V. 1973. Economic appraisal of the resident salmon and steelhead sport fishery of 1970. In: Michigan's Great Lakes Trout and Salmon Fishery. 1969-1970. Mich. Dept. Nat. Res. Fish. Mgmt., Rpt. No. 5.

Euston, E. T. and D. Mathur. 1979. Effects of heated discharges on the winter fishery in Conowingo Pond, Pennsylvania. Proc. Pa. Acad. Sci. 53(2).

Johnson, F. W. 1981. Acid precipitation: the ultimate threat to future fishing in Pennsylvania? Pa. Angler 50(5):16-18.

Johnson, M. W. and L. Wroblewski. 1962. Errors associated with a systematic sampling creel census. Trans. Amer. Fish. Soc. 91(2):201-207.

Lingenfelter, D. P. and R. C. Summerferfelt. 1972. Angler harvest in heated fishing docks on an Oklahoma reservoir. Proc. 26th Ann. Conf. S. E. Assoc. Game and Fish Commrs. 26:611-621.

Malvestuto, S. P., W. D. Davies and W. L. Shelton. 1978. An evaluation of the roving creel survey with non-uniform probability sampling. Trans. Am. Fish. Soc. 107(2):255-262.

Moore, C. J. and C. M. Frisbie. 1972. A winter sport fishing survey in a warm water discharge of a steam electric station on the Patuxent River, Maryland. *Chesapeake Sci.* 13(2):110-115.

Pfeiffer, P. W. 1966. The results of a non-uniform probability creel survey on a small state-owned lake. Proc. 20th Ann. Conf. S.E. Assoc. Game & Fish Comrs. 20:409-412.

Pfeiffer, C. B. 1982. Fishing the power plant lakes of the East. Outdoor Life. Jan(1982):40-41.

Poff, R. J. 1972. Social considerations of the Lake Michigan Fishery. 15th Conf. Great Lakes Res., Madison, WI.

Spigarelli, S. A. and M. M. Thommes. 1976. Sport fishing at a thermal discharge into Lake Michigan. *J. Great Lakes Res.* 2(1):99-110.

Sport Fisheries Institute. 1982. Recreational fishing is big business. Sport Fish. Inst. Bull. (336):2-4.

Stroud, R. H. 1981. Value of recreational fishing. Sport Fish. Inst. Bull. (323):1-2.

Tennessee Valley Authority. 1977. Fishing around TVA steam plants. Tennessee Valley Authority, Norris, TN. 10p.

Solid and Liquid Wastes: Management, Methods and Socioeconomic Considerations. Edited by
S. K. Majumdar and E. Willard Miller. © 1984, The Pennsylvania Academy of Science.

Chapter Sixteen

HEALTH EFFECTS RELATED TO SEWAGE EFFLUENTS DISCHARGED INTO FRESH WATER ENVIRONMENTS

Stanley J. Zagorski[1], Alfred P. Dufour[2], Richard A. Gammon[3], and Gerald A. Kraus[4]

[1]Professor of Biology,
Gannon University, Erie, PA 16541
[2]Chief Bacteriology Section, Health Effects Research Laboratory
Environmental Protection Agency, Cincinnati, OH 45268
[3]Professor of Microbiology
Gannon University, Erie, PA 16541
[4]Associate Professor of Mathematics
Gannon University, Erie, PA 16541

The practice of disposing domestic wastewaters into streams, lakes and coastal waters carries with it an element of health risk for those individuals who use the receiving waters for leisure time activities, such as swimming, scuba diving and water skiing. One of the earliest recorded outbreaks of enteric illness due to swimming in sewage-contaminated water took place in Walmer, England, in 1908! A group of young army recruits became ill with typhoid fever after a training exercise in an indoor seawater swimming pool which was usually filled on an incoming tide. Twenty of thirty-four cases were attributed to swimming in the pool. The source of the pathogen was sewage from two outfalls whose effluent was carried directly toward the pool intake with the rising tide. Eight of the first nine cases were members of a squad that had just completed their training in the swimming pool, a clear indication that it was a common source of the infections. Another well-documented outbreak of disease associated with swimming in sewage-contaminated water occurred a short distance downstream from Dubuque, Iowa, in 1975.[2] Thirty-one individuals were reported ill with shigellosis, and the only factor common to all the illnesses was swimming activity in the Mississippi River. Water samples from the river, examined some days after the peak of the outbreak, were found to contain high densities of coliforms. The pathogen which caused the illness, *Shigella sonnei*, was isolated from locations in the river where individuals who contracted shigellosis had been swimming. A sewage treatment plant 17 miles upstream was suspected as the source of the pathogen, but this suspicion was never confirmed. Table 1 lists the two outbreaks just described and some other well-known outbreaks of infectious disease where there was an apparent linkage between sewage-contaminated water and swimming-associated illness. It should be pointed out that while the disease outbreak information in the table has frequently been used as proof of the connection between illness in swimmers and the accidental ingestion of contaminated bathing water, it should be accepted with a degree of skepticism, since much of the evidence is clearly circumstantial. This is especially true of the New York and the New Haven outbreaks, where a sound relationship between swimming in polluted harbor water and typhoid fever was never clearly established.[3,4] The etiological agent of illness in swimmers was isolated in only two of the outbreaks. Coxsacki A virus was isolated from the bathing water in the Niort, France, outbreak[5] and *Salmonella paratyphi* B was recovered from the river which served as a source of water for a riverside pool, and from swimmers who used the pool near Beccles, England.[6] These two outbreaks present the clearest evidence that the accidental ingestion of sewage-polluted bathing water can result in gastrointestinal illness in swimmers.

Outbreaks of disease associated with swimming in sewage-contaminated water serve the very useful purpose of confirming what we intuitively believe to be true, i.e., if there are individuals in the community who have enteric disease, and their fecal wastes are discharged into aquatic environments without proper treatment, there is a probability that other individuals who use the receiving water for recreational purposes, such as swimming, will become ill if they

TABLE 1

Outbreaks of Disease Associated with Swiming Activity

Year	Reference	Type of Water	Disease or Agent	Water Quality
1908	1	Sea	Typhoid	Poor
1921	3	Sea	Typhoid	Poor
1932	4	Sea	Typhoid	Poor
1947	6	Fresh	Salmonellosis	Poor
1958	7	Sea	Typhoid	Poor
1973	8	Fresh	Coxsacki B Virus	Unknown
1974	5	Fresh	Coxsacki A Virus	Poor
1978	2	Fresh	Shigellosis	Poor
1979	9	Fresh	Viral Enteritis	Unknown

accidentally ingest some of the water. Disease outbreak information is also useful for identifying new pathogen reservoirs and in some cases, unrecognized diseases, but they provide little help for quantifying health hazards which result from the disposal of sewage into surface waters.

EPIDEMIOLOGICAL STUDIES

The use of epidemiological methods to establish possible relationships between swimming-associated illness and water quality was first attempted in the early 1950's. Stevenson[10] reported on a series of epidemiological studies which were conducted at marine and freshwater bathing beaches from 1948 to 1950. Significant swimming-associated health effects were not detected at marine bathing beaches, but an excess of gastrointestinal illnesses above what would be expected in the total study population was observed in swimmers at a freshwater beach on the Ohio River. Water samples taken during the course of the study were assayed for coliforms which at that time were the commonly accepted bacterial indicator for measuring the quality of the water. The geometric mean density of coliforms in the water samples was 2300 per 100 ml. Swimmers who used a public swimming pool containing high-quality water did not show an excess of gastrointestinal illness, as was observed in those who swam in a river. Stevenson concluded that there was a risk of enteric illness attributable to swimming in sewage-contaminated bathing water. This conclusion was tempered somewhat by the results reported by Moore in 1959.[11] He looked at the swimming activity of polio patients for the three weeks prior to the onset of the illness. Similar data was obtained from control subjects who had been matched to cases by age and sex. Moore did not find a higher frequency of swimming activity in the group that had contracted polio and he concluded that swimming in poor quality water was not a risk factor for poliomyelitis. His conclusion frequently has been used to justify the case against extensive treatment and disinfection of sewage. A similar retrospective study conducted by D'Alessio and his colleagues in 1981 determined that there was an increased risk of enterovirus-caused illness in children who swam in lake water.[12] Their results showed that children who contracted enteric illness swam significantly more often than those children who did not have enteric illness.

All of the epidemiological studies described above, though they were more elaborate than the disease outbreak investigations, were not very helpful for determining if a gradient existed between swimming-associated illness and decreasing water quality. In 1981, Cabelli, *et al,*[13] reported the results of a series of studies conducted at marine bathing beaches in New York City, Boston, and near New Orleans. The studies examined the differences in symptomatic illness rates between swimmers and nonswimmers on weekend days when the water quality was intensively monitored. The results of their studies clearly showed

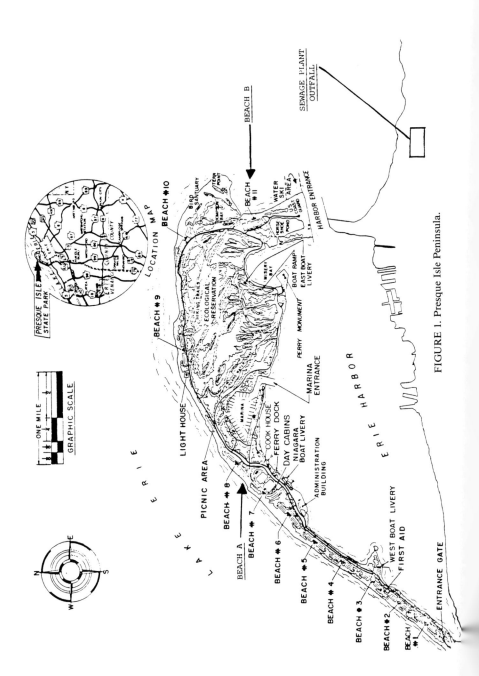

FIGURE 1. Presque Isle Peninsula.

that as the quality of the bathing water decreased, the rate of swimming-associated gastrointestinal illness increased. The indicator they used to measure the quality of the water was a group of bacteria called enterococci, which are commonly found in the feces of warm-blooded animals. These indicator bacteria were used because they showed the strongest relationship to swimming-associated illness. The rate of swimming-associated gastroenteritis ranged from zero to 34.5 per thousand swimmers, and the mean enterococci densities ranged from 3.6 to 495 colony-forming units per 100 ml. Paradoxically, significant illness rates were sometimes associated with very high quality bathing waters.

FRESHWATER STUDIES

The detectable excess of gastrointestinal illnesses observed in the freshwater studies reported by Stevenson[10] and the results of the marine bathing beach studies showing a direct linear relationship between water quality and swimming-associated gastrointestinal illness described by Cabelli *et al*[13] indicated that further freshwater studies should yield significant results. In 1979, a series of freshwater trials were begun at two beaches on Lake Erie at Erie, Pennsylvania. The beaches are located on a long peninsula which extends into Lake Erie from west to east. (See Figure 1.) One beach is located on the northern perimeter of the peninsula (designated Beach A), and its waters are normally of good quality, except in situations when pollutants are carried along the shore by westerly winds. The second beach is located on the southeast shore of the eastern tip of the peninsula (designated Beach B). Beach B is approximately three-fourths of a mile northwest of the sewage outfall of the Erie sewage treatment plant. The sewage treatment plant serves a population of about 170,000 people and treats an average of about 45 million gallons of sewage per day, using an activated sludge process. The sewage is chlorinated before it is discharged into Lake Erie.

The design of the prospective cohort study described here was the same as that used by Cabelli *et al*[13] at marine bathing beaches. The salient features of the design are as follows:

1. Beach surveys and water quality monitoring were conducted only on weekends. This restriction was imposed because maximum beach usage occurred on those days. Whenever feasible, family groups were recruited for the study. This was done so that information on a number of individuals could be obtained by questioning a group spokesperson, usually an adult member of the group, at the initial and follow-up interviews. Participants who swam in the five days before the trial weekend, or during the five days following the swimming experience were excluded from the study. Demographic data which included age, sex, ethnicity and socioeconomic status was obtained at the time of the beach interview.
2. Swimming was defined as completely immersing the head and body under water.

3. Eight to ten days following the initial interview the participants were contacted by telephone. After their eligibility was confirmed, they were questioned about the onset of any symptomatic illness that occurred after the swimming exposure. Information was obtained on gastrointestinal and respiratory symptomatology. Symptoms such as diarrhea, nausea, stomach ache and vomiting were considered as gastrointestinal, and respiratory symptoms included sore throat, bad cough and chest cold. Fever was categorized under the heading "other" and it indicated a body temperature greater than 100 degrees F. Gastroenteritis was defined as having any one of the following unmistakable symptoms or combination of symptoms: (1) vomiting; (2) diarrhea with fever or a disabling condition such as remaining at home or in bed or seeking medical advice; and (3) stomach ache or nausea accompanied by fever.

4. A swimming-associated illness rate was determined by subtracting the symptomatic illness rate in nonswimmers from the symptomatic illness rate in swimmers.

5. The water quality was measured by sampling the water at two-hour intervals during peak swimming activity at three locations on each beach. The samples were assayed for enterococci using the method of Levin et al.[14]

The results of the freshwater studies conducted on Lake Erie in 1979, 1980 and 1982 were very similar to those observed in the marine studies in three respects. First, swimmers appeared to have higher illness rates for all symptomatic illness compared to the rates in nonswimmers, although the rates were not statistically significant. This phenomenon also was observed in the marine recreational water studies[13] and in the freshwater studies report by Stevenson.[10] Second, there is a risk for swimming-associated gastroenteritis which is related to the quality of the water. This pollution-related effect is clearly shown in Table 2. At the relatively unpolluted Beach A, neither of the two summer trials yielded significant swimming-associated gastroenteritis symptom rates. However, at the more polluted Beach B, two of the three summer trials showed significant differences in the rate of gastroenteritis between swimmers and nonswimmers. The lack of corresponding significant swimming-associated rate differences for respiratory illness is indicative of the sewage-related nature of the illnesses observed in swimmers. These results clearly confirm the findings reported by Stevenson[10] in 1953, and they are very similar to the results obtained in the marine bathing beach studies described by Cabelli et al.[13] Third, if there is indeed a relationship between gastroenteritis and water quality, then there should be a high correlation between the density of bacteria used to indicate the presence of fecal contamination and the degree of risk to swimmers. In the freshwater studies on Lake Erie, there was a high correlation between gastroenteritis and the density of enterococci in beach water samples. The strength of the relationshp between enterococci and the relative risk for contracting gastroenteritis while swimming, or enterococci and the relative risk for

TABLE 2

Relative Risk Ratios and Microbial Indicator Densities
for Lake Erie Beaches, 1979, 1980 and 1982

Year	Beach	Relative Risk[1]		Mean[2] Enterococci Density/100 ml
		Gastroenteritis	*Respiratory Illness*	
1979	A	1.2	1.2	5
	B	1.3	1.3	13
1980	A	1.4	0.68[+]	25
	B	2.3*	1.3	71
1982	B	1.8*	1.3	20
Correlation Coefficient[3]		0.90	0.20	

[1]Illness rate in swimmers divided by the illness rate in nonswimmers.
[2]Geometric mean.
[3]Spearman's rank difference correlation coefficient—A measure of the strength of relationship of relative risk to indicator density.
*Swimmer illness rate statistically greater than the nonswimmer illness rate ($P < 0.05$).
[+]Nonwimmer illness rate statistically greater than the swimmer illness rate ($P < 0.05$).

respiratory illness was measured with the Spearman rank difference correlation coefficient.[15] The correlation coefficient for the enterococci, gastroenteritis variables was 0.90, which indicates that most of the increase in risk can be accounted for by the increase in indicator density. The correlation coefficient for the enterococci, respiratory illness, on the other hand, was very small, 0.20, indicating a weak relationship between these two variables, a result that was not unexpected since it is unlikely that respiratory pathogens would be transmissible by ingestion even if they were present in domestic sewage. This strong relationship between enterococci, a bacterium intimately associated with the feces of humans and other warm-blooded animals, and gastroenteritis clearly establishes the health significance of discharging sewage into waters used for recreational activities.

The results of the studies conducted at Lake Erie beaches indicated that significant swimming-associated gastroenteritis occurred in reasonably good quality fresh water, a phenomenon which was also observed at marine beaches. This phenomenon occurred when the mean enterococci density was as low as 20 per 100 ml. The mean fecal coliform density in the same water samples was 60 per 100 ml. This fecal coliform density, which is about one-third of the standard recommended limit used by most states, is considered to be indicative of good quality water and yet it was associated with a significant swimming-associated difference in illness rates. In the marine bathing beach studies, significant swimming-associated gastroenteritis rates occurred when the enterococci density was as low as 11 per 100 ml. or the E. coli density was 15 per 100 ml. The significant gastroenteritis rates associated with swimming in good quality water are not easily rationalized. Viruses are thought to be the etiologic agent responsible for the swimming-associated gastroenteritis in swimmers,[16] and this complicates the matter even more, since many viruses are not culturable and therefore their infectious doses are unknown. Although it was once assumed that one virus was sufficient to initiate an infection,[17,18] recent evidence indicates that multiple viral particles are required.[19] Furthermore, it also is generally assumed that the maximum amount of water swallowed during swimming activity is about 10 ml.[20] These two factors imply that the causative agent of the gastroenteritis in swimmers is either highly virulent and therefore only a few viruses are required to initiate an illness, or the viruses occur in very high densities in the water and therefore only a small volume of water has to be ingested to receive an infective dose. The former premise is not amendable to experimental testing since the causative agent has not been identified. The second premise, however, can be subjected to testing if it is assumed that the high densities of pathogen are associated with or encapsulated in particles of fecal material. Such a study has been conducted at Lake Erie beaches.

PARTICLES AND GASTROENTERITIS IN SWIMMERS

The particulate study was carried out using the same health survey design previously described. The protocol for monitoring the microbiological quality of the water was modified so that bacterial cell-containing particles in a water sample could be quantified and the number of viable units per particle could be estimated. Since viral and bacterial pathogens are difficult to measure and their occurrence is sporadic, a bacterial surrogate was used to examine the potential relationship between swimming-associated illness and particulates in bathing water. *Escherichia coli* was selected as a pathogen surrogate for three reasons. First, this organism is a common inhabitant of the gastrointestinal tract and therefore it is intimately associated with fecal material, just as a pathogen would be. Second, *E. coli,* as pointed out previously, is related to the incidence of gastrointestinal illness in swimmers, which implies that this organism is related to the presence of pathogens and may even physically occupy the same microenvironments as the pathogens. Enterococci, which also was closely related to the incidence of swimming-associated gastroenteritis, were not used as a surrogate because of their greater tendency to form chains and clumps. Third, a highly specific, facile enumeration technique was available for quantifying *E. coli.*[21]

The procedure for obtaining estimates of the number of particles containing *E. coli* and the number of *E. coli* per particle is given in Figure 2. A portion of a water sample was assayed for *E. coli* with the standard technique used throughout the study and another portion was filtered through a 3 micron pore size polycarbonate membrane filter. A 3 micron pore size was used because it

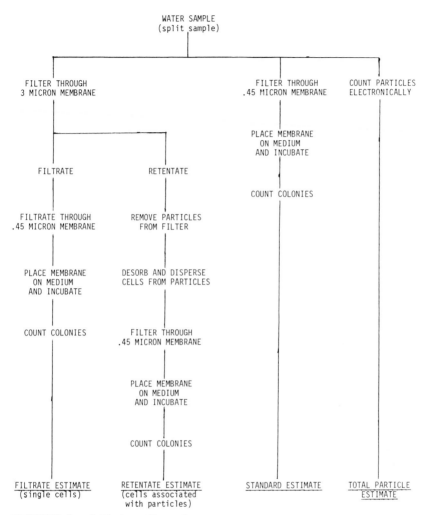

FIGURE 2. Sample Treatment Protocol.

retained particles that were larger than a single *E. coli* cell. The assumption was made that all single, non-particle associated *E. coli* cells passed through the membrane into the filtrate. The filtrate was subsequently assayed using the standard technique. The colony count resulting from the original sample was called the standard estimate and that from the filtrate samples was called the filtrate estimate. The number of particles 3 microns or greater in size, containing at least one *E. coli* cell, was estimated by subtracting the filtrate estimate from the standard estimate.

The density of *E. coli* cells in the retenate was estimated by breaking up the

TABLE 3

Gastroenteritis Illness Rates of Swimmers and Nonswimmers
by Trial Day at Beach B, 1980.

Trial Day	Swimmers		Nonswimmers	
	Number of Participants	% Ill	Number of Participants	% Ill
1	345	4.3	129	4.3
2	300	3.0	109	1.8
3	268	3.4	92	1.1
4	255	1.6	235	0.4

TABLE 4

Health Effects Risk, Water Qualtiy Level, and Particle-Related Measurements
for Beach B, 1980.

Relative Risk	E. coli Per 100 ml	Colony Associated Particles Per 100 ml	E. coli Density/Particle
1.0	138	30	1500
1.6	286	154	158
3.1	109	64	426
3.7	624	484	17

particles and then assaying the resulting dispersed bacterial cells with the standard technique. This was accomplished by immersing the polycarbonate filter in a buffer solution containing a surfactant and then subjecting the filter containing solution to a brief sonication and blending treatment to desorb and disperse the *E. coli* cells. The average number of *E. coli* per particle was estimated by dividing the total number of *E. coli* colonies obtained after the dispersion treatment by the number of *E. coli* associated particles.

Studies employing the particle enumeration procedures and the health effects survey methods were conducted during the summers of 1980 and 1982. The gastroenteritis illness rates observed in swimmers and nonswimmers in the summer of 1980 are given in Table 3. These data were used to calculate the relative risk in swimmers attributable to the water exposure. The relative risk was ordered from the smallest to the largest value and examined against the variables described above to determine if there was an association between increasing risk of illness and the measures of water quality. The values for the density of *E. coli*, the density of particles associated with *E. coli* colonies and the number of surrogate cells per particle, as well as the relative risk for each trial day are shown in Table 4. The Spearman rank difference correlation coefficient was used as a measure of the relative strength of association between the health effect parameter and each of the water quality variables.[15] Table 5 indicates that there is a relationship between relative risk and the number of *E. coli* cells per particle, but that it is an inverse relationship, i.e., as the risk of gastroenteritis increases, the number of cells per particle decreases, an unlikely event if the surrogate were indeed a pathogen. The correlation between the risk of illness

TABLE 5

Correlation of Swimmer-Associated Gastroenteritis with Water Quality Indicator Density and Particle Related Variables.

Comparison	Correlation[1] Coefficient
Gastroenteritis vs No. *E. coli*/Particle	−0.80
Gastroenteritis vs *E. coli* Density/100 ml	0.40
Gastroenteritis vs. Density of *E. coli* Associated Particles/100 ml	0.80

[1]Spearman's rank difference correlation coefficient.

TABLE 6

Gastrenteritis Illness Rates of Swimmers and Nonswimmers by Trial Day at Beach B, 1982.

Trial Day	Swimmers		Nonswimmers	
	Number of Participants	% Ill	Number of Participants	% Ill
1	335	1.8	147	1.4
2	422	2.1	203	1.5
3	352	2.3	137	1.5
4	451	6.9	141	4.3
5	460	3.3	174	1.1
6	495	4.0	128	0.7

and the *E. coli* density, as obtained using normal monitoring procedures, is moderately low, indicating a positive but slight association. The correlation between the health effect risk and the density of surrogate colony associated particles, on the other hand, is a relatively strong one. The order of the relative strength of the three variables to swimming-associated risk was not unexpected. It is what might be anticipated if particle associated pathogens are indeed responsible for swimming-associated gastroenteritis. Thus, the greater the density of particles containing multiple pathogens per volume of water, the greater would be the likelihood of a swimmer ingesting an infectious dose. This strong relationship between the density of *E. coli* associated particles and the risk of gastroenteritis would also explain the apparent lack of association between the number of surrogate bacteria per particle and the risk of illness. If the density of particles in the water is not great enough to insure a high probability of ingestion of at least a single particle by a swimmer, then it is irrelevant whether or not the particle contains a single pathogenic entity or a million.

The health effects and microbiological data collected during the summer of 1982 are shown in Tables 6 and 7. A new characteristic was substituted for the number of viable bacteria per particle in the second year of the study. The density of all particles less than 50 microns, those carrying viable bacteria as well as those not associated with microbial cells, was quantified using an electronic particle counter. This new parameter and two of the variables examined in 1980

TABLE 7

Health Effects Risk, Water Qualtiy Level, and Particle-Related Measurements
for Beach B, 1982.

Relative Risk	E. coli Per 100 ml	Colony Associated Particles Per 100 ml	E. coli Density/Particle
1.32	66	20	2030
1.44	60	32	1908
1.56	72	46	1521
1.62	65	24	1689
2.84	142	64	1435
5.17	135	104	1863

TABLE 8

Correlation of Swimmer-Associated Gastroenteritis with Water Quality Indicator
Density and Particle Related Variables.

Comparison	Correlation[1] Coefficient
Gastroenteritis vs Particle Density/100 ml	− 0.60
Gastroenteritis vs E. coli Density/100 ml	0.66
Gastroenteritis vs. Density of E. coli Associated Particles/100 ml	0.71

[1]Spearman's rank difference correlation coefficient.

were analyzed using the Spearman rank difference correlation coefficient described above. The results of this analysis are given in Table 8. Once again the
density of E. coli associated particles showed the strongest relationship to the
risk of gastrointestinal illness, as it did in the 1980 study. The relationship between E. coli density and swimming-associated risk of illness was quite strong
as compared to the slight association observed in 1980. The total particle count
showed a negative relationship to the illness risk very similar to the relationship observed between illness risk and E. coli associated particles in 1980. This
finding also points to the fact that risk of gastrointestinal illness is indeed
associated with particles of fecal material and not particulates in general.

The rationale for the trials which examined the relationship between health
effects in swimmers and particulates in the water was based on the hypothesis
that an infectious dose of a pathogen could be ingested in a small volume of
water if the pathogens were adsorbed to or encapsulated in fecal material. Thus,
a single particle might contain hundreds or even thousands of bacterial or viral
pathogens. This hypothesis was attractive because it could account for many
of the known factors which were incompatible with some of the observations
made from data collected in the epidemiological-microbiological studies at
marine and freshwater bathing beaches. For instance, it was known that many
virus particles were required for an infective dose, and that the amount of water
ingested during swimming activity was small, and yet a significant number of
illnesses were occurring in spite of the fact that the water was judged to be of

high quality by current standards. The hypothesis is also supported by available evidence which indicates that bacteria and viruses adsorbed to or encapsulated in particles are protected from environmental stresses and the effects of disinfection. Hejkal *et al*[22] was shown that polioviruses encapsulated in fecal material are more resistant to chlorination than nonencapsulated viruses. Similarly, a number of investigators[23,24,25] have presented evidence that particle associated bacteria and viruses survive longer than non-particle associated bacteria and viruses. These two factors could account for the occurrence of gastrointestinal illness at beaches affected by sewage treatment plant effluents that not only have been disinfected, but which sometimes must travel considerable distances before reaching a beach.

Although the research work on this problem is not complete, the preliminary evidence presented here strongly supports the hypothesis that swimming-associated gastrointestinal illness observed in individuals bathing in good quality water is at least in part the result of particles which contain an infective dose of a pathogen.

IMPLICATIONS OF PARTICULATE
ASSOCIATED ILLNESS IN SWIMMERS

The apparent relationship between swimming-associated illness and the density of bacteria laden particulates in bathing waters implies that some form of intervention might be practiced which could influence the gastrointestinal illness rate in swimmers. For instance, decreasing the density of microorganism associated particles in the water should lead to a corresponding decrease in the relative risk for gastroenteritis in swimmers. The breakup of particles in wastewater effluents would have a two-fold effect on any potential pathogens that might be contained in the particles. First, the exposed bacteria and viruses would be much more susceptible to disinfection, and, second, the pathogens discharged into receiving waters would be fully exposed to physical and biological factors which promote biological decay in aquatic environments. The results of this study also imply that incomplete intervention could have an adverse effect on the rate of swimming-associated health effects. Although the data are limited, they clearly indicate that breaking up large particles that contain high numbers of viable units per particle to smaller particles which still contain an infective dose of pathogen may actually increase the rate of gastroenteritis in swimmers. However, this conclusion must remain speculative until confirmed in future studies.

CONCLUSION

The result of the Lake Erie bathing beach studies on the relationship between swimming-associated illness and water quality lead to the following conclusions. The presence of sewage wastes in recreational waters, as measured with bacterial indicators, is a significant risk factor for gastrointestinal illness associated with swimming activity.

The rate of swimming-associated gastrointestinal illness is correlated with water quality; that is, as the water quality decreases, the rate of illness in swimmers increases.

Statistically significant swimming-associated illness rates occur in water considered to be high quality according to currently accepted standards. Preliminary evidence from studies examining the effect of particulates in water on the health of swimmers indicates that the swimming-associated illness in individuals bathing in good quality water may be the result of particulates laden with high numbers of pathogens. This finding, if confirmed in ongoing studies, could be useful in planning intervention measures that might appreciably lower the rate of gastrointestinal illness in swimmers.

REFERENCES

1. Moore, B. 1954. Sewage contamination of coastal bathing water. *Bulletin of Hygiene*. 29: 689-704.
2. Rosenberg, Mark L., M.D.; Kenneth K. Hazlet, M.D.; John Schaefer, M.S.; Joy G. Wells, M.S.; Rudy C. Pruneda, Ph.D. 1976. Shigellosis from swimming. *Journal of the American Medical Association*. 236: 1849-1852.
3. Ciampolini, Ettore. 1921. A study of the typhoid fever incidence in the health center district of New Haven. *Report to City of New Haven*.
4. ANON. 1932. Typhoid fever from bathing in polluted water. *City of New York Department of Health Weekly Bulletin*. 21: 257.
5. Denis, F. A., E. Blanchovin, A. DeLignieres and P. Flamen. 1974. Cosacki A_{16} infection from lake water. *Journal of the American Medical Association*. 228: 1370-1371.
6. Martin, P. H. 1947. Field investigations of paratyphoid fever with typing of *Salmonella paratyphi* by means of VI bacteriophae. *Bulletin of Hygiene*. 22: 754-755.
7. Anon. 1961. Typhoid traced to bathing at a polluted beach. *Public Works*. 92: 182-183.
8. Hawley, H. Bradford, M.D., David P. Morin, Margaret E. Geraghty, Jean Tomkow, C. Alan Phillips, M.D. 1973. Coxsackievirus B. epidemic at a boys' summer camp. Isolation of virus from swimming water. *The Journal of the American Medical Association*. 226: 33-36.
9. Koopman, James S., Edward A. Eckert, Harry B. Greenberg, Brian C. Strohm, Richard E. Isaacson, Arnold S. Monto. 1982. Norwalk virus enteric illness acquired by swimming exposure. *American Journal of Epidemiology*. 115: 173-177.

10. Stevenson, A. H. 1953. Studies of bathing water quality and health. *Journal of the American Public Health Association. 43: 529-538.*

11. Moore, B. 1959. The risk of infection through bathing in sewage-polluted water. *In: Proceedings of the First International Conference on Waste Disposal in the Marine Environment.* Pearson, E. A., Ed. Pergamon Press, New York. 27-37.

12. D'Alkessio, Donn J., Theodore E. Minor, Catherine I. Allen, Anastasios A. Tsiatis, Donald B. Nelson. 1981. A study on the proportions of swimmers among well controls and children with enterovirus-like illness shedding or not shedding an enterovirus. *American Journal of Epidemiology.* 113: 533-541.

13. Cabelli, V. J., A. P. Dufour, L. J. McCabe, M. A. Levin. 1982. Swimming-associated gastroenteritis and water quality. *American Journal of Epidemiology.* 115: 606-616.

14. Levin, M. A., J. R. Fischer, V. J. Cabelli. 1975. Membrane filter technique for enumeration of enterococci in marine waters. *Applied Microbiology.* 30: 66-77.

15. Tate, M. W., R. C. Clelland. 1957. *Nonparametric and Shortcut Statistics.* Interstate Publications, Danville, IL.

16. Cabell, V. J. 1981. Epidemiology of enteric viral infections. In: *Viruses and Wastewater.* M. Goddard and M. Butler, Eds. Pergamon Press, New York. 291-304.

17. Plotkin, S. A., M. Katz. 1965. Minimal infective doses of viruses for man by the oral route. In: *Transmission of Viruses by the Water Route.* G. Berg, Ed. John Wiley and Sons, New York. 151-166.

18. Westwood, J. C. N., S. A. Sattar. 1976. The minimal infective dose. In: *Viruses in Water.* G. Berg, *et. al.,* Eds. Interdisciplinary Books, Pamphlets and Periodicals for the Professional and Layman, American Public Health Association, Washington, DC. 61-69.

19. Minor, T. C., C. I. Allen, A. A. Tsiatis, D. B. Nelson, D. J. D'Alkessio. 1981. Human infective dose determination for oral poliovirus type 1 vaccine in infants. *Journal of Clinical Microbiology.* 2: 388-389.

20. Shuval, H. I. 1975. The case for microbial standards for bathing beaches. In: *Proceedings of the International Symposium on the Discharge of Sewage from Sea Outfalls.* H. Gameson, Ed. Pergamon Press, London. 95.

21. Dufour, A. P., E. R. Strickland, V. J. Cabelli. 1981. Membrane filter method for enumerating *Escherichia coli. Applied and Environmental Microbiology.* 41: 1152-1158.

22. Hejkal. T. W., F. M. Wellings, F. A. LaRock, A. L. Lewis, 1979. Survival of poliovirus within organic solids during chlorination. *Applied and Environmental Microbiology.* 38: 114-118.

23. Bitton, G., R. Mitchell. 1973. Effect of colloids on the survival of bacteriophages in seawater. *Water Resources.* 8: 227-229.

24. Bitton, G., R. Mitchell. 1974. Protection of *E. coli* by montmorillonite in seawater. *Journal of the Environmental Engineering Division, ASCE.* 100: 1310-1320.
25. Gerba, C. P., G. E. Schaiberger. 1975. Effect of particulates on virus survival in seawater. *Journal of Water Pollution Cont. Fed.* 47: 93-103.

Solid and Liquid Wastes: Management, Methods and Socioeconomic Considerations. Edited by
S. K. Majumdar and E. Willard Miller. © 1984, The Pennsylvania Academy of Science.

Chapter Seventeen

Fluorides, Water Fluoridation and Environmental Quality: A Retrospective Study

J. -B. Bundock, O.B.E., M.D., Frsh.,[1] J. R. Graham, BA., LLB,[2] and P. J. Morin, Ph.D.[3]

[1]Scientific Adviser for the Minister of Environment for Quebec
2540 Carre Pijart, Sainte-Foy,
Quebec, Canada G1V1H8
[2]Member of the Minnesota Bar
[3]Scientific Consultant of the Ministry of the Province of Quebec

"For the first time in the history of the world, every human being is now subjected to contact with dangerous chemicals from the moment of conception until death."

Rachel Carson *Silent Spring*[1]

In June 1975, the Quebec Government enacted Bill 88, compelling municipalities which were equipped with their own water treatment plants to fluoride their water supplies to a fluoride concentration of 1.2 (parts per million). The purpose of this legislation was to reduce tooth decay.[2]

Since that time, it has been difficult to enforce this new act. A number of major scientific and political problems which had not been foreseen or had not been thoroughly investigated before the Bill was adopted have arisen, and pressure has been continuously brought to bear on the Government to reexamine the effects of fluoridation on the environment and on public health before proceeding further.

As early as 1975, the environmental council for Quebec had voiced concern over the increase in fluoride sources in the environment, and the failure on the part of responsible authorities to recognize the problem. The council recommended a general study of the matter to reevaluate all the consequences, i.e., social, medical and environmental, before any new source of fluoride was introduced into the environment.[3]

To clarify the issues and develop a satisfactory solution, the minister of Environment instructed the *advisory committee on fluoridation* to make a thorough analysis of the situation and report to him accordingly.

In carrying out its mandate, the committee used a comprehensive and multidisciplinary approach. A first analysis of the situation revealed that many questions raised on the subject of fluoridation were only part of a wide range of much deeper problems linked to the increase of fluorides and their effects on public health and the environment. Artificial fluoridation of water supplies is in fact only one of the many sources of fluorides to which people, animals and plants are increasingly exposed, fluorides are present in water, air, and in various foods, and there are other very specific sources such as toothpastes and teflon cooking utensils which have a polyvinyl fluoride base. But all are potention environmental contaminants.

The committee conducted a study of some of the most pressing problems at hand and submitted its findings and recommendations to the minister of the Environment in November 1979.[4]

In the first place, I will start this address with a summary of the principal findings of the *advisory committee on fluoridation.*

The second part deals with the sources of fluorides originating from water, food and air, including industrial emissions and the ecological effects of fluoridation.

I will conclude with a summary of the principal recommendations made by the committee.

SUMMARY OF FINDINGS

Fluoride in the Air

Fluoride emissions in the atmosphere are as a rule related to certain types of industrial activities. In the United States, in 1970, the annual fluoride emissions into the atmosphere by industry were estimated at some 120,000 tons. It is believed that this figure has doubled during the 1971-1980 period, despite the fact that 90 percent of all emissions are intercepted at the source by various devices.

However, new technology in aluminum processing may curb fluoride emissions into the atmosphere substantially in the coming years and it is hoped that by year 2000 this problem will be confined to history.

Fluoride in Water and Food

A number of controlled studies have shown that since 1940, changes in agricultural processes have brought about major increases in the fluoride content of food. This increase, coupled with increase in the fluoride content of water, now exposes humans as well as domestic animals to what appears to be a dangerously high level of fluorides. The margin of safety between harmless and toxic dose would appear to be slight. On the other hand, opinions expressed on the subject by various experts or groups may differ by a factor of ten or more. This controversy appears impossible to resolve, at least for the present time, since the supporting data for the opinions expressed is either incomplete or missing.

Maximum Safe Dosage

Although American health authorities consider a 1.2 PPM fluoride concentration a safe and effective dose for public water supplies, no scientific consensus has ever been reached on a safe maximum absorption dose in absolute terms (MG/DAY).

Synergetic Effects

The synergetic effects of fluorides and the threat they pose to human and animal health and to the environment are poorly understood. There is some evidence that fluorides can accumulate along specific food chains, leading to major increases in the food consumed by man and animals. Whatever data is available is particularly worrying since so little is known on this subject. Little is known also about the capacity of fluorides to form new substances in the presence of other chemicals found in the environment. The development of organic fluoro compounds by plants is a good example of the formation of such unexpected but highly toxic new substances.

Toxic Effects of Fluorides

Fluorides exert their toxic effects on various component parts of the body. Dental fluorosis appears to be a first sign of intoxication in humans. It exerts its action on the ameloblasts which are component cells of teeth. Fluorides also have been reported to react with the genetic material of cells where they appear capable of producing mutations or congenital malformations. Strong documentation supports the thesis that fluorides can interefere with normal enzyme reactions in living cells.

SOURCES OF FLUORIDES

The chemical element *fluorine* is among the most violently active substance known. Because of this high reactivity, it does not exist in nature in a free state. The compounds of fluorine number more than six hundreds.[5,6] Water, food, and air are the three major sources of fluorides and the contribution of each may vary from person to person depending on weather and climatic conditions. This last point is particularly important. It should be remembered that in Quebec, for example, the daily temperature varies by as much as sixty-one degrees celsius, and changes may be abrupt. This could certainly contribute to large changes in individual daily water consumption.

FLOURIDE INTAKE FROM WATER

The quantity of fluorides ingested from water supplies depends of course on the fluoride concentration of the water and the amount consumed. Fluoride concentration can easily be determined but water consumption is extremely difficult to assess.

In a study conducted on adults, Marier and Rose showed that water containing 1 PPM of fluoride gives a daily fluoride intake of between one and three MG.

Following an exhaustive review of the literature on the subject, Groth estimates that adults consume between one and five liters of water daily while children drink from 200 to 500 ML. He pointed out that heavy tea drinkers may ingest between 2 and 3 MG/day of fluorides from this source alone. With beer drinkers, the fluoride quantities ingested vary greatly from one individual to another and can exceed 6 MG per day.[7]

FLUORIDE INTAKE FROM FOOD

As a rule, all foods contain a certain amount of fluoride. For example, beets contain 17.70 PPM dry base, celery 7.92 PPM, spinach 1.11 PPM, salmon 19.3 PPM, etc.[8]

The fluoride increase during food processing must be taken into account. When fluoridation was first introduced, it was accepted without question that this would add little or nothing to fluoride intake. But today the facts are better known. The consumption of processed foods has outstripped that of fresh foods and most of the North American food-processing industry is supplied with fluoridated water.[9] Martin showed that when foods are cooked in water containing one PPM of fluoride, their fluoride content is increased three to five times. This demonstrates the multiplier effect of water fluoridation.[10]

The study of Marier and Rose of the *National Research Council* of Canada shows that commercial foods and beverages prepared with fluoridated water contain on the average about three and a half times or more fluorides than the rate established by Hodge and Smith for non-fluoridated regions. Their results again demonstrate that the fluoride content is increased when fluoridated water is used in the industrial or domestic preparation of foods.

Their work enables one to predict that an adult exposed to water containing 1 PPM of fluoride will consume on the average between two and five MG of fluoride from food only.[11]

According to Groth, ingestion of fluorides from food has increased significantly since fluoridation became generalized on the North American continent. He adds that a comparison of the McClure estimates and of those arrived at during the last seven years reveal that flouride quantities in the human diet have increased from three to ten times since fluoridation. The total quantity of fluorides ingested from different other sources has presumably doubled or even tripled during the same period.[12,13]

Considering the absence of data for "a safe daily dosage of fluorides," the committee could only take notice of the large recent increase in the total amount absorbed in the fluoridated areas.

FLUORIDE FROM THE AIR

The fluoride ion concentration in the air in rural or urban residential areas is generally very low, ranging between 0.04 and 1.20 PPB (part per billion), which makes an analytical determination difficult. But this concentration can be increased appreciably by industrial activity. In the vicinity of aluminum plants, for instance, fluoride concentrations in the air have been found to range from 3 to 18 PPB and sometimes even as high as 80 PPB.

EFFECTS ON ANIMALS[14]

Domestic animals fed on fodder containing fluorides ultimately show signs of the poisoning known as fluorosis. In addition to its own cycotoxic properties, fluorine, because of its affinities for calcium, disturbs the ossification process. Taken in excessive quantities, it causes fluorosis, symptoms of which appear in various disorders of increasing severity. The effects of the fluorine vary according to the intensity of the poisoning. Where the emission of fluorine is greatest, the animals' teeth decay and wear out completely, they are no longer white but yellow or brown. The animals become incapable of grinding food. The teeth work loose and finally fall out. As a result, chewing becomes difficult if not impossible, causing digestive ailments, a rapid loss of weight, a drop in production, and finally, complete loss of life. The ministry of Environment has now passed a law limiting the flouride content of animal fodder.[15]

Ecological Effects of Fluorides

All aquatic animals ingest fluorides in water directly. Moreover, herbivorous and carnivorous animals can be affected in their food.

In the vertebrates, it should be noted that frog's eggs undergo a delay in their embryonic development when they are subjected to a concentration of one PPM of fluorides. This is also true for tadpoles.[16]

Wiber reports that water with a concentration of fluorides of 1.5 PPM gave visible harmful effects on the embryogenesis of fish eggs. Following an extensive review of literature on the effects of fluorides, he considers that it is necessary to suggest a maximum concentration of 1.5 PPM of fluorides in water if the Aquatic life is to be maintained and preserved.[17]

Also, from studies conducted by H.L. Richardson, pathologist at the University of Oregon, it has been shown conclusively that fluorides in a concentration of 1 PPM can sterilize chinchillas on a farm.[18]

Effectiveness of Fluorides on Dental Caries

Fluoridation of drinking water dates back a little more than thirty years. It was first tried out after epidemiological research by Dean[19] to isolate the causes of dental fluorosis observed in part of the population of about 25 American states. The studies by Dean and several others, which showed that fluorides can cause dental fluorosis, were later confirmed by experiments under controlled laboratory conditions. Subsequent research, in fact, revealed that fluorides can cause poisoning of the ameloblasts, resulting in the formation of dental enamel with obvious signs of hypomineralization. The anomaly is commonly called dental fluorosis.[20,21,22,23,24]

Depending on the amount of fluoride consumed, the enamel will show changes ranging from irregular whitish opaque areas to severe brown discoloration of the entire crown, with a roughened surface.[25,26,27]

Dean, in his studies, also noted that there seemed to be a correlation between the presence of dental fluorosis and a reduction in caries. His proposed dosage (1.2 PPM), seems to be a compromise between "an acceptable rate of dental fluorosis" and some reduction in caries. Calculations made at the time estimated that an adult would absorb from one to two MG per day.[28]

When the time came to confirm the effectiveness in preventing tooth decay by adding fluorides at this level, matters became quite complicated, and even today the only valid experience is that obtained from studies made on human population. We must recognize that in this respect we are witnessing the most extensive toxicological study ever made on the human race, and that this study is usually being carried out without the unanimous consent of the people involved.

The difficulties in carrying out the experiments stem from the very large number of factors that can influence the rate of dental caries in the population. Among them, mention may be made of dietary habits,[29,30] the hardness of food,[31] the presence in drinking water of such elements as vanadium, strontium , calcium, magnesium, phosphorus, fluorine, and copper.[32,33]

Also on the type of bacterial growth in the mouth,[34,35] the presence or absence of antibodies against "streptococcus mutans" in the saliva,[36] dental hygiene,[37] genetic factors,[38] the use of dentrifices with a fluorine base or those containing antiseptics, etc.[39]

The presence of such a large number of variables, all of which may have important effects on the results, creates serious problems in interpreting the collected epidemiological data. It is therefore not surprising that many scientists question the interpretation given to these studies by public health services.[40] The very large differences in percentages of reduction in dental caries found in the various studies appear to be further confirmation of the doubts expressed, and in general it can be said that even today, it is impossible to predict the results for any given population.[41,42]

In a situation like this, the data assembled in epidemiological studies are normally confirmed by laboratory tests under controlled conditions. But in this instance, while it would seem a priori that experiments with animals could easily be carried out, this is not the case. Fluorides are so prevalent in nature that it is almost impossible to prepare a diet for laboratory animals that does not contain excessive amounts of these substances. The few experiments that have been made with diets that are chemically pure are so far removed from normal experience that any extrapolation to the human conditions seems irrelevant. This is a problem that modern technology has not resolved.

It is not surprising that works published about the prevention of tooth decay have not yet established any consensus within the scientific community. For the present, it seems that research efforts should be directed to an understanding of the basic mechanisms of dental caries, since only more complete knowledge in this field will enable us to improve the present situation.

Considering all the data available at the time the review was made, the *committee* members were somewhat surprised at the gratuitous assurances of efficacy made by various bodies and associations. They could not concur with them on the basis of available data.

Secondary Effects of Fluoridation

Rose and Marier and several others[43,44,45] state that, whether we like it or not, our systems absorb larger quantities of fluorides than the 1 to 2 MG the *World Health Organization* (WHO) considers a safe daily dose.[46] Furthermore, according to Rose and Marier and others,[47,48,49] fluoridation of drinking water has extraordinary consequences when food is prepared in factories and when food concentrates are reconstituted. It therefore seems that at present we should be more concerned with possible intoxication than with fluorides deficiency.

Fluorine has a very toxic effect on the human organism, and there is a very fine line between the acceptable level and the toxic level. Dean[50,51] had noted that with a fluoride level of 0.9 PPM in drinking water, close to twelve per cent of children suffer from dental fluorosis. At the level of 1.2 PPM, this disease affects 20 to 30 per cent of the child population. Dean's work was partially confirmed by a recent publication which states that nearly 70 per cent of young children who take fluoridated vitamins drops suffer from dental fluorosis.[52]

Dental fluorosis seems to be a first indication of fluorine intoxication of the population.[53,54,55] The seriousness of disorders and diseases caused by fluorine seems to increase with the degree of intoxication. This is true, for example, for persons who absorb a higher than normal amount of fluorides, such as those with kidney deficiencies, sufferers of polydipsia, diabetics and those on dialysis.[56,57,58] A number of cases of osteomalacia and osteosclerosis have also been reported. In regions where fluorosis is enemic, Singh[59] has noted calcification of the joints, tendons and spinal column, with fusion of vertebrae and major malformations of different parts of the skeleton. In these regions there are also a number of children with a form of fluorosis that can cause major malformation of the knee (genu valgum).[60]

The examples Singh documented can be considered advanced cases of intoxication. But they also provide evidence of certain processes of chronic fluorine intoxication. Some patients with fluorosis also suffer from hyperparathyroidism and severe neurological disturbances.[60,61] Finally, it seems that fluorosis can cause a kidney disease called insipid diabetes.[64,65]

Down's Syndrome

Two reports, among the very extensive literature on fluorine, deserve particular attention. The first deals with studies published by Rapaport, who relates an increase in the frequency of the Down's Syndrome (mongolims) which is proportional to the level of fluorides in drinking water in certain cities in Minnesota. The genetic components of this disease and the incidence of its concentration in the population are well known. This study would not be particularly significant if it were not indirectly confirmed by recent studies showing that even a low level of fluorides can cause modifications in the genetic material of the cell.[66,67,68]

(We have just recently completed a review of the world literature on the subject of Down's disease and fluorine intoxication, and we have come to the conclusion of a positive assocation. Our work on this topic will be ready for publication during the course of the next few months.)

Cancer Deaths

A second point of concern is found in an article by Burk and Yiamouyianis,[69] which is essentially an epidemiological study showing the fluoridation—cancer experience in 18 million Americans over a period of thirty years.

The basic data showed that two geographically balanced groups of ten large central cities of the United States, each having virtually identical average crude cancer death rates for a period of 10 - 12 years, these rates had been gradually rising together year by year before the introduciton of artificial fluoridation of public water supplies in one group of cities.

Thereafter, the crude cancer death rate in the fluoridated group began to accelerate much faster than in the control group. After a period of 13 - 18 years, there was a relative increase exceeding 10 per cent in the experimental cities.

The cancer sites involved were essentially nonrespiratory: hence, the results were not contrary to what would have been normally expected in water fluoridation-caused carcinogenesis.

The major conclusion of the study by Burke and Yiamouyiannis is that fluoridation of pubic water supplies is causally related to a very sizeable increase in human cancer deaths. The Yiamouyiannis and Burk report is not completely surprising since several other scientists have published studies showing that inorganic fluoride compounds do seem to have carcinogenic effects.[70-79]

Intolerance to Fluoridated Water

Finally, there is also a whole series of disorders which could be classified as intoxications ranging from light to severe, and which are caused by an intolerance to fluorides. Walbott reports more than four hundred cases of intolerance with symptoms that are either minor or major. Many others have also documented clincial cases of intolerance to fluorides.

RECOMMENDATIONS

Considering all the available data, the *Advisory Committee* on the fluoridation of water supplies made a large number of recommendations which may be summarized as follows:[81]

— Steps must be taken in order to determine with accuracy and on a continuing basis the various sources of fluorides in the environment, their mode of diffusion, their location and their effects on the environment and public health.

— Fluoride emissions into the environment must be controlled and curbed as part of the total emissions into the environment. The committee therefore recommended that the moratorium imposed on Bill 88 be sustained as part of the control of emissions.

— Research must be carried out on the effects of fluorides on ecosystems so as to allow some much needed understanding of the interactions of fluorides with the environment. The research carried out under this program should include synergetic effects and accumulation of fluorides along food chains.

— Large scale epidemiological studies should also be carried out on human populations exposed to fluorides and remedial measures taken to prevent chronic fluoride intoxication and carcinogenic diseases.[82]

— Finally, citizens must be made aware that fluorides are increasing in the environment. Steps must be taken to involve the population in the setting and continuous re-evaluation of the norms imposed to curb fluoride increases.

ACKNOWLEDGEMENTS

The authors are grateful to the following persons for their precious collaboration in the preparation of this paper: Mr. Robert Poisson, Me. Pierre Chatillon, Mr. Gerard Nogrega, Dr. Marcel Ouellette, Mr. Clement Audet, Mr. André Couillard, Mr. Robert Boudreau, Mr. Conrad Anctil, Dr. Henri Saint-Martin, Mr. Jean Roy, Dr. Philip R. N. Sutton, Mr. John R. Marier and Dr. H. C. Moolenburgh.

BIBLIOGRAPHY

1. Carson, Rachel, Silent Spring, p. 15, Houghton Mifflin Company, Boston, 1962.
2. L Q Chapitre 63, Loi modifiant la Loi de la protection de la santé publique, Section IV, A, "Fluoration des eaux de consommation."
3. Conseil consultatif de l'environnemont. Conséquences écologiques de la fluoruration de l'eau au Québec. Gouvernement de Québec, août 1975.
4. Report prepared for the Minister of the Environment by the Advisory Committee on the fluoridation of Water Supplies. "Fluorides, Fluoridation and Environmental Quality", November 1979.
5. Maier, F. J. Fluoridation, p. 9. CRC Press, Cleveland, Ohio, U.S.A.
6. Gabovich, R.E. and Ovrutskiy, G.D., Fluorine in Stomatology and Hygiene, DHEW Publ. No. (NIH) 890785, Bethesda, Maryland, 1977.
7. Groth, E., 1973. Two issues of science and public policy: air pollution . . . Stanford University. Ph.D. Biology Thesis, 534 pp.
8. World Health Organization, 1970. Fluorides and human health. Genvea.
9. Rose, Dyson and Marier, John R. Environmental Fluoride 1977. Associate Committee on Scientific Criteria for Environmental Quality. National Research Council, Canada, 1977.
10. Martin, D. J., 1951. Fluoride content of vegetables cooked in fluoride containing waters. J. Dent. Res., 30P 676.
11. Anonymous. Rapport sur la situation de la qualité de l'air au Québec. S.P.E.Q., 1977.
12. Groth, E., 1973. Two issues of science and public policy: air pollution . . . Stanford University. Ph.D. Biology Thesis, 534 pp.
13. Ibid.
14. Report prepared for the Minister of the Environment by the Advisory Committee on the fluoridation of Water Supplies. "Fluorides, Fluoridation and Environmental Quality," November 1979.
15. Article 7 relating to air quality prohibits the emission of fluorides into the environment exceeding the prescribed standards under the terms of the air quality regulations for the Province of Quebec (R.R.Q., c. Q-2, r. 20).

16. Cameron, J.A., 1940. The effect of fluorine on the hatching time and hatching stage in Rana pipiens. Ecology, 21: 288.
17. Wiber, C.G., 1969. The biological aspects of water pollution. Charles C. Thomas Publisher, Illinois, U.S.A. 296 pp.
18. National Health Federation. Fluoridation. U.S.A. pp. 40 and 199-203.
19. Dean, H.T. and Elvove, E., 1937. Further studies on the minimal threshold of chronic endemic dental fluorosis. Publ. Health Rep. Vol. 52: 1249.
20. Ibid.
21. Bundock, J.B., Graham, J.R. and Morin, P.J., June 1982. Fluorides, water fluoridation and environmental quality. Science and Public Policy, p. 138.
22. Kempf, G.A. and Mackay, F.S., 1930. Mottled enamel in a segregated population. Publ. Health Rep. 45: 2923.
23. Smith, M.C., Lantz, E.M. and Smith, H.V., 1931. The cause of mottled enamel, a defect of human teeth. Univ. of Arizona, Agr. Exp. Sta. Bull. No. 32.
24. McKay, F.S., 1929. The establishment of a definite relation between enamel that is defective in its structrue, as mottled enamel, and liability to decay, II Dent. Cosmos, 71, 747.
25. Thylstrup, A. and Fejerskov, O., 1978. Clinical appearance of dental fluorosis in permanent teeth in relation to histologic changes. Community Dent. Oral Epideomiol. 6: 315.
26. Fejerskov, O., Thylstrup, A. and Lorsen, M.J., 1977. Clinical and structural features and possible pathogenic mechanisms of dental fluorosis. Scand. J. Dent. Res. 85: 510-534.
27. The "Merck Manual," Merck Sharp and Dohme Research Laboratories, 11th Edition, 1966.
28. Dean, H.T., 1938. Endemic fluorosis and its relation to dental caries. Publ. Health Rep. 53: 1443.
29. Jenkins, G.N., 1968. Art and Science of dental caries research Ed. Harris, R.S. Academic Press. N.Y. p. 331.
30. Donahue, J.J., Kestenbaum, R.C. and King, N.J., 1966. The utilization of sugar by selected strains of oral streptococci. IADR. Program and abstracts of papers, no. 58.
31. Egelberg, J., 1969. Local effect of diet on plaque formation and development of gingivitis in dogs. Part I. Effect of hard and soft diets. Odont. Revy. 16:31.
32. Adkins, B.L. and Losee, F.L., 1970. A study of the covariation of dental caries prevalence and multiple trace element content of water supplies. N.Y. State Dent. J. 36, 618.
33. Losee, F.L., Cadell, P.B. and Davies, G.N., 1961. Caries, enamel defect and soil: Owaka-Cheviot districts. New Zealand Den. J. 57, 135.
34. Gibbons, R.J. and Banghart, S., 1967. Cariogenicity of a human levan forming streptococcus and streptococcus isolated from subacute bacterial en-

docarditis. IADR Program and abstracts of papers, No. 137.
35. Carlsson, J., 1968. Plaque formation and streptococcal colonization on teeth. Ondontal Revy. 19:1.
36. Lehner, T., Challacombe, S.H., Wilton, J.M.A. and Caldwell, J., 4 Nov. 1976. Cellular and humoral immune response in vaccination against dental caries in monkeys. Nature 264 (5581) 69-72.
37. Fosdick, L.S., 1950. The reduction of the incidence of dental caries Part I. Immediate tooth-brushing with a neutral dentifrice. J.A.D.A. 40:133.
38. Hunt, H.R., Hoppert, C.A. and Rosen, S., 1950. The distribution of carious cavities in the lower molars of caries susceptible and caries resistant albino rats. J. Dent. Res. 29, 157.
39. Loe, H. and Schiott, C.R., 1970. The effect of mouth rinses and topical application of chlorhexidine on the development of dental plaque and gingivitis in man. J. Periodont. Res. 5:79.
40. World Health Organization, 1970. Fluorides and human health. Geneva.
41. Imai, Y., 1972. Relationship between fluoride concentration in drinking water and dental caries in Japan. Koku Eisi, Gakkai Zasshi, 22:144-196.
42. Hilleboe, H.E., Schlesinger, E.R., Chase, H.C., Cantrell, K., Ast, D.E., Smith, D. J., Wachs, B., Overton, D.E., Hodge, H.C., 1956. Newburgh Kingston caries-fluorine study: final report. Journal of the American Dental Assocation, 52:3. pp. 290-325.
43. Greenberg, L.W., Nelson, C.E. and Kramer, N., 1974. Nephrogenic diabetes insipidus with fluorosis. Pediatrics 54:320.
44. Posen, G.R., Marier, J.R. and Jaworski, Z.F., 1971. Renal osteodystrophy in patients on long term hemodialysis with fluoridated water. Fluoride 4 (3) 114.
45. Cordy, P.E., Gagnon, R., Taves, D.R. and Kaye, M., 1974. Bone disease in hemodialysis patients with particular reference to the effect of fluoride. C.M.A.J. 110,1349.
46. Rose, Dyson and Marier, John R. Environmental Fluoride 1977. Associate Committee on Scientific Criteria for Environmental Quality. National Research Council, Canada, 1977.
47. World Health Organization, 1970. Fluorides and human health. Geneva.
48. Ibid.
49. Martin, D.J., 1951. Fluoride content of vegetables cooked in fluoride containing waters. J. Dent. Res., 30—676.
50. Dean, H.T. and Elvove, E., 1937. Further studies on the minimal threshold of chronic endemic dental fluorosis. Publ. Health Rep. Vol. 52:1249.
51. Dean, H.T., 1938. Endemic fluorosis and its relation to dental caries. Publ. Health Rep. 53:1443.
52. Aasenden, R. and Peebles, T.C., 1974. Effects of fluoride supplementation from birth on human deciduous and permanent teeth. Arch. Oral Biol. 19:321.

53. Zipkin, I., Posner, A.S. and Eanes, E.C., 1962. The effect of fluoride on the X-ray diffraction pattern of the apatite of human bone. Biochem. Biophys. Acta. 59,255.

54. Yeager, J.A., 1966. The effect of high fluoride diet on developing enamel and dentine in the incisors of rats. Am. J. Anat. 118,665.

55. Walton, R.E. and Eisenmann, D.R., 1974. Ultrastructural examination of various stages of amelogenesis in the rat following parenteral fluoride administration. Arch. Oral. Biol. 19,171.

56. Greenberg, L.W., Nelson, C.E. And Kramer, N., 1974. Nephrogenic diabetes insipidus with fluorosis. Pediatrics 54:320.

57. Cordy, P.E., Gagnon, R., Taves, D.R. and Kaye, M., 1974. Bone disease in hemodialysis patients with particular reference to the effect of fluoride. C.M.A.J. 110,1349.

58. Ibid.

59. Singh, A., Jolly, S.S., Bansal, B.G. and Mathur, C.C., 1963. Endemic fluorosis. Medicine 42:229.

60. Krishnamachari, K.A.V.R. and Krishnaswamy, K., 1973. Genu valgum and osteoporosis in an area of endemic fluorosis. The Lancet Oct. 20, 887.

61. Teotia, S.P.S., Teotia, M., Burns, R.R. and Heels, S., 1974. Circulating plasma immunoreactive parathyroid hormone levels in endemic skeletal fluorosis with secondary hyperparathyroidism. Fluoride 7 (4) 200.

62. Greenberg, L.W., Nelson, C.E. and Kramer, N., 1974. Nephrogenic diabetes insipidus with fluorosis. Pediatrics 54:320.

63. Singer, I. and Forrest, J.N., 1976. Drug induced states of nephrogenic diabetes insipidus. Kidney Internat. 10, 82-95.

64. Cousins, J.J. and Mazze, R.I., 1974. Methoxyfluorane Nephrotoxicity—a study of dose response in man. J. Amer. Med. Assoc. 225, 1611-1616.

65. Manocha, S.L., Warner, H. and Olkowski, Z.L., 1975. Cytochemical response of kidney, liver and nervous system to fluoride ions in drinking water. Histochem, J. 7,343.

66. Mohamed, A.H. and Weitzenkamp-Chandler, M.E., Sept. 3, 1976, Cytological effects of sodium fluoride on mitotic and meiotic chromosomes of mice. Pres. at Amer. Chem. Soc. meeting, San Francisco.

67. Mitchell, B. and Gerdes, R.A., 1973. Mutagenic effects of sodium and stannous fluoride upon drosophila melanogaster. Fluoride 6 (12) 113.

68. Mohamed, A.H., 1971. Induced recessive lethals in second chromosomes of Drosophila Melanogaster by hydrogen fluoride. Proc. 2nd clean air Congr. Ed. H.M. England and W.T. Beery. Academic Press, London, p. 158.

69. Yiamouyiannis, J.A. and Burk, D., 1977. Fluoridation and cancer: age dependence of cancer mortality related to artificial fluoridation. Fluoride 10 (3): 100-123.

70. Mohamed, A.H. and Weitzenkam-Chandler, M.E., Sept. 3, 1976. Cytological effects of sodium fluoride on mitotic and meiotic chomosomes

of mice. Pres. at Amer. Chem. Soc. meeting, San Francisco.

71. Mitchell, B. and Gerdes, R.A., 1973. Mutagenic effects of sodium and stannous fluoride upon drosophila melanogaster. Fluoride 6)2) 113.

72. Mohamed. A.H., 1971. Induced recessive lethals in second chromosomes of Drosophila Melanogaster by hydrogen fluoride. Proc. 2nd clean air Congr. Ed. H. M. England and W.T. Beery. Academic Press, London, p. 158.

73. Yiamouyiannis, J.A. and Burk, D., 1977. Fluoridation and cancer; age dependence of cancer mortality related to artificial fluoridation. Fluoride 10 (3): 100-123.

74. Court of common please of Allegheny County, Pennsylvania, Civil division. Paul W. Atikenhead et al, plaintiffs v. Borough of West View, defendants. Judge Flaherty, Presiding. No GD 4585, 78.

75. Little, J.B., Radford, E.P., McCombs, L., and Hunt, V.R., 1965. Distribution of polonium[210] in pulmonary tissues of cigarette smokers. New England. J. Med. 273, 1343-1354.

76. De Villiers, A.J. and Windish, J.P., 1964. Lung cancer in a fluorspar mining community I. Radiation, dust mortality experience. Br. J. Indust. Med. 21, 94-109.

77. Hirayama, T., 1975. Epedemiology of cancer of the stomach with special reference to its recent decrease in Japan. Cancer Res. 35:3460-3463.

78. Cecilioni, V.A., 1972. Lung cancer in a steel city. Its possible relation to fluoride emissions. Fluoride 5:172-181.

79. Taylor, A., 1954. Sodium fluoride in the drinking water of mice. Dent. Digest. 60:170-172.

80. Taylor, A. and Taylor, N.C., 1965. Effect of sodium fluoride on tumor growth. Proc. Soc. Exp. Biol. Med. 119:252-255.

81. Waldbott, G.L., Fluoridation—The Great Dilemma, p. 131. Coronado, 1978.

82. Leverett, H., 1982. Fluorides and the Changing Prevalence of Dental Caries, Science, Vol. 217, p. 26.

Solid and Liquid Wastes: Management, Methods and Socioeconomic Considerations. Edited by
S. K. Majumdar and E. Willard Miller. © 1984, The Pennsylvania Academy of Science.

Chapter Eighteen

The Ecological Effects of Acid Deposition in Eastern North America

Patricia T. Bradt and Judith L. Dudley

Department of Biology
Lehigh University
Bethlehem, PA 18015

Acid deposition is falling over a large portion of eastern North America. It may be defined as the wet and dry deposition of acidic substances. Most current research has focused on wet deposition, e.g. precipitation including rain and snow, hence the popular name "acid rain." Acidic substances also come down in dry form, but techniques for measuring dry deposition are not yet perfected so most current data concern wet deposition. Precipitation is labeled "acid" if the pH is less than 5.6, which is the pH of carbon dioxide in equilibrium with distilled water. "Unpolluted" precipitation, therefore should have a pH close to 5.6, but precipitation falling over much of North America in 1980 registered mean weighted pH values from 4.1 to 4.4 (Figure 1) which are much more acidic than the "ideal" pH of 5.6, especially considering that the pH scale is logarithmic. The pH of precipitation over eastern North America has decreased over the last 25 years and the areas receiving acid precipitation have increased (Likens et al. 1979; Likens and Butler 1981). The northeastern United States receives the most acidic precipitation in the country (Wisniewski and Keitz 1983). Parts of Europe, especially southern Norway and Sweden, are also experiencing acid precipitation (Barnes 1979). Our knowledge of the pH of precipitation comes from a number of monitoring networks which measure not only the pH of the precipitation but also some of the chemical parameters such as sulfate and nitrate concentrations. There are several such networks which measure precipitation in both North America and Europe. The location of the North American sampling stations are shown in Figure 1.

The probable cause of the increase in acid precipitation over the last 25 years is the increase in the discharge of sulfur dioxide and oxides of nitrogen into the atmosphere. These pollutants interact with atmospheric water vapor and fall to earth as solutions of sulfuric and nitric acids. Pollutants discharged into the air to the west are carried eastward, as the prevailing winds blow from west to east, and fall as acid precipitation, or as dry deposition, over eastern North America. Such long range transport of air pollutants (LRTAP) has been demonstrated to occur over hundreds of kilometers both in North America (National Academy of Sciences 1981) and in Europe (Barnes 1979). In the United States the annual release of these pollutants into the atmosphere has increased from 22×10^{12}g sulfur dioxide in 1940 to 36×10^{12}g in 1972 (National Research Council 1978) and from 8.1×10^{12}g oxides of nitrogen in 1950 to 22.2×10^{12}g in 1975 (U.S.-Canada Res. Grp. 1979). In 1980-82, nitrate deposition was higher in Eastern Pennsylvania precipitation in the winter and sulfate was higher in

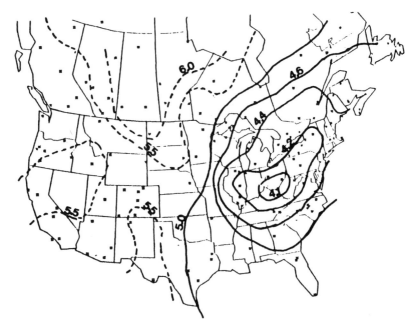

FIGURE 1. Isopleth map of annual mean weighted pH of precipitation falling on North America—1980. Location of sampling stations of U. S. National Atmospheric Deposition Program and Canadian Network for Sampling Precipitation (CANSAP) are designated by asterisks. Reproduced by permission of the National Atmospheric Deposition Program and J. H. Gibson, Natural Resource Ecology Laboratory, Colorado State University, Fort Collins, CO.

the summer (EPRI 1983). In 1980 sulfate deposition was 1671 mg/m^2 and nitrate deposition was 1116 mg/m^2 so sulfate was the major anion in precipitation (EPRI 1983). In the next 15 years emissions of nitrogen are expected to increase and, concurrently, nitrate in precipitation will increase. Emissions of oxides of sulfur, however, are projected to remain relatively unchanged (Mitre Corp. 1979). The National Academy of Sciences (1981) states that the circumstantial evidence for the role of sulfur and nitrogen oxides in the production of acid precipitation is "overwhelming".

AREAS SENSITIVE TO ADVERSE EFFECTS FROM ACID PRECIPITATION

Knowing that acid precipitation is falling over eastern North America, the question which should be asked is, "Is it doing any harm?" The answer depends on the bedrock and the surficial geology of the areas receiving the precipitation. If the precipitation falls on areas underlain by limestone bedrock and thick, fertile soils, the impact on the land and waters will be minimized because carbonates leaching from the relatively soluble limestone can effectively neutralize the acid. A severe depression in pH is thereby prevented. Much of eastern North America, especially the southeast, is underlain by limestone (Figure 2), but other sections (e.g. New England—1 & 2, eastern New York—3, parts of Pennsylvania—4 & 5, the Blue Ridge and the Great Smoky Mountains—6 & 7) are underlain by insoluble bedrocks like granite, sandstones and shales. These regions are labeled as sensitive to damage from acid precipitation, and are also referred to as sensitive to acidification (OTA 1982). Similar sensitive areas are found in eastern Canada, especially Ontario and Nova Scotia, and in the northern midwestern areas (Figure 2—11-14) of the U.S. (Galloway and Cowling 1978).

The surface waters of these sensitive areas are often: acidic because of leaching from acid soils and vegetation; very "soft" (i.e. low in dissolved minerals); and consequently lacking in the chemical ability to neutralize the additional acidic input entering from precipitation. Following a heavy rain or snowmelt, the pH of sensitive surface waters may fall to a level at which susceptible aquatic organisms may be adversely affected. Some of these adverse effects include: impairment of reproductive or other physiological processes; and/or cessation of processes vital to the organism's survival. Populations of organisms sensitive to low pH may decline and eventually die out completely.

The disappearance of certain sensitive species of fish like the Atlantic salmon, brook trout, brown trout and rainbow trout has been documented in the Adirondac Mountains of New York state (Schofield 1976), in the LaCloche mountains of Ontario (Beamish and Harvey 1972), in Nova Scotia (Farmer et al. 1981), in Norway (Overrein et al. 1980) and in Sweden (Almer et al. 1974). The disappearances of these sensitive game fish were some of the first widely publicized

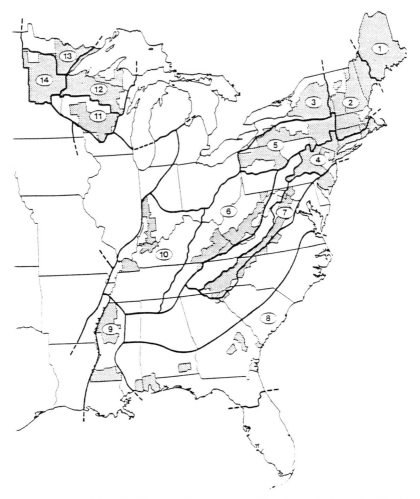

FiGURE 2. Areas of the United States sensitive to adverse effects from acid disposition. Sensitive regions are stippled. Sensitive areas are determined by soil depths, bedrock characteristics and water chemistry. Courtesy of Office of Technology Assessment (1982), Washington, D.C. p. G-5.

effects of acid precipitation.

TERRESTRIAL EFFECTS: PLANTS, SOILS, MAN-MADE STRUCTURES

Acid precipitation also affects the terrestrial environment. Forest growth may initially be stimulated by the increased inputs of the nutrients sulfate and nitrate from precipitation. This stimulation, however, is thought to be short term with deleterious long term effects. Red spruce in high elevations (> 3000m) in New England are experiencing large-scale die back and mortality (Vogelman 1982). Acid deposition is suspected to be a contributing factor. A decline in the growth rate of many forest trees, as measured by the width of the growth rings, has been observed in the northeastern U.S. (Johnson 1983). It is suspected that acid precipitation contributes to this growth decline in combination with other stress factors such as drought and exposure to other air pollutants (Tomlinson 1983). The forests of West Germany are being affected by acid deposition (Steinen and Rademacher 1983) as are those of Norway (Overrein et al. 1980) and Sweden (Ministry of Agriculture 1982).

When acid precipitation falls on forest soils, major elements and trace metals may be leached away. When the receiving soils are naturally acid (e.g. podzol-a soil type found in mountainous coniferous forests) with a pH below 5.5, the essential plant nutrients of calcium, potassium and magnesium are mobilized in the soil by the acidic input and leach out into the surface waters. If the soil pH is less acid, i.e. > 5.5, toxic metals like aluminum become mobilized in the soil and may damage sensitive tree roots and/or leach into the surface waters and cause damage to aquatic organisms. In addition, if both the chemical composition and the ability of the soils to neutralize acidic inputs are adversely affected, then the soils may become increasingly acid and nutrient deficient. The neutralizing capacity and plant nutrients are leached away.

Acid precipitation may cause direct damage to both forest trees and farm crops as it falls on them before reaching the soils. The deposition of acidic substances, both dry and wet, on sensitive plant leaves may damage the foliage, decrease photosynthetic area and therefore decrease the yields of timber or cultivated crops. Additional effects on vegetation have been documented (Krupa 1982; Lee 1981). Limestone is added to many agricultural soils to increase soil pH and the availability of nutrients. Such soils do not appear to be adversely affected by acid deposition at this time, but the amount of limestone added to the soils may have to be increased as the neutralizing ability is washed away. At this time the long term effects on limestoned soils are not known.

The corrosion rate of man-made structures such as buildings, bridges and statuary is greater in urban areas than in rural regions because air pollution is greater in cities. Sulfur dioxide, particular sulfates, and acid precipitation are known to cause corrosion (Ministry of Agriculture 1982). Many types of steel, especially galvanized and painted, are subject to corrosion in areas where the air is polluted by sulfur dioxide. Limestone, sandstone and marble have been widely used for buildings and sculptures. These structures show greater evidence of deterioration in sulfur dioxide laden urban and industrial areas (Gauri 1980; Luckat 1981).

Aquatic Effects

The impact of acid deposition on the aquatic ecosystem is well documented in North America and Europe. The loss of fish from sensitive rivers, streams and lakes in susceptible areas has received wide coverage by the news media. Subtle changes occurring in the aquatic ecosystem as a whole, while less widely publicized than the fish losses, are of greater potential ecological significance. Changes occur not only in water chemistry, but also in the plant and animal communities.

Water Chemistry

The ability of a body of water to neutralize acid is measured by its total alkalinity. This parameter is determined by measuring the amount of weak acid required to decrease the pH of the water being tested to a predetermined pH, usually 4.6-4.4. If the sample requires a large amount of weak acid to lower its pH the water is high in total alkalinity and, therefore, also high in its ability to maintain a fairly constant pH, even if there is some acidic input. It will require a large amount of acid precipitation to reduce the pH of such well buffered (high alkalinity) waters to a critical level (< 5.0). However, if the tested water requires only a small amount of acid to reduce its pH, the water is labeled as low in total alkalinity and low in neutralizing or buffering capacity. For those waters, continued acid precipitation represents a potential source of damage because those waters do not require much acid to reduce their pH to a critical level for sensitive species of plants and animals (< 5.0). Some surface waters have no alkalinity and therefore no buffering capacity. Such waters are labeled "acidified" and their pH may be severely depressed following a heavy rain or snowmelt. The pH of "acidified" waters may approximate the pH of incoming precipitation following a heavy influx of precipitation.

Total alkalinity has become one of the most important criteria by which many investigators evaluate the susceptibility of surface waters to acidification. A reduction in total alkalinity over time is considered indicative of damage from acid precipitation because the buffering capacity is being exhausted. Waters low in alkalinity are shown in Figure 2. These waters are located over large areas of New England, eastern New York, parts of Pennsylvania, West Virginia, North Carolina, Virginia, Kentucky and eastern Canada, the same areas mentioned previously as being underlain by insoluble bedrocks (Figure 2).

A survey of 226 surface waters in New England (Haines and Akielaszek 1983) found that 53% of the waters were vulnerable to acidification, i.e. total alkalinity < 200 ueq l⁻¹. In nine middle Atlantic states a survey of 278 lakes and streams indicated that 45% of these waters were sensitive to acidification, based on total alkalinity and underlying bedrock (Arnold et al. 1984). When alkalinity in these waters was compared to historical records, 77% of the waters registered a decline in total alkalinity. In the sensitive regions of Pennsylvania (the Poconos, the anthracite region and the Allegheny Plateau) 36% of the 197 streams sampled showed evidence of decreasing alkalinity. Many of these streams also showed a decrease in the number of fish species (Arnold et al. 1980). F. Johnson (1983) considers 30% of Pennsylvania's 8000 kilometers of stocked trout streams and 40% of the 14400 kilometers of unstocked streams to be sensitive to acidification because these streams have total alkalinities less that 200 ueq l^{-1}. Cold water fisheries in Pennsylvania are valued at $516 million, based on 12,000,000 angler days at $43 per day (F. Johnson 1983). These studies confirm that many surface waters in the northeastern United States are vulnerable to acidification (i.e. loss of total alkalinity and subsequent depression of pH). As these waters become increasingly acidified the pH may fall to a level incompatible with the continued existence of many valued game fish and the algae, zooplankton, aquatic macrophytes and invertebrates which support them. There are many varied responses of the aquatic ecosystem to acid precipitation (Haines 1981).

A survey of 10 high risk lakes in the Pocono mountains area of northeastern Pennsylvania (Bradt and Berg 1983) showed that two lakes were already acidified (total alkalinity = 0.0 ueq l^{-1}), three were very sensitive to acidification (< 40 ueq l^{-1}) and five were moderately sensitive (< 200 ueq l^{-1}), based on the total alkalinity criteria of 200 ueq l^{-1} (Glass and Brydges 1981). The Pocono mountains area, as would be expected, is underlain by insoluble sandstone and shale bedrock. From this initial survey three lakes were chosen to study intensively biologically and chemically for three years. In the study, Deep Lake was already acidified (total alkalinity = 0 ueq l^{-1}) and the mean pH of this lake dropped from a high of 7.0 in the summer of 1981 to a low of 3.98 in the spring of 1982. The pH remained below 4.6 for the remainder of the study. The second study lake, Lacawac, had a mean total alkalinity of 35.8 ueq l^{-1} and a mean pH of 6.45 (range 5.1 to 7.2). The third study lake, Long Pond, had a mean total alkalinity of 170 ueq l^{-1} with a mean pH of 7.42 (range 6.4 to 8.2). The algae, macrophytes, zooplankton, and benthic invertebrates of these lakes were sampled seasonally from summer 1981 to summer 1983. The vegetation, surficial geology and the hydrology of the drainage basins were also studied. The fish community of each lake was sampled annually. Precipitation chemistry was also analyzed. Mean weighted pH of pecipitation in the area was 4.17 in 1982 (Lynch 1983). The three lakes are apparently reacting differently to chemically similar precipitation. The reasons for the varying responses are based on complex interactions between the vegetation, surficial geology and hydrology of the drainage basins. Pennsylvania receives some of the most acidic precipitation in the country (EPRI 1983) and the sensitive areas such as the Poconos and parts of central Pennsylvania should be carefully monitored for both biological and chemical impacts.

Many waters in the sensitive areas of North America (Figure 2) are of great recreational value and are located in popular, mountainous vacation areas which are heavily forested and valued as wilderness areas. Waters in such remote areas are being damaged by pollution produced hundreds of kilometers to the west, though in some cases locally produced pollutants may also be involved.

Additional changes may occur in water chemistry, often in the major ions. The major cations in circumneutral water are usually calcium and magnesium and the major anion, the bicarbonate ion (HCO_3^-). In waters affected by acid precipitation the hydrogen ion may become the dominant cation and sulfate and nitrate the major anions. The important buffering capacity of the calcium/magnesium/bicarbonate system is greatly reduced and the pH of the water decreases.

Bacteria

Acidification of a freshwater lake causes changes in the bacterial community. Bacterial populations (especially nitrifying and some sulfur cycle bacteria) are lower in acid stressed waters than in non-acid stressed waters (Rao and Dutka, 1983). The rate of decomposition is retarded which leads to an increase in the accumulation of organic debris and a decrease in the rate of nutrient recycling. This decrease in decomposition may affect the whole lake ecosystem because the higher trophic levels are dependent upon the recylcing of those nutrients.

Phytoplankton and Macrophytes

The single-celled algae which float in the water column are called phytoplankton. These microscopic organisms are at the base of the food pyramid in most aquatic systems. They convert incoming solar energy into chemical energy by the process of photosynthesis. Chemical energy in the form of carbohydrates, proteins, etc. is thereby made available to the microscopic invertebrates (zooplankton) which feed upon the algae.

The larger invertebrates, often insect larvae or crustaceans, feed on the zooplankton, and small vertebrates such as young fish and amphibians devour the insect larvae in turn. Larger fish and fish eating birds are often present especially in large bodies of water and they prey upon the smaller fish. The phytoplankton, therefore, are vital to the functioning of the entire aquatic ecosystem because they provide the initial energy to the system.

Acid precipitation can severly depress the pH of a susceptible body of water to the point where the numbers of algae are greatly reduced and only certain acid tolerant species remain. The waters become exceptionally clear as the phytoplankton numbers are reduced and sunlight penetrates deeply into the water column, more deeply than it does in most circumneutral waters. Large numbers of filamentous acid tolerant algae may develop on the lake bottom because the sediment is not usually quite as acidic as the waters above.

In the Pocono mountain study the acidified lake, Deep Lake, had fewer species of algae than the other two lakes and the moss *Sphagnum* was growing on the bottom. Golden-brown algae (Chrysophyeae) dominated numerically at Deep Lake and the blue-green algae (Cyanobacteria) dominated at both Lacawac and Long Pond. There are additional effects of acid deposition on both algae and macrophytes (Conway and Hendrey 1982). The effects of acidification upon

the macrophytes (rooted aquatic plants) appear to occur in three important ways.
1. *Sphagnum* increases in abundance as a benthic macrophyte. Dense mats of this bryophyte may cover the lake bottom (Grahn 1976). This mat inhibits the recylcing of nutrients between the sediment and the overlying water.
2. The decrease and/or disappearance of species in the littoral zone of the lake. Stands of acid-intolerant macrophytes will disappear as acidification progresses (Hendrey and Vertucci 1980). At Deep Lake the emergent macrophytes in the littoral zone decreased in both density and diversity during the study, until in the last year (1983) there were only a few struggling *Nuphar* plants in the sampling quadrats and *Sphagnum* was evident on the sediment.
3. Increase in the numbers of acidophilic plants, in addition to the benthic *Sphagnum*. Some of these plants include *(Lobelia dortmanna),* bullrushes *(Scirpus subterminalis),* pondweed *(Potamogeton confervoides),* water milfoil *(Myriophyllum tenellum),* water lily *(Nuphar leuteum),* bladderwort *(Utricularia* sp.*)* and pipewort *(Eriocaulon septangulare).* At Deep Lake a species of *Myriophyllum, M. humile,* was found in abundance in 1983.

Zooplankton

An ecosystem is considered healthy when it has a large number of different kinds of species, a property called diversity. The diversity of the zooplankton community in acid lakes is usually lower than that in circumneutral lakes. Only a few acid tolerant species may persist, the community is simplified and the numbers of organisms may also decrease, thereby making less food available for the next level of the food pyramid. Larger invertebrates such as clams, snails, and insect larvae feed upon the zooplankton and if their preferred zooplankton food decreases, less preferred zooplankton species may survive.

Hobaek and Raddum (1980) and Malley et al. (1982) have proposed several ways in which a zooplankton community may be affected by acid precipitation. These mechanisms include:
1. The hydrogen ion concentration and/or concentration of toxic metals (like aluminum) may increase to a level at which they become lethal to some sensitive zooplankton species.
2. These microscopic organisms must use a large amount of energy in order to maintain the delicate ionic balance within their bodies as the pH of the water surrounding them drops to a stressful level.
3. The quantity of the surviving acid tolerant algae may not be adequate to sustain the zooplankton or the algae may not be palatable as food for the zooplankton.
4. A change in the predator/prey dynamics in an acid lake may determine the presence or absence of zooplankton species.
5. As sunlight penetrates deeper into the water column, larger portions of the water column are warmed and the heat budget of the entire lake is altered. The zooplankton are exposed to warmer temperatures which further stress

the community.

In the Pocono lake study five species of zooplankton were found in acidified Deep Lake and only one of these species was a rotifer (Rotatoria). Three of the Deep Lake species have been previously reported as acid tolerant. In Lacawac 14 species were found including 7 species of rotifers. In Long Pond there were 15 zooplankton species found during the study including 9 species of rotifers. Only one species, *Mesocyclops edax* (Copepoda), was found in all three lakes.

Benthic Invertebrates

Most of the larger invertebrates in a body of water live on the bottom, under rocks or in the sediment, or on submerged plant stems. This habitat is called the benthos, hence the adjective "benthic". In circumneutral waters this ecological niche is usually occupied by snails (Gastropoda), freshwater mussels and clams (Pelecypoda), tiny freshwater shrimp or scuds (Amphipoda) and immature forms of insects such as dragonflies (Odonata), caddisflies (Trichoptera), mayflies (Ephemeroptera) and true flies (Diptera).

As a body of water becomes acidified the diversity of the inverterbrate bottom dwellers decreases. Snails, mussels and tiny clams are often no longer found because they cannot tolerate the low pH and the calcium which they need to build their shells is not in adequate supply or cannot be utilized (Okland and Okland 1980). The minute scuds and the mayflies also disappear at low pH values, but the immature dragonflies are acid tolerant as are some caddisflies and many species of midge larvae (Chironomidae, order Dipera) (Singer 1982). These latter three groups along with several other acid tolerant organisms like the water louse (Isopoda) and the alderfly larva (Megaloptera), proliferate. The composition of the benthic community changes as the tolerant species increase and the sensitive species decrease. An increase in the number of the larger predators (e.g. immature dragonflies and alderflies, and certain midge larvae) may be noticed as more of these organisms grow to maturity because the number of fish which prey upon them may be severely reduced by the low pH.

In the Pocono lake study the midge larvae (Chironomidae) dominated (70% of total numbers) the benthic community in the acidified lake (Deep Lake), and there were no snails or clams found and only one mayfly and one scud (Figure 3). In the least acid lake (Long Pond) clams, snails, mussels and scuds made up a large portion (54%) of the benthic population while in the intermediate lake (Lacawac) the Chironomidae comprised 43% of the population of benthic invertebrates, scuds 30% and snails and clams only 7%. There was little difference in the numbers or diversity in the three lakes. The major difference was in the types of organisms composing the community, a phenomenon that has occured at the three levels of the aquatic ecosystem (phytoplankton, zooplankton and benthic invertebrates) examined so far.

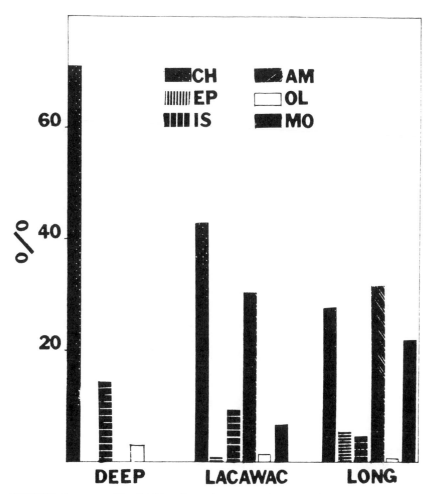

FIGURE 3. Frequency of Benthic Macroinvertebrate groups in each study lake (1981-1983), based on percentage of total numbers. CH = Chironomidae (Diptera), EP = Ephemeroptera, IS = Isopoda, AM = Amphipoda,   OL = Oligochaeta,   MO = Mollusca.

Fish, Amphibians and Waterfowl

Like the other aquatic organisms discussed to this point, different species of fish vary greatly in their ability to tolerate low pH values. The desirable game fish from the salmon family (Salmonidae) such as the Atlantic salmon, lake trout, brook trout, brown trout, and rainbow trout, in addition to the walleye, bass and pickerel are all acid sensitive. The brook trout is probably the most acid tolerant and has been stocked for years in the Adirondacks where lakes have historically been naturally acid (although they are now artificially acidified as well). It was the disappearance of the brook trout from many Adirondack ponds and lakes that first signaled the problems of decreasing pH in that mountainous area (Schofield 1976; Pfeiffer and Festa 1980). Problems with fish populations had also surfaced in Canada where fish were disappearing from the LaCloche mountain area of Ontario, many kilometers east of a large sulfur dioxide emitting smelter (Beamish and Harvye 1972), and Atlantic salmon are disappearing from rivers in Nova Scotia (Wall et al. 1983). From Europe came reports of massive fish kills in the Torval River in Norway during the spring snow melt (Leivestad and Muniz 1976) and trout were disappearing from over 700 Norwegian lakes (Jensen and Snekvik 1972). Populations of the roach, an acid sensitive fish, have declined in over 2000 Swedish Lakes (Ministry of Agriculture 1982).

In most of the above cases the young of the year disappeared first and the fish population became increasingly older because the adult fish were more acid resistant. The young fail to hatch or to survive immediately following hatching. The reproductive stage is the most sensitive to stresses of all kinds (this applies to eggs, sperm and the survival of the immature and very young). Increasing levels of toxic aluminum, leached from the surrounding soils into the water, stress both immature and adult fish.

The number of fish species declines as the pH declines and the acidified waters have only a few species of fish, or no species at all. The fish community becomes composed of aging acid tolerant fish such as pumpkinseed, yellow perch or white sucker. Even the white sucker is disappearing from some acidified Ontario lakes. As the fish population decreases the larger, acid tolerant invertebrates, which were former fish prey, may assume the role of predator upon the smaller benthic invertebrates. The changes in the fish population are subtle, like the other changes in the biota, and are only apparent after careful monitoring of the population for many years.

In the Pocono Mountain study the acidified lake (Deep) was populated almost completely by stunted pumpkinseeds with an occasional golden shiner, which may have been introduced as bait. The pumpkinseed is the most acid tolerant fish, being able to tolerate a pH as low as 4.2 (Harvey 1979). The least acid lake (Long Pond) supported a diverse fish fauna, but it was regularly stocked throughout the study making it difficult to ascertain how diverse the natural population might have been. However, the pumpkinseeds were not stocked in Long Pond and they are growing much faster there than in acidified Deep Lake.

Adverse effects of acid precipitation on some amphibians have also been documented. Tome and Pough (1982) reported a high mortality rate for embroyos of 13 frog species and two salamander species at pHs below 5.0. Other species of amphibians, however, appear to be acid tolerant, and have been observed in acid lakes by Singer (1928) and also in Deep Lake during the Pocono mountain study.

The effects of acid precipitation upon waterfowl have not, as yet, been documented extensively. It is suspected, however that waterfowl must avoid acidified waters because those waters are devoid of most prey, especially fish and invertebrates. Carnivorous birds which feed upon the insects and fish contaminated with heavy metals mobilized by acid precipitation, may experience impaired deposition of their eggshells (Nyholm and Myrberg 1977).

CONCLUSIONS

Acid precipitation is occurring over eastern North America. Much of this area is valued as recreation area. There are lakes and streams in the Pocono Mountains which are sensitive to acidification. At least two lakes are already acidified and there is unpublished information on at least three other lakes and one stream which are acidified. Anecdotal information from the Pocono area indicates that many more lakes and streams are very sensitive. Soils, forest, and food crops are affected by acid precipitation, but knowledge about the extent of the problem is quite limited not only in the Pocono Mountain region but also in eastern North America as a whole. Future research should focus not only on monitoring possible effects of acid precipitation on potentially sensitive lands and waters but also on methods of mitigating the damage from acid precipitation.

LITERATURE CITED

Almer, B., W. Dickson, C. Ekstrom, E. Hornstrom and U. Miller. 1974. Effects of acidification of Swedish lakes. Ambio. 3:30-36.

Arnold, D. E., R. W. Light and V. J. Dymond. 1980. Probable effects of acid precipitation on Pennsylvania waters. Report EPA-600/3-80-012. U. S. Environmental Protection Agency, Corvallis, Oregon. 24 pp.

Arnold, D. E., R. W. Light and E. A. Paul. 1984. Vulnerability of selected lakes and streams in the middle Atlantic states to acidification. U. S. Fish and Wildlife Service. Eastern Energy and Land Use Team. FWS/DBS-80/40.17 (In Press).

Barnes, R. A. 1979. The long range transport of air pollution. A review of the European experience. J. Air. Pollut. Cont. Assoc. 29(12):1219-1235.

Beamish, R. J. and H. H. Harvey. 1972. Acidification of the La Cloche Mountain Lakes, Ontario and resulting fish mortalities. J. Fish. Res. Bd. Canada 29:1131-1143.

Bradt, P. T. and M. B. Berg. 1983. Preliminary survey of Pocono Mountain lakes to determine sensitivity to acid deposition. Proceed. PA Acad. Sci. 57: 190-194.

Conway, H. L. and G. R. Hendrey. 1982. Ecological effects of acid precipitation on primary producers. In F. M. D'Itri (Ed.), Acid precipitation effects on ecological systems. Ann Arbor Science, Ann Arbor, MI pp 277-295.

Electric Power Research Institute (EPRI). Precipitation data displays for January 1, 1979 - June 30, 1982. Vols. 1 and 2. UAPSP 103. Vol. 1 204 pp. Vol. 2 184 pp.

Farmer, G., T. Goff, D. Ashfield and H. Samant. 1981. Some effects of the acidification of Atlantic salmon rivers in Nova Scotia. Intern. Atlantic Salmon Found. Spec. Public. 10:73-91.

Galloway, J. N. and E. B. Cowling. 1978. The effects of precipitation on aquatic and terrestrial ecosystems—a proposed precipitation chemistry network. J. Air. Pollut. Control Assoc. 28:229-235.

Gauri, K. Lal. 1980. Deterioration of architectural structures and monuments. In T. V. Toribara, M. W. Miller and P. E. Morrow (Eds.), Polluted rain. Plenum Press, New York. pp. 125-144.

Glass, G. E. and T. G. Brydges (Eds.). 1981. United States-Canada Memorandum of Intent on Transboundary Air Pollution. Aquatic Impact Assessment (Final Draft Copy). USEPA, Environmental Research Laboratory, Duluth, Minn. 150 pp.

Grahn, O. 1976. Macrophyte succession in Swedish lakes caused by deposition of airborne acid substances. In Proceed. Intern. Symp. on Acid Precipit. and the Forest Ecosystem. pp. 516-530. Columbus, Ohio, 1975. NTIS PB 258-645, Springfield, VA.

Haines, T. A. 1981. Acidic precipitation and its consequence for aquatic ecosystems. Trans. Amer. Fish Soc. 110:669-707.

Haines, T. A. and J. J. Akielaszek. 1983. Acidification of headwater lakes and streams in New England. In Lake Restoration, Protection and Management. USEPA 440/5-83-001. pp. 83-87.

Harvey H. H. 1979. The acid deposition problem and emerging research needs in the toxicology of fishes. Proc. 5th Annual Aquatic Toxicity Workshop, Hamilton, Ont., Nov. 1978. Fish. Mar. Serv. Tech. Rep. 862:115-128.

Hendry, G. K. and F. A. Vertucci. 1980. Benthic plant communities in acidic Lake Colden, N.Y. *Sphagnum* and the algal mat. In D. Drablos and A. Tolan (Eds.), Ecological impact of acid precipitation. SNSF Project, Oslo, Norway. pp. 314-315.

Hobaek, A. and G. G. Raddum. 1980. Zooplankton communities in acidified lakes in south Norway. SNSF Project IR 75/80. Internal Report IR 75/80. SNSF Project, As. Norway. 132 pp.

Jensen, K. W. and E. Snekvik. 1972. Low pH levels wipe out salmon and trout populations in Southernmost Norway. Ambio 1(6):223-224.

Johnson, A. H. 1983. Red spruce decline in the northeastern U. S.: hypotheses regarding the role of acid rain. J. Air. Pollut. Control Assoc. 33(1):1049-1054.

Johnson, F. 1983. Trends of alkalinity and pH in Pennsylvania's low order stocked mountain trout streams, and potential economic implications. Presented at Northeast Div., Amer. Fish. Soc., Dover, Vermont, May 1983.

Krupa, S. V. 1982. Effects of dry deposition components of acidic precipitation on vegetation. In F. D'Itri (Ed.), Acid Precipitation. Effects on ecological systems. Ann Arbor Press, Ann Arbor, MI. pp. 469-482.

Lee, J. J. 1982. The effects of acid precipitation on crops. In F. D'Itri (Ed.), Acid precipitation. Effects on ecological systems. Ann Arbor Press, Ann Arbor, MI. pp. 453-468.

Leivestad, H. and I. P. Muniz. 1976. Fish kill at low pH in a Norwegian river. Nature 259:391-392.

Likens, G. E. and T. J. Butler. 1981. Recent acidification of precipitation in North America. Atmos. Environ. 15(7):1103-1109.

Likens, G. E., R. F. Wright, J. N. Galloway and T. J. Butler. 1979. Acid Rain. Sci. Am. 241(4):43-51.

Luckat, S. 1981. Quantitative Untersuchung des Einflusses von Luftverunreinigungen bei der Zerstorung von Naturstein. Staub-Reinhalt-Luft 41:440-442.

Lynch, James. 1983. Personal communication. PSU/PADER Atmospheric Deposition Monitoring Program.

Malley, D. F., D. L. Findlay and P. S. S. Chang. 1982. Ecological effects of acid precipitation on zooplankton. In F. D'Itri (Ed.), Acid Precipitation. Ann Arbor Science, Ann Arbor, Michigan. pp. 297-327.

Ministry of Agricultural Environment Committee. 1982. Acidification today and tomorrow. Risbergs Tryckeri AB, Uddevalla, Sweden. 231 pp.

Mitre Corporation. 1979. National Environmental Impact Projection No. 1. Report No. HCP/P-6119. Washington, D.C.: U.S. Dept. of Energy.

National Academy of Sciences. 1981. Atmosphere-Biosphere interactions: Toward a better understanding of the sociological consequences of fossil fuel combustion. National Academy Press, Washington, D.C. 263 pp.

National Research Council. 1978. Sulfur Oxide's. Comm. on Sulfur Oxides, Board on Toxicology and Environmental Health Hazards. Assembly of Life Sciences. Washington, D.C. National Academy of Sciences.

Nyholm, N. E. I. and H. E. Myhrberg. 1977. Severe eggshell defects and impaired reproductive capacity in small passerines in Swedish Lapland. Oikos, 29:336:341.

Office of Technology Assessment (OTA). 1982. The Regional Implications of Transported Air Pollutants: An Assessment of Acidic Deposition and Ozone. Interim Draft. Congress of the United States, Washington, D.C. 193 pp.

Økland, J. and K. A. Økland. 1980. pH level and food oganisms for fish: Studies of 1000 lakes in Norway. In D. Drablos and A. Tollan (Eds.), Ecological Impact of Acid Precipitation. SNSF Project, Oslo, Norway. pp. 326-327.

Overrein, L. N., H. M. Seip and A. Tollan. 1980. Acid precipitation—effects on forest and fish. Final Report SNSF Project 1972-1980. Oslo, Norway. 175 pp.

Pfeiffer, M. and P. Festa. 1980. Acidity status of lakes in the Adirondack region of New York in relation to fish resources. N. Y. Dept. Environ. Conserv. Report

FW-P168. Albany, New York.

Rao, S. S. and B. J. Dutka. 1983. Influence of acid precipitation on bacterial populations in lakes. Hydrobiologia 98: 153-157.

Schofield, C. L. 1976. Acid precipitation: effects on fish. Ambio 5: 228-230.

Singer, R. 1982. Effects of Acidic Precipitation on Benthos. In F. D'Itri (Ed.), Acid Precipitation. Ann Arbor Science, Ann Arbor, Michigan. pp. 329-363.

Stienen, H. and P. Rademacher. 1983. Abiotische und biologische Aspekte des Waldsterbens. Nat. Mus. 113:6.

Tome, M. A. and F. H. Pough. 1982. Responses of amphibians to acid precipitation. In R. E. Johnson (Ed.), Acid Rain/Fisheries. Amer. Fish. Soc., Bethesda, MD. pp. 245-254.

Tomlinson, G. H. 1983. Air pollutants and forest decline. Environ. Sci. Technol. 17(6): 246a-256a.

United States-Canada Research Consultation Group on the Long-Range Transport of Air Pollutants. 1979. The LRTAP problem in North America: A preliminary overview: Downsview ON: Information Directorate, Atmospheric Environment Service.

Vogelmann, H. W. 1982. Catastrophe on Camels Hump. Nat. Hist. 91: 8-14.

Watt, W. D., C. D. Scott and W. J. White. 1983. Evidence of acidification in some Nova Scotia rivers and its impact on Atlantic salmon, *Salmo Salar*. Can. J. Fish. Aquat. Sci. 40: 462-473.

Wisniewski, J. and E. L. Keitz. 1983. Acid rain deposition patterns in the continental United States. Water, Air, Soil Pollut. 19: 327-339.

RECOMMENDED READING

D'Itri, F. M. (Ed.). 1982. Acid precipitation. Effects on ecological systems. Ann Arbor Science. Ann Arbor, MI. 506 pp.

Drablos, D. and A. Tollan (Eds.). 1980. Ecological impact of acid precipitation. SNSF Project. Oslo, Norway. 383 pp.

Johnson, R. E. (Ed.). 1982. Acid Rain/Fisheries. Amer. Fish. Co. Bethesda, MD. 357 pp.

Lihach, N. 1983. Discerning the change in waters and woodlands. EPRI Journal. November, 1983: 29-39.

NRCC. 1981. Acidification in the Canadian Aquatic Environment. NRCC #18475. National Research Council of Canada, Ottawa. 369 pp.

NSEPB. 1983. National Swedish Environment Protection Board. Ecological effects of acid deposition. Report 1636. Stockholm. 340 pp.

Peterson, M. A. 1982. The effects of air pollution and acid rain on fish, wildlife, and their habitats—introduction. U. S. Fish and Wildlife Service. Biological Services Program, Eastern Energy Land Use Team, FWS/OBS-8040.3. 181 pp.

Part 4
Disposal, Recycling and Energy Recovery

Because modern society is creating wastes in such huge quantities, efforts are now being directed to develop not only better disposal technology, but also to recycle and reuse waste products. This is a worldwide phenomenon.

The initial paper presents a basic analysis of the relationship between soil chemistry and minerology and the disposal of solid wastes. When wastes are applied to the land as fertilizer there must be a continuing monitoring program to protect water quality and plant uptake of elements and compounds in quantities that could be detrimental to plants and/or animals.

As the amount of wastes increases, new disposal areas are continually being sought. As a consequence more and more waste has been disposed of in the oceans. While it was once thought the oceans could absorb any possible amount of waste dumped into them, it is now recognized that this is not correct. Many countries now control waste disposal in coastal waters. Increased research is needed to gain a better understanding of the marine ecosystem and how it responds to conventional and exotic waste materials.

As the traditional options for the disposal of wastes in the atmosphere, waterways and land diminish, other options must be investigated. Of these, bio-conversion, the controlled decomposition of organic matter, is of growing interest. In the future there must be a massive recycling of the nutrients that have been in the past considered waste.

A number of pilot plants are now being developed to utilize waste materials. The Organic Processing System of Erie, Pennsylvania is an example of a pilot operation. It is recognized that for such a program to be successful there must be education of the public of its value, followed by the motivation to consume the resources, and finally, the cooperation of all participants to reduce costs, not only in dollars, but in environmental damage.

A specific example of waste recovery and utilization is provided in the chapter on potential economic benefits by using the vast coal waste piles associated with coal mining. The study reveals that reusing coal wastes has not only produced benefits in cost savings in disposing of coal wastes, but also consumer benefits in terms of the increase in the quantity of resources and the resultant reduction in market price.

The last two papers develop recycling and energy recovery from waste materials. While these systems are complex and costly, they can yield stable revenues when properly sited and constructed.

Solid and Liquid Wastes: Management, Methods and Socioeconomic Considerations. Edited by
S. K. Majumdar and E. Willard Miller. © 1984, The Pennsylvania Academy of Science.

Chapter Nineteen

Soil Chemistry, Soil Mineralogy and the Disposal of Solid Wastes*

Dale E. Baker[1] and Ann M. Wolf[2]

[1]Professor of Soil Chemistry and [2]Research Assistant in Soil Chemistry
Department of Agronomy
221 Tyson Building
The Pennsylvania State University
University Park, PA 16802

Soils as the outer part of the earth's crust or regolith have been subjected to biological influences as well as to chemical and physical weathering. A generalized theory of soil development as presented by Simonson (1959) emphasizes a combination and balance among physical, chemical and biological processes acting on parent material. Jenny (1941) proposed soil formation to be a function of numerous environmental factors acting on parent material over time. For the purpose of this presentation, it may be satisfactory to consider soil development as the result of climate, relief, gravity and the biosphere acting on a parent material in the course of time. With respect to disposal of solid waste on land, we must modify our concepts of soils and consider the solid waste as a component of soil parent material, a fertilizer, and/or a biocide in soils. Since soil development is a function of time, we should also expect the soil reactions with solid waste to change over time. Soil development processes never stop; soils continue to change; likewise, we should not use a short-time approach to evaluating the effects of land application of solid wastes.

For a comprehensive review of pedology, the reader is urged to study the book

*Contribution of the Pennsylvania Agri. Exp. Sta., University Park, Pa. Authorized for publication on December 8, 1983 as paper number 6837 in the Journal Series.

by Marshall (1977). The title of the book is *The Physical Chemistry and Mineralogy of Soils, Vol. II: Soils in Place.* Beginning with the development of pedological concepts, Marshall provides a comprehensive treatment of soil development processes and ends with a chapter on "Obstacles and Vistas." Marshall's book serves as a tribute to him and historically great scientists from whom he learned and with whom he studied—L. D. Baver, F. E. Bear, R. Bradfield, G. Brindley, E. Buckingham, N. T. Coleman, N. M. Comber, F. G. Donnan, K. K. Gedroiz, C. D. Jeffries, H. Jenny, J. S. Joffe, C. F. Marbut, S. Mattson, G. Wiegner and many others.

The objective of this presentation is to summarize the historically important concepts of soils as they relate to results of more recent investigations which can provide a framework for using and interpreting data from soil assay methods. The goal is to have these concepts and methods applied in the field for solid waste management to provide acceptable pollution of the soil environment because over time soil pollution will affect water quality and the composition of the food chain.

Soil Composition

From a geochemical perspective, soils can be thought of as an "excited skin of the subaerial segment of the earth's crust" (Nikijoroff, 1959). Soil processes consist of transactions in matter and energy between the soil and its surroundings. Soils comprise the outermost layer of the silicate shell which makes up the land resources which are managed by man. Soil has also been defined as "rock on its way to the ocean" (Lindsay, 1979). Since soils comprise the residual weathering products of rocks and minerals left after soluble components have leached away, solid wastes must be compared with soil parent material to determine which constituents will leach and which constituents will be incorporated into the soil or absorbed by soil organisms and higher plants.

Soils, then, consist of minerals from igneous rocks or their derivatives. Ninety-eight percent of igneous rocks consists of oxygen, silicon, aluminum, iron, calcium, potassium, sodium and magnesium. From the results of Clark (1924), it was found that the weight percentage ratios of Ca, K, Mg and Na in igneous rocks to the respective elements in the ocean are 72.6, 64.8, 14.9 and 2.5. Thus with time the rate of loss of Na has been greater than for the other macro cations which weather from igneous rocks (Baker, 1972). The weighted average composition of igneous rocks is used in Table 1 for comparing soils. For a prairie soil of South Dakota the igneous rock:soil ratio of K_2O was 1.73 while the ratio for MgO was 3.83. For the respective soils the elements rank as follows with respect to igneous rock:soil ratios:

Barnes loam
$$MgO > Na_2O > CaO > K_2O > P_2O_5 > Al_2O_3 > SiO_2 > Fe_2O_3$$
Caribou loam
$$MgO > CaO > Na_2O > K_2O > Al_2O_3 > P_2O_5 > Fe_2O_3 > SiO_2$$

Cecil sa. cl. 1.

$$MgO > CaO > Na_2O > K_2O > P_2O_5 > Al_2O_3 > Si_2O_3 > Fe_2O_3$$

Columbiana cl.

$$K_2O > CaO > Na_2O > MgO > Si_2O_3 > P_2O_5 > Al_2O_3 > Fe_2O_3$$

The rankings are important because they are predictable from concepts of soil development. Of the above soils, the Barnes loam from South Dakota is younger and less weathered than the other soils; therefore, the abundance of Si and Fe are near that of igneous rock. The Caribou soil of Maine is a residual soil developed under a cool climate with rainfall greater than evapotranspiration. Under these conditions Si in the form of quartz sand accumulates and the other macro cations tend to leach to lower horizons of the soil profile (Fe and Al) or be leached into ground water (Mg, Ca, Na, K). The Cecil soil of Georgia developed under similar rainfall as the soil of Maine, but the average temperature is higher. Under these conditions and the more extreme situation for the soil from Costa Rica, Fe_2O_3 and Al_2O_3 tend to accumulate. Just as the more abundant elements are lost at different rates and indifferent proportions under different climates, solid wastes can be predicted to behave differently in different locations because the soils differ and the climatic effects will also be different.

Studies of soils of the northeastern United States indicate substantial ranges in composition for surface (A) and subsoil (B) with respect to most elements (Tables 2 and 3). These soils have been used for many investigations as a part of NE-96, a regional research project supported by the Agricultural Experiment Stations in the respective states. The sorption of trace metals has not been found predictable from these results or from mineralogical data published by Johnson and Chu (1983). For example, the retention of Cu may be compared among soils using γ_{cu} of the equation,

$$\gamma_{cu} = A_{cu}/C_{cu} \tag{1}$$

TABLE 1

Soils[1] compared to the weighted average for the igneous rocks[2].

	Igneous Rocks (average)	Barnes Loam (S. Dak.)	Caribou Loam (Maine)	Cecil Sandy Clay Loam (N. C.)	Columbiana Clay (Costa Rica)
	---------------------------------- Igneous Rock/Soil ----------------------------------				
SiO_2	59.1	0.85	1.03	0.79	2.98
Al_2O_3	15.3	1.34	1.95	1.24	0.41
Fe_2O_3	7.3	0.82	1.24	0.63	0.20
CaO	5.1	3.19	4.25	25.50	25.50
Na_2O	3.8	3.37	3.71	18.55	18.55
K_2O	3.1	1.73	3.46	4.18	31.10
MgO	3.5	3.83	5.75	>34	6.90
P_2O_5	0.3	1.50	1.50	1.50	1.00

[1]Reported from Byers, Alexander and Holmes (1935) by Black (1968).
[2]Clark (1924).

where γ_{cu} is defined as the total activity coefficient, A_{cu} is the total activity and C_{cu} is total labile Cu in the soil (Baker and Low, 1970). As γ approaches 1 the binding of Cu by the soil would approach zero. Dragun and Baker (1982) found that γ_{cu} for the soils of Tables 2 and 3 was negatively correlated at the 1% level with pH (r = -0.84), percent silt (r = -0.62) and the percent Na_2O (r = -0.69). The correlation was not significant for organic matter, cation exchange capacity (CEC) or percent clay for the soils. These data agree with results for 64 soils of India where the availability of Cu was negatively correlated with pH (Prabhakaran and Cotteni, 1971). Korte et al. (1976) also concluded that available Cu was not related to clay content nor CEC. However, the results for soils of the northeastern United States do not agree with those of McLaren and Crawford (1973). They reported that soil organic matter was important in controlling the concentration or activity of Cu in the soil solution. The correlations relating Cu retention with silt and Na_2O suggest that Cu retention would be greater for soils which contain more silt which remains high in plagioclase feldspars and other sodium-bearing minerals. These minerals tend to be present in the early to intermediate stages of soil weathering following the calcic plagioclases and preceeding the potassium feldspars (Jackson and Sherman, 1953). These resutls for Cu in soils of the Northeast are in agreement with numerous studies of Cd retention in soils. Retention of Cd is very high for high pH, less weathered soils of the western and midwestern United States, and decreases for highly weathered soils of eastern United States (CAST, 1976).

Soil Organic Matter and Microbial Degredation of Wastes

Soil organic matter includes all organic substances in soils, whether living or dead, fresh or decomposed, or simple or complex. Therefore, all organic compounds added to soils with solid or liquid wastes become a part of soil organic matter. William A. Albrecht when teaching soil microbiology would define soil humus as "an intermediate product in the cycle in which living tissue returns to dust." His definition not only indicates the dynamic nature of soil organic matter, but is also stipulates that soil organic matter in nature is derived from "living tissues." For any inorganic or organic material added to soil which is phytotoxic or potentially harmful to animal or human health, it is important that the effects of the material on soil microorganisms be determined. Much of this type of work has been done to determine the effects of Cd in wastes (Babich and Stotzky, 1978; Bollag and Duzota, 1983; Kurek, et al., 1982). It has been found that microorganisms, dead or alive, will sorb Cd in the presence of soil. The toxic effect of Cd on growth of mircoroganims is usually greater under anaerobic than aerobic conditions. Under the different soil aeration conditions, different soil organisms were capable of sorbing Cd. While soil organisms are capable of removing Cd from solution and thereby reducing its availability to higher plants and the concentrations leaching into ground water, the dynamic nature of soil organic matter suggests that the cycle of microbial

TABLE 2

Macro element composition of soils of the Northeast. Both the surface soil and sub-soil were analyzed.

Soil and State		SiO₂	Al₂O₃	Fe₂O₃	CaO	MgO	Na₂O	K₂O	P₂O₅	Ignition Loss	Total
						%					
Caribou	A	68.13	13.92	6.94	0.20	1.44	1.55	1.56	0.25	6.70	100.69
ME	B	68.75	14.45	6.90	0.46	2.12	1.83	2.25	0.16	3.91	100.83
Groveton	A	69.94	10.16	5.18	0.57	1.10	1.38	1.23	0.34	10.96	100.56
NH	B	67.35	11.34	8.69	0.60	1.43	1.63	1.30	0.17	6.80	99.31
Paxton	A	70.12	12.19	4.25	1.74	1.18	1.51	1.48	0.20	8.39	101.06
CT	B	73.31	11.91	4.18	1.75	1.35	1.62	1.47	0.11	5.30	101.00
Mardin	A	75.49	10.44	4.58	0.10	0.90	0.68	1.76	0.19	5.91	100.05
NJ	B	74.84	11.81	5.51	0.09	1.23	0.84	2.31	0.11	3.46	100.20
Lima	A	80.22	7.91	3.04	0.55	0.89	0.92	1.82	0.11	4.65	100.11
NY	B	77.50	8.64	3.86	0.66	1.22	0.95	1.96	0.12	4.37	99.28
Fauquier	A	54.61	15.45	12.37	5.24	1.79	1.65	0.45	0.26	8.37	100.19
MD	B	52.42	18.71	14.48	3.51	2.07	2.07	0.41	0.24	7.62	101.53
Gilpin	A	76.51	10.16	4.07	0.07	0.70	0.57	1.58	0.16	6.10	99.92
PA	B	75.54	11.24	5.90	0.01	0.98	0.51	1.96	0.09	4.89	101.12
Dekalb	A	80.80	6.33	2.04	0.06	0.25	0.20	0.70	0.08	9.07	99.53
WV	B	83.11	8.32	2.83	0.01	0.42	0.20	0.98	0.08	5.15	101.10
Minesoil	A	71.33	12.66	4.04	0.00	0.47	0.18	2.11	0.08	8.35	99.22
WV	B	71.03	14.93	3.47	0.00	0.52	0.20	2.43	0.10	7.66	100.34
Evesboro	A	94.46	1.89	0.75	0.00	0.07	0.03	0.27	0.11	1.53	99.11
NJ	B	96.06	1.70	0.43	0.00	0.02	0.02	0.13	0.07	0.74	99.17
Sassafras	A	88.44	5.53	1.43	0.15	0.16	0.57	1.45	0.10	1.87	99.70
NJ	B	87.73	5.81	1.50	0.17	0.15	0.56	1.53	0.05	2.93	100.43
Pocomoke	A	90.51	2.27	0.57	0.02	0.07	0.22	0.70	0.06	5.28	99.70
NJ	B	93.44	2.82	0.82	0.05	0.06	0.33	0.98	0.03	1.40	99.93
Vergennes	A	65.64	15.40	4.86	0.60	1.50	1.50	2.54	0.19	8.04	100.27
VT	B	63.20	16.58	6.76	0.49	2.13	1.28	2.98	0.13	6.13	99.68
Hagerstown	A	77.03	9.50	4.72	0.10	0.61	0.40	1.64	0.23	6.13	100.36
PA	B	71.10	12.28	6.87	0.01	1.15	0.15	2.18	0.07	5.53	99.34
Christiana	A	91.21	3.45	2.43	0.00	0.09	0.02	0.38	0.05	3.18	100.81
MD	B	75.99	11.95	6.43	0.00	0.29	0.05	1.28	0.14	4.51	100.64
Marlton	A	70.58	4.30	12.94	0.00	1.72	0.11	3.34	0.26	6.21	99.46
NJ	B	58.34	6.00	22.26	0.00	2.89	0.03	3.68	0.26	6.29	99.75

sorption incorporation into soil humus and mineral adsorption with a gradual change in the soil solution concentration could continue for several years. Studies of ion adsorption and organic chemical incorporation into the humic and fulvic acids of soils are very important. The organic geochemistry approach of R. L. Malcolm and co-workers is extremely important (Malcolm, 1972; Malcolm, et al., 1977; Aiken, et al., 1979).

While Cd and other metals added to soils with solid wastes have received much attention, the mutagenic organics in these materials have been assumed to be rendered "harmless" by microbial degradation or by selective adsorption by soils. Such assumptions must be critically questioned in view of the results from research conducted at the University of Illinois (Hopke et al., 1982). Sewage

TABLE 3

Trace element composition of soils of the Northeast. Both surface soil
and sub-soil were analyzed.

Soil		Cr	Mn	Co	Zn	Ni	Cu	Pb	Cd
					ppm				
Caribou	A	99.2	246.8	6.2	97.0	23.1	126.9	77.5	0.28
	B	118.3	388.0	49.0	97.0	61.9	45.1	70.0	1.34
Groveton	A	93.2	321.5	0.0	87.4	56.4	36.5	25.0	0.12
	B	96.6	445.5	47.4	93.0	53.3	2.2	37.5	0.54
Paxton	A	65.6	406.0	12.0	73.1	41.6	109.4	25.0	0.60
	B	67.2	388.0	85.0	81.1	46.3	18.8	50.0	0.48
Mardin	A	69.4	418.2	0.0	97.0	46.3	52.1	43.8	0.12
	B	74.4	229.7	0.0	85.1	49.4	43.1	85.0	0.30
Lima	A	59.2	243.3	50.5	77.1	30.8	37.9	38.8	0.24
	B	62.4	305.8	55.2	85.1	30.8	27.8	37.5	0.44
Fauquier	A	205.0	1042.6	90.8	99.4	77.5	127.5	30.0	1.59
	B	184.0	805.5	84.9	97.0	61.9	196.4	0.0	1.47
Gilpin	A	68.1	633.2	45.5	127.2	52.7	24.6	46.3	0.43
	B	75.3	239.7	24.0	105.7	61.9	60.9	58.8	0.52
Dekalb	A	43.4	49.5	0.0	46.8	0.0	29.9	5.0	1.54
	B	46.7	60.7	73.7	73.1	48.6	41.8	38.8	0.93
Minesoil	A	59.3	65.5	52.7	45.2	39.3	76.8	0.0	0.67
	B	60.8	90.1	1.3	54.8	54.1	66.3	0.0	0.89
Evesboro	A	7.6	35.6	35.6	28.4	29.2	34.8	87.5	0.06
	B	7.6	21.6	21.6	42.0	28.5	46.7	107.5	0.0
Sassafras	A	28.9	192.2	42.0	49.2	30.8	52.8	67.5	0.34
	B	34.7	161.8	38.6	43.6	20.7	55.4	65.0	0.70
Pocomoke	A	13.0	70.5	14.0	5.3	3.8	39.6	125.0	0.06
	B	6.6	54.4	26.3	5.3	2.3	49.0	87.5	0.06
Vergennes	A	91.0	447.0	57.0	79.5	40.8	32.2	0.0	0.51
	B	118.6	325.1	29.2	95.4	48.7	55.1	83.8	0.85
Hagerstown	A	110.1	3129.7	94.3	60.4	25.4	28.0	40.0	0.23
	B	130.7	461.3	91.7	60.4	65.8	40.8	40.0	0.31
Christiana	A	41.6	43.9	0.0	43.6	28.5	84.9	68.8	0.10
	B	93.8	28.6	0.0	71.5	35.4	57.4	77.5	0.19
Marlton	A	275.3	87.3	13.2	120.9	69.7	46.8	30.0	0.13
	B	483.4	30.0	15.4	140.7	71.2	46.7	0.0	0.04

sludge added to soil led to mutagenic effects in field corn (*Zea mays* L.). The presence of polyhalogenated organics, organochlorine pesticides and polyaromatic hydrocarbons including the known mutagenic agents benzo(a)pyrene and benzofluroanthene have been noted in different sludges (Erickson and Pellizzari, 1979). Telford et al. (1982) extracted sludge from Syracuse, New York with methanol and the extract solicited a mutagenic response from *Salmonella typtrimurium* strains TA98 and TA100 in the Ames test (Babish et al., 1983). Similarly, 33 of 34 dichloromethane extracts from municpal sludges were shown to be mutagenic to one or more of five strains of the microorganism used in the Ames test. Stoewsand, Babish and Lisk (1977) fed cabbage grown on sludge-treated soil to guinea pigs for 100 days and observed no gross changes other than an increased intestinal aryl hydrocarbon hydroxylase and greater levels of polychlorinated biphenol (PCB) levels in the liver (Babish et al., 1979). While much more experimentation will be required to separate organic chemical effects from inorganic chemical effects from sludges on land, the existing data prompted Baker, et al., 1984b, to recommend that growers of food crops avoid the use of solid wastes and sludges on food crops. In addition, it was recommended that food crops not be grown on soils treated with sludges in the previous five years.

From the relationships among soil composition, soil development processes and the biological effects of wastes from non-living systems, all materials and effluents to be land applied should be carefully monitored using water quality standards where possible (Baker and Chesnin, 1975). Since this is not generally possible, monitoring of soils using physical chemistry concepts has been found helpful and will be considered in a subsequent section of this presentation.

Soil Structure and Air-Water Relations

The interactions among the soil formation factors including topography, parent material and organisms lead to heterogeneous properties for soil within and among soil orders, soil series and even within fields where the climatic and other factors are not greatly different. All soils should be considered as polydispersed systems composed of solid, liquid and gaseous phases in various proportions (Baver, 1956). The particles in soils are in a fine state of subdivision and are dispersed by the soil water which acts as the dispersion medium. Regardless of the size of particles and their aggregation, each particle is dispersed within the soil solution. The composition of soil particles, their arrangement into structural units and the composition of the soil solution all affect the percent by volume occupied by the soil solution and soil air at a given water tension. For natural soils, the composition of the soil solution is determined by the solubility of minerals remaining within the solid phase. However, when solid wastes, lime, manures, fertilizers and other chemicals are added to soil by man, the composition of the soil solution may not reflect the solubility of soil minerals.

Soil management with erosion control has little effect on soil texture which reflects particle size distribution and specific surface area. However, the arrangement of soil particles into desirable structural units can be rapidly degraded but very slowly improved by soil management. In addition to cultivation, exposure of soil to the impact of raindrops, removal of crop residues, and compaction by machinery, the addition of chemicals can also effect the degree of dispersion of soil particles over time.

With optimum control of erosion which must be a number one national priority, the loss of ions and compounds from soils will result from crop removal and by leaching losses from soils exposed to rainfall greater than evapotranspiration during the year. With respect to waste disposal on land, the "living filter" relates to (1) the filtration of solid material from waste water, manure, sludges, etc.; (2) the exposure of the material to the soil environment for microbial degradation of organic compounds; and (3) exchange or specific adsorption of ionic and molecular species by the soil solid phase (Kardos and Sopper, 1973; Sopper and Kerr, 1979). While of substantial value for land disposal of wastes, the living filter has limitations. If loading rates are excessive, the properties of the waste can substantially modify the soil air-water relationships. The waste can occupy the large pores which are responsible for water infiltration and percolation into ground water or for some wastes the soil may become more dispersed which will also lead to fewer large pores needed for the diffusion of air into the soil. Under these conditions the beneficial filtration, degradation effects are lost. Because of these limitations, the "living filter" process should generally be limited to waste additions for nutrient utilization in crop production.

When wastes are applied to land over time, the soil should be monitored continuously with respect to soil aeration, water infiltration and percolation in addition to chemical monitoring to protect water quality and plant uptake of elements and compounds in quantities which are harmful to plants and/or animals. Chemical monitoring of soils is accomplished with prepared samples within the laboratory and will be considered in a subsequent section of this presentation. The air-water relationships must be determined by field observations and techniques which can be supplemented with laboratory measurements on undisturbed core samples taken from the field. Most of these methods have been published by Black et al. (1965).

Waste water and sludges high in Na or Li especially as the hydroxides can be expected to have detrimental effects on soil aeration and water infiltration because of the effectiveness of these highly hydrated monovalent cations in dispersing soils. As soils become more dispersed, the structure changes from various forms of aggregates toward a single grain or massive structure which eliminates large pores in all soils except those classified as sands. The adverse effects of Na can be predicted from the sodium adsorption ratio (SAR) of the waste water or from a water extract of solid wastes:

$$SAR = [Na^+] / \sqrt{\frac{[Ca^{++}] + [Mg^{++}]}{2}}$$

The SAR is the ratio of the Na^+ concentration to the square root of the average concentration of divalent Ca^{++} and Mg^{++} in the solution extracted from a waste or from a soil. If the SAR is greater than 10 and the solution concentration is low (less than 10^{-3} M), soil dispersion and clogging of pores can occur (Oster and Schroer, 1979).

Intensive and Extensive Properties and Interrelationships

Generally, biological systems respond directly to the intensive properties of nature. Intensive properties include temperature, relative humidity, and pressure.

They are not additives; for example, the temperature of any small part of a system in equilibrium is the same as the temperature of the whole (Moore, 1972). The chemical potential, μ, of an ion, i, in soil-water systems is also an intensive property:

$$\mu_i = \mu_i^\circ + RT \ln A_i/A_i^\circ$$

where μ_i° and A_i° refer to the standard state chemical potential and activity, respectively; R is the universal gas constant; T is temperature in $^\circ$K; ln is the natural logarithm; and A_i is the activity of ion, i:

$$A_i = \gamma_i C_i$$

where γ_i is the activity coefficient which corrects for non-ideal solutions and is determined from the ionic strength of the solution and some form of the Debye-Huckel equation (Lindsay, 1979). The familiar solution pH is a measure of μ_{H+}:

$$\mu_{H+} = \mu^\circ_{H+} + RT \ln A_{H+}/1.0 = -1364 \text{ pH}$$

where the units for μ_H is calories per mole at 25°C and the standard state is an ideal 1.0 molar soluton of H^+. The concept of chemical potential and ionic activities for aqueous systems is discussed by Stumm and Morgan (1981).

The application of chemical potentials to soils and plant systems has been the subject of several investigations (Baker and Low, 1970; Baker, 1971, 1973, 1974, 1976a, 1977; Baker and Eshelman, 1979; Stout and Baker, 1978). For most biologically essential elements, living systems will show deficiency symptoms—normal growth, reduced growth, and in some cases toxicities—as the chemical potential decreases (becomes less negative relative to a standard state of unit activity) as a result of increasing activity of an ion in solution. For most productive soils the abundant ions (NO_3^-, K^+, $SO_4^=$, Ca^{++}, Mg^{++}, Cl^-, N^+, HCO_3^-) will have activities in the range of 10^{-2} to 10^{-4} M. Other essential elements (P, Fe, Mn, Cu, Zn, Mo, and B for plants plus F, V, Cr, Co, Ni, Si, and I for animals) will range from 10^{-5} to 10^{-9} M or less in the soil solution. While the quantities of these elements in soils (extensive property) is usually sufficient to support plant growth, the culture of crop plants and water quality require the control of chemical potentials within acceptable limits. Ratios of ionic activities or the difference between the chemical potentials of ions can be important where a high activity of one ion will depress the availability of another ion. For example, high levels of K^+ can depress plant uptake of Mg^{++} (Stout and Baker, 1978; Rahmatullah and Baker, 1981).

The importance of chemical potentials for ions and compounds in the soil solution and water in general cannot be over-emphasized. It is not only theoretically valid, but it also enables comparisons among soil water, surface water and groundwater. It has been predicted from experimental data reported in the literature that the normal range and values for excessive activities for various elements expressed as the negative logarithm of the ion activity (analogous to pH) should be expressed as follows:

Pi	Normal Range for Soil Water	Excessive Level
	------------------ ($l$log A_i) ------------------	
pH	6.0-7.0	<5.0
pCa	2.3-2.7	NA*
pMg	2.8-3.3	NA
pNa	3.0-4.0	NA
pK	3.0-4.0	NA
pFe^{+3}	16-18	NA
pMn^{++}	5-8	NA
pPb	8.5-10	NA
pCu	10-12	<6.0
pNi	11-13	NA
pZn	6.5-7.5	NA
pCd	7-8	<6.0
pMo	9-10	<8.0
pH$_2$PO$_4$	4-6	NA

*NA = not available

The values above were predicted from data presented by Allaway (1969), Baker (1973), Hornick et al. (1977), Lagerwerff (1972), Lindsay (1979), and Dragun (1977). The assay methods of Baker and Amacher (1981) are used routinely to evaluate intensive-extensive property relationships for soils of Pennsylvania.

As important as these values are, only Cd and Cu have been studied with respect to waste disposal on land (Baker, 1974; Baker, et al. 1979). More recently, Baker, Kotuby and Rasmussen (1984a) and Rasmussen (1984) have developed methods for measuring the loading capacity of different soils (extensive property) at a given pCd of 6.5 in the equilibrium solution. The results of their investigations agree with the conclusion reached by Baker et al. (1984b) that for soils of the northeastern United States the maximum amount of Cd to be applied to soils ranges from 1 to 2 ppm depending on soil texture and assuming that the pH may reach values below 6.0. Investigations of the intensive/extensive properties and kinetics of all potentially toxic elements and compounds added to land from waste should be a goal for agencies and individuals interested in land application of wastes.

The Kinetic or Replenishment Factor in Soil-Plant Relations

Historically, soil chemistry, soil fertility and soil management priorities for research involved studies to relate extensive properties or the quantities and proportions of various essential ions in soils with plant growth. For example, Miller and Krusekopf (1919, 1929) published "The Soils of Missouri." In addition to providing detailed descriptions of the soils with respect to parent material, geology, topography, altitude, climate and native vegetation, they reported the total pounds of N, P, K and lime requirements for the acre furrow slice (2,000,000 pounds of soil) for both the surface soil and the subsoil. The productivity of the soils at that time was related to these results which could be predicted from the composition of the parent material:

Soil	Parent Material	N	P	K	Lime Req. as CaCO₃
		(pounds/acre)			
Marshall	Loess in N. W. Mo.	3,685	1,675	33,720	3,960
Cherokee	Shale in S. W. Mo.	1,950	810	24,845	2,215
Lebanon	Limestone of Ozarks	1,470	705	30,730	1,505
Tilsit	Sandstone of Ozarks	1,320	620	23,990	1,580

All of these soils are silt loams, so the composition reflects soil formation factors independent of the distribution of particle sizes.

The reported higher land values for these soils in 1929 ranged from $250.00 per acre for the Marshall silt loam to $5.00 to $50.00 per acre for the Tilsit soils. Because the water-holding capacity is also a direct and indirect function of the nitrogen content of the original soils, the relative productivity of the above soils probably still ranks in about the same order, but the use of commercial fertilizers and lime has greatly improved the productivity of all agricultural soils.

Total composition of soils is no longer considered an important parameter for use in predicting soil productivity. The concepts currently used to assay soils for predicting nutrient requirements of crops were included in the proceedings of a symposium on "Soil Testing: Correlating and Interpreting the Analytical Results" (Peck et al., 1977). The complexity of the soil-water system has been illustrated by Wolf and Baker (1984). Each mechanism which inactivates or sorbs soluble ions from the soil solution will gradually release a fraction of them into the soil solution as the activities of the ions are lowered by crop removal or leaching.

Commercial fertilizers, lime, manure, sewage sludge and other sources of essential and non-essential elements for plants, animals and human beings should be applied for the benefit of society as well as for the farmer (Baker, 1977). The amount of an element to be added to a soil for a crop growth is a function of crop removal in relation to the following:

a. Plant availability of each essential plant nutrient which is a function of its chemical potential in the soil solution as affected by adsorption and dissolution of soil minerals and/or man-induced changes from applications of fertilizers and other materials (manure, sewage sludges and other solid and liquid wastes).

b. The effect that other ions have on the uptake of the specific ion under consideration. As indicated above, ionic activity ratios can be of substantial importance.

c. Changes in (a) and (b) to be expected over the growing season.

d. Differential soil-plant interactions such as soil volume per plant (Baker and Woodruff, 1963) and plant species or variety (Baker, 1977).

The approach to soil testing currently available to growers of Pennsylvania through the Soil and Environmental Chemistry Laboratory involves measurement of the change in soil amounts (extensive properties) of adsorbed ions which are required to make all soils equal with respect to chemical potentials and differences among chemical potentials:

pH, pK, pCa, pMg, pFe, pMn, pZn, pCu, pAl, pPb, pNi, and pCd.

The method has been described by Baker and Amacher (1981). The approach is considered excellent, but the method has limitations because the solution activities of the trace metals are usually very low and often below the detection limits of most instrumental methods. For the trace metals, the ionic activities are calculated from the formation constants for the chelator DTPA (diethylene triamine penta acetic acid) at a pH just above the pH of the soil. Since the DTPA removes the metals from soils to form metal-chelate complexes in solution, the equilibrium chemical potentials represent lower solution activities than would be present in the soil initially. The test or assay results correlate well with crop uptake of the elements, however. It is postulated that the method works because it combines chemical potentials, adsorbed amounts, and other parameters important in ionic diffusion (Baker and Low, 1970):

$$J_i = \frac{dQ}{dt} = -A \, \beta \, \bar{\gamma}_i \, \bar{C}_i \, \frac{d\bar{\mu}_i}{dx}$$

where J_i is the diffusive flux (change in quantity, dQ, with a change in time, dt); A is the cross-sectional area of the diffusion path (proportional) to root area where absorption occurs); β is the geometry factor that corrects for porosity or tortuosity of the soil medium; $\bar{\gamma}_i$ is the average mobility of the ion, i; $\bar{C}_i$ is the average concentration of the ion, i, in equilibrium with ion, i, in solution; and $d\bar{\mu}_i/dx$ is the change in electrochemical potential with distance, x, in the direction of diffusion. It will be noted that the chemical potentials by this method are higher (more negative) than those presented above because of the DTPA effect.

For sustained use of land for disposal of solid and liquid wastes, experimentation to improve the approach to soil testing of Baker is recommended (Baker, 1971, 1973, 1976a, 1976b, 1977; Baker and Amacher, 1981; Baker, et al. 1984d; and Rasmussen, 1984). In addition, it will be necessary to extend the approach to potentially harmful organics if they are found in soil water and in the food chain.

REFERENCES

Aiken, G. R., E. M. Thurman, R. L. Malcolm, and H. Walton. 1979. Comparison of XAD macroporous resins for the concentration of fulvic acid from aqueous solution. Anal. Chem. 52:1799-1803.

Allaway, W. H. 1968. Adv. in Agron. 20:235-272.

Babich, H. and G. Stotzky. 1978. Effects of cadmium on the biota: Influence of environmental factors. Adv. Appl. Microbiol. 23:55-117.

Babish, J. G., B. E. Johnson and D. J. Lisk. 1983. Mutagenicity of Municipal Sewage Sludges of American Cities. Environ. Sci. Technol. 17:272.

Babish, J. G., G. S. Stoewsand, A. Furr, T. F. Parkinso, C. A. Bache, W. H. Gutenmann, P. C. Wszolek and D. J. Lisk. 1979. Elemental and Polychlorinated Biphenyl Content of Tissues and Intestinal Aryl Hydrocarbon Hydroxylase Activity of Guinea Pigs Fed Cabbage Grown on Municipal Sewage Sludge. J. Agric. Food Chem. 27:399.

Baker, D. E. 1971. A new approach to soil testing. Soil Sci. 112:381-391.

Baker, D. E. 1972. Soil Chemistry of Magnesium. IN: Magnesium in the Environemnt. pp. 1-39. Taylor County Printing, Reynolds, GA.

Baker, D. E. 1973. A new approach to soil testing: II. Ionic equilibria involving H, K, Ca, Mg, Mn, Fe, Cu, Zn, Na, P and S. Soil Sci. Soc. Am. Proc. 37:537-541.

Baker, D. E. 1974. Copper: Soil, water, plant relationships. Symposium on Ecological Problems of High Level Nutrient Feeding. Federation Proc. 33:1188-1193.

Baker, D. E. 1976a. Ion activities and ratios in relation to corrective treatments of soils. IN: Soil Testing: Correlating and Interpreting the Analytical Results. pp. 55-74. Am. Soc. of Agron., Madison, WI.

Baker, D. E. 1976b. Soil chemical monitoring for environmental quality and animal and human health. Yearbook of Science and Technology. McGraw Hill Book Company, New York.

Baker, D. E. 1977. Soil chemical constraints in tailoring plants to fit problem soils. 1. Acid soils. IN: M. J. Wright (ed.), Plant Adaptation to Mineral Stress in Problem Soils. Cornell Univ. Agric. Exp. Stn., Ithaca, NY. pp. 127-140.

Baker, D. E. and M. C. Amacher. 1981. Development and Interpretation of a Diagnostic Soil Testing Program. Bul. No. 826, Pennsylvania Agric. Exp. Stn., University Park, PA. 18 pp.

Baker, D. E., M. C. Amacher and R. M. Leach. 1979. Sewage sludge as a source of cadmium in soil-plant-animal systems. Environ. Health Perspectives 28:45-49.

Baker, D. E. and L. Chesnin. 1975. Chemical monitoring of soils for environmental quality and animal and human health. Adv. Agron. 27:305-375.

Baker, D. E. and R. M. Eshelman. 1979. Ionic activities. IN: The Encyclopedia of Soil Science. Part I. Physics, Chemistry, Biology, Fertility, and Technology. Dowden, Hutchinson & Ross, Inc., Stroudsburg, PA. pp. 241-245.

Baker, D. E. and P. F. Low. 1970. Effect of the sol-gel transformation in clay-water systems on biological activity: II. Sodium uptake by corn seedlings. Soil Sci. Soc. Am. Proc. 34:49-56.

Baker, D. E., D. S. Rasmussen and J. Kotuby. 1984. Trace metal interactions affecting soil loading capacities for cadmium. IN: Industrial and Hazardous Waste, Proc. Int. Symposium, ASTM Committee D-34, Philadelphia, PA (in press).

Baker, D. E. and C. M. Woodruff. 1963. Influence of volume of soil per plant upon growth and uptake of phosphorus by corn from soils treated with different amounts of phosphorus. Soil Sci. 94:409-412.

Baker, D. E., et al. 1984b. Criteria and Recommendations for Land Application of Sludges in the Northeast. Pa. Agri. Exp. Sta. Bul. 851 (in press).

Baver, L. D. 1956. Soil Physics. Third Ed. John Wiley and sons, Inc., NY. 489 pp.

Black, C. A. 1968. Soil Plant Relationships. Second Ed. John Wiley and Sons., Inc., NY. 792 pp.

Black, C. A. et al. (eds.). 1965. Method of Soil Analysis. Chemical and Microbiological Properties. Agronomy No. 9, part 2, pp. 771-1572. Amer. Soc. Agron., Madison, WI.

Bollag, J. M. and M. Duzota. 1983. Effect of the physiological state of microbial cells on cadmium sorption. Arch. Environ. Contam. Toxicol. (in press).

CAST. 1976. Application of sewage sludge to cropland. Appraisals of potential hazards of the heavy metals to plants and animals. Council for Agri. Sci. & Tech. Rept. 64, Ames. IA.

Clark, F. W. 1924. The data of geochemistry. Fifth Ed. U.S. Geological Survey Bul. 770. 841 pp.

Dragun, J. 1977. Copper availability in soils and montmorillonite suspensions. Ph.D. thesis in Agronomy. The Pennsylvania State University, University Park, PA. 91 pp.

Dragun, J. and D. E. Baker. 1982. Characterization of copper availability and corn seedling growth by a DTPA soil test. Soil Sci. Soc. Amer. J. 46:921-925.

Erickson, M. D. and E. D. Pellizzari. 1979. Analysis of Municipal Sewage Sludge Samples by GC/MS Computer for Polychlorinated Biphenyls and Other Chlorinated Organics. Bul. Environ. Contam. Toxicol. 22:688.

Hopke, P. H., M. J. Plewa, J. B. Johnston, D. Weaver, S. G. Wood, R. A. Larson, T. Henesly. 1982. Multitechnique Screening of Chicago Municipal Sewage Sludge for Mutagenic Activity. Environ. Sci. Technol. 18:140.

Hornick, S. B., D. E. Baker, and S. B. Guss. 1976. Crop production and animal health problems associated with high soil molybdenum. IN: Molybdenum in the Environment.

Jackson, M. L. and G. D. Sherman. 1953. Chemical weathering of minerals in soils. Adv. in Agron. V:219-318.

Jenny, Hans. 1941. Factors of Soil Formation. McGraw-Hill, NY. 281 pp.

Johnson, L. J. and P. Chu. 1983. Mineral Characterization of Selected Soils from Northeastern United States. Pa. Agri. Exp. Sta. Bul. 847.

Kardos, L. T. and W. E. Sopper. 1973. Renovation of municipal wastewater through land disposal by spray irrigation. IN: Recycling Treated Municipal Wastewater and Sludge through Forest and Crop Land. W. E. Sopper and

L. T. Kardos (eds.). The Pennsylvania State University Press, University Park, PA. pp. 148-163.

Korte, N. E., J. Skopp, W. H. Fuller, E. E. Niebla and B. A. Alesh. 1976. Trace element movement in soils: Influence of soil physical and chemical properties. Soil Sci. 122:350-359.

Kurek, E., J. Czaban, and J. M. Bollag. 1982. Sorption of cadmium by microorganisms in competition with other soil constituents. Appl. Environ, Microbiol. 43:1011-1015.

Lagerwerff, J. V. 1972. Lead, mercury and cadmium as environmental contaminants. IN: Micronutrients in Agriculture. Soil Sci. Soc. Am., Inc., Madison, WI. pp. 593-636.

Lindsay, W. L. 1979. Chemical Equilibria in Soils. Wiley-Interscience, NY. 449 pp.

Malcolm, R. L. 1972. Comparison of conditional stability constants of North Carolina humic and fulvic acids with Co. (II) and Fe(III). The Geological Society of America, Inc., Memoir 133, pp. 79-83.

Malcolm, R. L., E. M. Thurman, and G. R. Aiken, 1977. The concentration and fractionation of trace organic solutes from natural and polluted waters using XAD-8, a methylmethacrylate resin. 11th Annual Conf. on Trace Substances in Environmental Health, Columbia, MO. pp. 307-314.

Marshall, C. E. 1977. The Physical Chemistry and Mineralogy of Soils. Vol. II: Soils in Place. Wiley-Interscience, NY. 313 pp.

McLaren, R. G. and D. V. Crawford. 1973. Studies on soil Cu: I. The fractionation of Cu in soils. J. Soil Sci. 24:174-181.

McLean, E. O. 1977. Contrasting concepts in soil test interpretation: Sufficiency levels of available nutrients versus basic cation saturation ratios. IN: Soil Testing: Correlating and Interpreting the Analytical Results. pp. 39-54.

Miller, M. F. and H. H. Krusekopf. 1918. The Soils of Missouri. Univ. of Mo. Agri. Exp. Sta. Bul. 153. 130 pp.

Moore, W. J. 1971. Physical Chemistry. Fourth Ed. Prentice Hall, Inc., NJ. 977 pp.

Nikiforoff, C. C. 1959. Reappraisal of the soil. Science 129:186-196.

Oliver, S. and S. A. Barber, 1966. An Evaluation of the mechanism governing the supply of Ca, Mg, K and Na to soybean roots (*Glycine max L.*). Soil Sci. Am. Proc. 30: 82-86.

Olsen, S.R. and S.A. Barber. 1977. Effect of waste applications on soil phosphorus and potassium. IN: L.F. Elliott and F.J. Stevenson eds. Soils for management of organic wastes and waste waters. Amer. Soc. of Agron., Madison, WI.

Oster, J. D. and F. W. Schroer. 1979. Infiltration as influenced by irrigation water quality. Soil Sci. Soc. Am. J. 43:444-447.

Peck, T.R., et. al. (eds.), 1977. Soil Testing: Correlating and Interpreting the Analytical Results. Spec. Publ. 29, Amer. Soc. Agron., Madison, WI 117 pp.

Prabhakaran, N.K.D. and A. Cotteni. 1971. A statistical evaluation of the interrelationships between particle size fractions, free iron oxides and trace elements. J. Soil Sci. 22:203-209.

Rahmatullah, and D. E. Baker. 1981. Magnesium accumulation by corn (*Zea mays* L.) as a function of potassium-magnesium exchange in soils. Soil Sci. Soc. Am. J. 45:899-903.

Rasmussen, D. S. 1984. Laboratory measurement of cadmium loading capacities for soils of the Northeast. M. S. Thesis, Dept. of Agronomy, The Pennsylvania State University, University Park, PA. 75 pp.

Shaffer, K. A., D. D. Fritton, and D. E. Baker. 1979. Drainage water sampling in a wet, dual-pore soil system. J. Environ. Qual. 8(2):241-246.

Simonson, R. W. 1959. Outline of a generalized theory of soil genesis. Soil Sci. Soc. Amer. Prod. 23:152-156.

Sopper, W. E. and S. N. Kerr. 1979. Utilization of Municipal Sewage Effluent and Sludge on Forest and Disturbed Land. The Pennsylvania State University Press, University Park, PA, 537 pp.

Stoewsand, G. W., J. G. Babish and D. J. Lisk. 1977. Activity of Intestinal Aryl Hydracarbon Hydroxylase in Guinea Pig Fed High Element Containing Sludge-Grown Cabbage. Fed. Proc. 36:1146 (abs.).

Stout, W. L. and D. E. Baker. 1978. A new approach to soil testing: III. Differential adsorption of potassium. Soil Sci. Soc. Am. J. 42:307-310.

Stumm, W. and J. J. Morgan, 1981. Aquatic Chemistry. Second Ed. John Wiley and Sons, Inc., NY. 780 pp.

Telford, J. N., M. L. Thonney, D. E. Hogue, J. R. Stouffer, C. A. Bache, W. H. Gutenmann, D. J. Lisk, J. G. Babish, and G. S. Stoewsand. 1982. Toxicological Studies in Growing Sheep Fed Silage Corn Cultured on Municipal Sludge-Amended Acid Subsoil. J. Tox. and Environ. Health 10:73.

Wolf, A.M. and D. E. Baker. 1984. Chapter 13. This publication.

Solid and Liquid Wastes: Management, Methods and Socioeconomic Considerations. Edited by
S. K. Majumdar and E. Willard Miller. © 1984, The Pennsylvania Academy of Science.

Chapter Twenty

The Marine Repository - A Vital Waste Management Resource

Robert F. Schmalz, Ph.D.

Department of Geosciences
503 Deike Building
The Pennsylvania State University
University Park, PA 16802

From the earliest times, man has disposed of his wastes in the ocean. In the past, the quantity of dumped waste was relatively small, and the materials largely innocuous. In the years following World War II, however, the United States, like most industrialized nations, experienced a dramatic increase in the generation of domestic and industrial waste. With more than twenty percent of the population concentrated in coastal cities, it was natural to turn to the ocean as a repository for waste materials. The increase in ocean dumping which resulted is startling: in the years after the war and extending into the middle 1950's, the United States dumped an average of 1.7 million tons of solid waste (exclusive of dredge spoil) into the sea each year. Ten years later, in the middle 1960's, this figure had grown to 7.4 million tons annually, and in 1973 the quantity of solid waste dumped into the ocean adjacent to the shores of the United States was in excess of 10.9 million tons.[2]

Increased Federal regulation during the 1970's has put an end to uncontrolled ocean dumping, and the tonnage dumped in 1973 has not yet been exceeded. However, it is estimated that the *per capita* production of solid domestic waste in the United States will increase from the present level of 3.6 pounds per day to between 4.0 and 4.5 pounds per day in 1990, when seventy-five percent of the population will live and work within fifty miles of the ocean coastline or the Great Lakes.[3,4] Industrial waste production will nearly double during the same period, and the bulk of this waste material will also be generated in the coastal zone. Even if only partially realized, these estimates make it quite clear that during the coming decades there will be increased economic and practical benefits to be derived from ocean dumping and a corresponding increase in efforts to relax existing restrictions on waste disposal at sea. It seems unlikely that the level of dumping set in 1973 will not soon be exceeded.

The problem of ocean dumping is complicated by the fact that we are not only producing more waste than ever before, but by the change in character of the waste generated. Radioactive materials and toxic or hazardous chemicals make up an increasingly significant part of the waste burden, and these materials cannot always be isolated for special treatment. Chemical wastes afford an example. In 1975 the United States produced approximately 2,302,000 tons of organic chemical waste and an additional 440,000 tons of toxic organochlorine waste. The generation of these materials nearly doubled by 1980 when there were produced 4,249,000 tons of organic chemical waste and 606,000 tons of organochlorine wastes, and projected 1990 waste production figures indicate a further increase of 50% above the 1980 level.[5] Although production wastes which may include dioxin, for example, can be disposed of under controlled conditions, insecticides or herbicides once applied cannot be recovered, and excess quantities or breakdown products (which may also include dioxin) may enter the biosphere, contaminate groundwater or be carried off in storm water, ultimately to enter the sea. The suggestion that such hazardous materials might be deliberately dumped into the ocean provokes an immediate and usually vigorous response from environmentalists.

If we are to deal rationally and wisely with the problem of growing quantities of increasingly hazardous waste, we must begin by examining the nature of the waste materials and their potential impact upon the environment, and consider the economic and environmental consequences of alternative methods of disposal. It is the purpose of this paper to review the practice of ocean dumping and to assess its future role in waste management.

OCEAN DUMPING—MEANS AND MATERIALS

Waste is dumped into the oceans either in the form of liquids or suspended solids delivered directly to the nearshore region by pipeline ('outfalls') or as solid, sludge or packaged liquids dumped at sea from barges or ships.

Though often overlooked, pipeline outfalls constitute a major source of pollution in the sea, and may include industrial waste, raw sewage or treated sewage effluent, and stormwater runoff. These materials are likely to be particularly damaging to the environment because generally outfalls extend only a few miles (sometimes only a few hundred yards) from shore and the wastes are in effect introduced into the especially fragile coastal wetlands zone.[6] Toxic materials from outfalls may adversely affect the organisms of the littoral zone directly, they may be concentrated by bioaccumulation, or they may affect the base of the marine food web and so have vast and far-reaching effects which cannot be anticipated. Insecticides and PCB's as well as lead from stormwater runoff have been identified over wide areas off the California coast, and mercury from an industrial outfall was concentrated by bioaccumulation in marine invertebrates and ultimately posed a serious health hazard for humans at Minimata, Japan.[7,8]

Today in the United States, all such outfall discharges are regulated under the Federal Water Pollution Control Act (FWPCA) and must be licensed by the U.S. Army Corps of Engineers; the waste is monitored by the Environmental Protection Agency. Significant environmentally as outfall discharges may be, however, it is toward the dumping of solid or packaged waste from ships or barges at sea that this review is primarily directed.

The principal forms of waste dumped at sea can be grouped into seven major categories, within which different specific waste materials share similar properties and thus pose similar problems of disposal. The forms generally recognized are defined and discussed below.

Sewage sludge. Sewage sludge is the solid material remaining after municipal water treatment, composed of residual human wastes and other organic and inorganic wastes.[9] Though primarily of organic origin, sewage sludge may include a wide range of inorganic materials, many of which must be viewed as contaminants. Silt and other mineral matter pose few problems, but concentrations of heavy metals, notable cadmium, zinc, lead, molybdenum, copper, barium and manganese may reach several parts per thousand, and nutrients (notably phosphate and nitrate) are often found in concentrations sufficient to significantly affect organisms at the disposal site.[10,11] Unoxidized organic matter in sludge may impose a high biological oxygen demand (BOD) at the site, and can substantially reduce or totally deplete the available supply of dissolved oxygen in ocean bottom waters.[12] Finally, sewage sludge may contain pathogenic bacteria and protozoa, including varieties resistant to antibiotic drugs or genetically adapted to elevated concentrations of heavy metals.[13,14] Many ecologists believe that sludge dumps form 'breeding grounds' for such pathogenic organisms and may thereby pose a threat to human health.[15]

Secondary treatment of sewage prior to dumping reduces the biological and chemical oxygen demand (BOD and COD, respectively) of the sludge, but by oxidizing the heavy metals (if any) present it may render them more assimilable and therefore more hazardous to many marine organisms.[16]

Several fish kills in the waters of New York Bight in the late 1960's and again in 1976 were attributed in part to depletion of oxygen in the waters round the

Apex sludge dumping site, and together with repeated pollution of beaches by waste matter, these kills helped to trigger public concern and to stimulate the investigation by the Council on Environmental Quality which led ultimately to the current Federal regulation and international conventions relating to ocean dumping.[17,9]

Industrial waste. Industrial wastes include acids, refinery, pesticide and paper-mill wastes as well as various liquid wastes.[9] Industrial waste consistently makes up between 3% and 5% of the waste dumped at sea by the United States, second only in tonnage to dredge spoil, and comparable in amount to sewage sludge (see Table 1).

Papermill waste, largely "black liquor" and pump suspensions, like mine and mill tailings are commonly discharged directly by coastal outfalls. Other industrial wastes, however, are commonly packaged in steel drums and carried to offshore dump sites by ship or barge. Such packaged waste may include acids, refinery waste, insecticide and herbicide residues and byproducts, and a broad spectrum of solid and semi-solid materials. Packages of waste are usually punctured on the surface at the disposal site, by axe blows or small arms fire, to ensure sinking; unpackaged wastes are simply dumped over the sides. In either case, the waste is brought into immediate contact with the biosphere. Although some chemicals will be neutralized by reaction with seawater or the bottom sediment, a few, particulary PCB's and many herbicide and insecticide residues (dioxin, DDT, kepone, etc.) may persist for long periods and accumulate in amounts

TABLE 1

Changing patterns of ocean dumping of waste in the United States, 1968-1980.
(Thousands of tons.)

Waste Type	1968[1]	1973[2]	1975[2]	1977[2]	1978[2]	1979[2]	1980[2]
Industrial waste	4,690.5	5,050.8	3,441.9	1,843.8	2,548.2	2,577.0	2,928.0
Sewage sludge	4,477.0	4,898.9	5,039.6	5,134.0	5,535.0	5.932.0	7,309.0
Construction/ Demolition debris	574.0	973.7	395.9	379.0	0	107.0	89.0
Solid (domestic) waste	26.0	0.2	0	<0.1	0	0	0
Explosives	15.2	0	0	tr[4]	0	0	0
Incinerated wood	—[3]	10.8	6.2	15.1	18.0	36.0	10.5
Incinerated chemicals	—	0	4.1	29.7	0	0	0
TOTAL[5]	9,782.7	10,934.4	8,887.7	7,401.2	8,101.2	8,652.0	10,336.5

Notes & Sources: 1 U.S. Army Corps of Engineers, *in* CEQ, 1970 (ref. 9)
2 EPA Annual Reports
3 Dash indicates that data were not reported.
4 Actual quantity dumped was approximately 850 lbs.
5 Excluding dredge spoil.

which are potentially harmful to humans or to marine organisms.[7] Such concentrations have been identified as causes of or contributors to diseases, tumors and fatalities in a wide range of higher marine organisms.[19]

Domestic waste. This category includes all forms of domestic solid waste (refuse, garbage, or trash) generated by residences, commercial, agricultural or industrial establishments, hospitals and other institutions, and municipal operations. It is made up chiefly of paper, food wastes, garden wastes, metal and glass containers and other miscellaneous materials. As used here, the term corresponds to the general category of "solid waste" in Federal statistical studies.[9]

Domestic waste is generally innocuous, and its disposal at sea poses little threat to the environment except insofar as the bulk of the material is buoyant and if not compacted and baled, it will float and be dispersed by currents, eventually to wash ashore as flotsam. Incineration or conventional landfill operations afford more economical means of disposing of domestic waste in most situations, so such waste rarely makes up as much as 1% of the waste dumped at sea. Experiments have shown, however, that when baled domestic waste can provide slow-release nutrients and a solid substrate for a variety of bottom-dwelling marine organisms.[20]

Construction and demolition debris. This category includes all forms of debris and rubble produced in the course of demolition and construction—masonry, tile, stone, plastics, wiring, piping, shingles, glass, cinderblock, tar, tarpaper, plaster, vegetation, metal, timber and excavation dirt.[9] It is probably the most benign of all waste dumped at sea, and is likely to be least restricted by government regulation. The greatest threat posed by ocean dumping of such materials is that of physical burial of bottom-dwelling organisms or alteration of the substrate character (edaphic environment). These effects are so limited in extent as to be regarded as trivial except in certain nearshore (fishing or shellfishing) areas.[17]

Explosives and munitions. Obsolete or superannuated explosive materials, ammunition, rocket fuel, small arms ammunition, bombs and other explosives, as well as certain chemical munitions, included in this category have been disposed of by ocean dumping. The average quantities involved are rather small (less than 5,000 tons a year) and the materials are regarded as too sensitive to dispose of in any other fashion. Small quantities of munitions may be dumped in designated dumping areas, while larger quantities are generally packed in overage hulks and towed to the disposal site where the hulk is scuttled or the explosives deliberately detonated.[9] The principal adverse effect of such disposal of explosives in the sea is the shock trauma to fish and other marine organisms; it has been estimated that the detonation of 1,000 tons of explosive in the seas will be fatal to most organisms within one mile of the blast and will cause death among fish with swim bladder at a distance of up to four miles.[9] Less than one-half ton of explosive has been disposed of at sea by the United States since 1973 according to published reports by the Environmental Protection Agency (Table 1).

Radioactive waste. Though included in one or more of the foregoing categories, radioactive wastes pose such special problems of disposl as to justify separate discussion. The category includes all solid and liquid wastes that result from the processing of radioactive ores or materials, fuel elements, nuclear reactor operations, medical and research use of radioactive isotopes, and from equipment and containment vessels made radioactive by induction.[9]

Prior to 1970 radioactive wastes were dumped at sea by the United States though on a rather limited scale.[9] European nations have continued to dispose of radioactive materials at sea, and though the tonnages have declined since 1967, the total radioactivity of the dumped materials has increased.[21] Some radioactive material was formerly discharged directly into the Irish Sea through the outfall of the British 'Windscale' fuel reprocessing plant, but most of the European waste has been packaged in 55-gallon mild steel drums and dumped onto the seafloor at designated locations.[21] By either method, the radioactive isotopes are, in effect introduced directly into the biosphere since the isotopes in the waste have half-lives on the order of thousands of years and the containers used are unlikely to resist corrosion on the seafloor for more than a few score years.[22]

The United States stopped all disposal of radioactive material at sea after 1970, and under the terms of the London Ocean Dumping Convention ratified by the Senate in 1974, the disposal of all high-level radioactive material at sea is prohibited (Annex I) and disposal of intermediate- and low-level radioactive waste in the ocean is regulated by permit.[23]

Dredge spoil. Dredge spoil comprises all solid materials removed from the water bodies generally for the purpose of improving navigation, including sand, silt, clay, rock and pollutants that have been deposited from municipal and industrial

TABLE 2

Comparison of dredge spoil and all other forms of waste dumped at sea, 1968-1978.
(All figures in thousands of tons.)

	1968	1973/74	1977	1978
Dredge spoil:	54,900	144,100	59,100	75,600
All other:	9,783	10,934	7,401	8,101
TOTAL	64,683	155,034	66,501	83,701
% Spoils:	84.7%	92.9%	88.9%	90.3%

Sources: Dredge spoil; EPA, 1980 (ref. #2)
Other; EPA Annual Reports

discharges.[9] The quantities of dredge spoil dumped at sea range widely from year to year, but are consistently three to five times as great as the combined tonnages of all other waste forms disposed of in the ocean combined (see Table 2). Though at first sight the dumping of dredge spoil in the ocean would appear relatively innocuous—sediment dredged from the sea floor near shore is simply being returned to the sea floor offshore—its dumping causes a range of problems which reflect the complex and often subtle interactions of the various components of the marine ecosystem.

Approximately one-third of all dredge spoil is considered 'polluted' by reason of its biological or chemical oxygen demand, bacteriological activity (chlorine demand), or contained concentrations of nitrogen, phosphorous, heavy metals,

TABLE 3

Media Jurisdiction of Key Environmental Laws of the 1970's.

Statue	Medium Regulated	Geographic Constraints
Clean Air Act	Air	None
Marine Protection, Research and Sanctuaries Act	Marine waters	Applies only seaward of the baseline of the territorial sea.
Safe Drinking Water Act	Underground water	None
Federal Water Pollution Control Act	Internal waters	"Navigable" waters and territorial sea*
Resource Conservation and Recovery Act	Land	None

* *Under the Federal Water Pollution Control Act, "navigable" is broadly defined and includes waters innundated by the tide, and waters that presently are, historically were, or could, in fact, be made navigable. The Act also regulates pipeline 'outfalls' regardless of the distance of the mouth of the pipeline from the shore.*

(from N.A.C.O.A., 1981, page 11) (ref. #41)

volatile solids or oil and grease.[9] Dredge spoil also comprises a range of mineral species and particle sizes. When dumped at sea, the coarser particles settle rapidly to the bottom, forming a deposit of sand or silt, often in a location where the natural deposits are much finer grained. The accumulating sediment may bury some bottom-dwelling organisms, and may so alter the bottom conditions as to force the survivors to emigrate. Such changes can substantially alter the bottom cummunity and the ecosystem in the overlying water column. The accumulating sediment also alters the bottom topography and may thereby affect the local wave and current regimen. Off New York, for example, dumping of dredge spoil at one site built a mound more than ten meters (33 feet) high during the period from 1936 to 1973; the mound was observed to rise an additional three meters (10 feet) between 1973 and 1978.[24] In extreme cases, such sediment build-ups may become hazards to navigation or even break the surface as islands.

The finer particles in the dredge spoil include clay materials on which a range of toxic metals or organic chemicals may be absorbed.[25,26] These particles may require hours or days to settle to the sea floor and may be distributed over a very large area as a down-current 'plume' of suspended sediment.[24] The sediment may smother filter-feeding organisms (clams or oysters in particular), or it may reduce the light level below the minimum necessary to support photosynthesis. There is also evidence that the suspended sediment may be fatal to some swimming organisms (hekton) as well as to bottom-dwelling forms.[27] The basemetals and toxic chemicals adsorbed on the suspended particles may have additional adverse effects, though there is evidence that some micro-organisms develop genetic resistance to such material in the environment.[14] Many of the larger marine organisms absorb these substances and the level of such toxins in their tissue may reach levels considered hazardous to human health.[14,17,22,24,28,29,30,31,32]

GOVERNMENT REGULATIONS

Fish kills at sea, and recurrent accumulations of 'foreign' matter on beaches, particularly those adjacent to New York Bight in the late 1960's, focused public attention on the potential hazards of uncontrolled ocean dumping. Federal acknowledgement of a growing environmental problem in the form of the National Environmental Protection Act (1970) was followed by the report of the Council on Environmental Quality entitled, *"Ocean Dumping:A* National Policy."[9] This report is an outstanding summary of the problem as it was then perceived. The report notes that,

"Ocean-dumped wastes are highly concentrated and contain materials that have a number of adverse effects. Many are toxic to human and marine life, deplete oxygen necessary to maintain the marine ecosystem, reduce populations of fish and other economic resources, and damage esthetic values. . . communities are looking to the ocean as a dumping ground for their wastes . . . (and) . . . faced with higher water quality standards, industries may also look to the ocean for disposal. The result could be a massive increase in the already growing level of ocean dumping. *Current regulatory activities and authorities are not adequate to handle the problem . . ."*
(9,page v. Emphasis added.)

The report proposes that the Environmental Protection Agency be given authority and responsibility to regulate ocean dumping by a system of permits, in the issuance of which the EPA Administrator should be guided by two overriding principles,

"Ocean dumping of materials clearly identified as harmful to the marine environment or man should be stopped."

"When existing information on the effects of ocean dumping are inconclusive, yet the best indicators are that such materials could create adverse conditions if dumped, such dumping should be phased out. When further information *conclusively proves that such dumping does not damage the environment, including cumulative and long-term damage,* ocean dumping could be conducted under regulation."

(9, page vi. Emphasis added.)

The report identifies areas where research is badly needed to evalute the effects of ocean dumping, then concludes that,

"Unilateral action by the United States can deal with only part—though an important part—of the problem. Effective international action will be necessary if damage to the marine environment from ocean dumping is to be averted."

(9, page v.)

Congressional response was swift. Legislation to control ocean dumping was introduced in February, 1971, and in early 1972 the Congress approved and the President signed the Marine Protection, Research and Sanctuaries Act (MPRSA, P. L. 92-532). Together with the Federal Water Pollution Control Act (FWPCA, especially section 404) this act established federal regulatory authority over waste disposal in coastal waters, the territorial sea and beyond. The MPRSA also provided inspiration and guidance for the United Nations Environmental Conference's International Working Group on Marine Pollution which agreed upon a convention regulating international use of the sea for waste disposal in London later that year. The London "Ocean Dumping Convention" of 1972 was based largely upon the MPRSA, and was ratified by the Senate after amending the MPRSA to fully conform to it. The recommendations of the National Environmental Policy Act and the Council on Environmental Quality were thus largely enacted and implemented on an international and national level in just four years. The despatch with which goal was achieved must set some sort of international record, and affords some measure of the level of concern about ocean pollution around the world.

After enactment of the MPRSA, ocean dumping of waste in the United States shows a sharp decline (Table 1). Dumping of radioactive waste had ceased in 1970, and from 1973 to 1977 the quantities of all classes of waste dropped substantially. Incineration of wood at sea (largely waterfront demolition debris) continued at a relatively constant average rate. The experimental incineration of chemical waste at sea (mainly hazardous chemicals and the defoliant, "Agent Orange") is responsible for the sharp increase of incineration in the period 1975-1977.[33,34] During the last years of the decade, however, the trend toward reduced ocean dumping is reversed. Total tonnages of waste dumped at sea rose by nearly 40% from 1977 to 1980 (the latest year for which official EPA statistics are available). Though a small increase in the quantity of industrial waste dumped at sea contributed to this change, the primary factor was increased dumping of sewage sludge, which reached an all-time high of 7,309,000 tons in 1980.[35] Virtually all of this sludge was generated by three municipalities; Philadelphia, Pennsylvania, Camden, New Jersey, and the New York City-New Jersey metropolitan area. The increase in sludge dumping is surprising in the light of the EPA's intention to phase out all disposal of sewage sludge at sea by December 31, 1981 a goal which was given statutory authority by the (1977) "Hughes Amendment" to the MPRSA (P. L. 95-153).

Although the EPA Administrator adopted the strictest possible interpretation of the Hughes Amendment, New York City, anticipating that it would be unable to comply with its provision, challenged the Administrator's interpretation in court (The City of New York *vs.* the EPA, No. 80, Civ. 1677, S.D.N.Y.). Judge Sofaer's decision in the case (see 543 F. Supp. 1084) emphasized that the law permitted sludge dumping to continue indefinitely provided that it caused no "unreasonable degradation" of the marine environment and provided it could be shown that there was reason to believe that the available alternatives might cause greater environmental damage or hazard. This decision is a landmark and represents a formal turning point in the regulatory attitude toward ocean dumping. Freed of statuatory restraints, New York slowed efforts to find economical alternatives to ocean dumping of sludge. It is not surprising to find that the quanitity of sludge alone dumped into the Atlantic Ocean in 1980 was nearly as great as the total quantity of waste of all classes dumped into the ocean just three years earlier.[35] The practice of dumping sewage sludge at sea is clearly likely to continue for many years, though the quantity may not increase dramatically over the present level as alternative methods of disposal(landspreading, composting, etc.) come into wider use.

Except in the case of sewage sludge, however, the anti-pollution legislation of the early 1970's appears to have been effective in controlling ocean dumping of waste. Disposal of solid waste (domestic rubbish) at sea has always been a trivial contributor to the total, and has been essentially nil since 1975. Similarly, the disposal of explosive materials and conventional munitions has been negligible since the mid-1970's and is likely to remain so. The disposal of chemical and biological weapons at sea is specifically prohibited by Annex I of the London Convention. We may anticipate a small rise in the practice of incinerating certain forms of waste (notable hazardous chemical wastes) at sea; the increase may be substantial if at-sea incineration proves to be the most satisfactory method of disposing of toxic or carcinogenic waste such as dioxin. The most benign waste form, construction and demolition debris, has been dumped at sea in diminishing quantities since 1973, and there seems to be no reason to anticipate a long-term reversal of this trend. All waste dumping in the Gulf of Mexico ceased in 1978, and there has been no significant dumping in the Pacific Ocean for several years.[2]

FUTURE PROSPECTS

In 1980 the Environmental Protection Agency published three formal goals for its plan to curtail ocean dumping of waste:[2]
• Elimination of all ocean dumping of sewage sludge by December 31, 1981.
• Strict regulation and limitation of industrial waste dumping by 1985.
• Elimination of *all* ocean dumping of waste by 1990.
The first goal was announced by the EPA as early as 1976 and was confirmed by the Hughes amendment in 1977. Unless Judge Sofaer's decision is challenged and reversed, it will postpone indefinitely the accomplishment of this objective, however.

The second objective was to limit the dumping of industrial waste to the "most benign" forms of waste, and would impose strict regulation by individual permits for each dump. This objective may be attainable, but the identification of appropriately "benign" waste forms will almost certainly prove somewhat controversial.

The third objective, the elimination of all ocean dumping by 1990 would not include either dredge spoil which is regulated by the Federal Water Pollution Control Act (Section 404) or direct discharge by pipeline 'outfall', both of which are under the jurisdiction of the Corps of Engineers, though monitored by the EPA.

Despite notable successes already achieved, it is evident that it will be difficult (if not impossible) to fully attain all these objectives, and there is a growing belief that ocean dumping may prove to be the most environmentally sound method of disposing of certain forms of waste. This belief reflects a gradual change in the philosophy and attitude of government regulatory and legislative agencies, and of the scientific community toward ocean dumping which began to appear in the middle-to late-1970's.[51] This change in attitude was stimulated by two factors, the first was a growing body of research data that indicated the environmental hazards of ocean dumping, of at least some classes of waste, may have been overstated.

priority" to research, development and monitoring of (i) outfalls of municipal sewage, (ii) ocean disposal of industrial waste, (iii) ocean disposal of radioactive waste, and (iv) ocean disposal of dredged material. Research on sewage dumping. The first two needs were satisfied (promptly) by the MPRSA and the London Dumping Convention. It was not until 1978, however, that Congress passed the National Ocean Pollution Research and Development and Monitoring Act (P.L. 95-273) to establish a plan for a systematic program of ocean pollution research and to coordinate the ongoing research effort sponsored by the National Science Foundation, Office of Naval Research, National Oceanographic and Atmospheric Administration, Geological Survey and other agencies. The five-year plan developed by the Interagency Committee on Ocean Pollution Research (ICOPRDM) under the terms of this Act assigned "high priority" to research, development and monitoring of (i) outfalls of municipal sewage, (ii) ocean disposal of industrial waste, (iii) ocean disposal of radioactive waste, and (iv) ocean disposal of dredged material. Research on sewage sludge dumping was assigned a "low priority" because it as understood at that time (1979) that ". . . all ocean dumping of sewage sludge (will) be phased out by 1981 . . . ."[36] A higher priority would almost certainly be assigned to research in this area today, in the light of the New York City case, cited earlier.

A vast body of research conducted under the auspices of this Act together with research programs undertaken by several Federal agencies and private institutions has now provided much greater understanding of the problems associated with ocean dumping than was possible in 1970. Much of this research confirmed what was already known (or suspected) about the characteristics of waste materials and their effects when dumped in the sea, but some of the results were unexpected and some confusing or contradictory. While some studies indicated that toxic materials associated with dumped wastes had little or no adverse effect on either nekton or plankton at the dump site[37,38] other investigations developed convincing evidence of substantial effects on micro-organisms, invertebrates and vertebrates.[14,24,28,30,39] In many instances, however, the research results were ambiguous, and though concentrations of some pollutants were found to be unusually high in the tissues of some organisms inhabiting the area of the dump sites, it was not evident that the pollutants were directly related to the dumped materials, that they were harmful to the organisms or that they were potentially harmful to man.[14,28,40] The specific nature of the waste, the location of the disposal site, the organisms present and even the method of dumping all were found to affect the environmental impact of the ocean dumping.[17] The environmental extremism which held that all dumping was necessarily bad was no longer tenable. The policy which in 1970 permitted ocean dumping only when, " . . . information *conclusively proves* that such dumping does not damage the environment . . ?"[9] began to yield to a more moderate, and perhaps more realistic view that ocean dumping might be permitted under suitable regulation when it could be shown to cause no "unreasonable degradation" of the marine environment and to cause less environmental damage than the available alternative(s).[51]

The second factor which contributed to this change of attitude was the growing realization that the five major environmental protection laws passed by the Congress in the early 1970's were often conflicting, sometimes contradictory and occasionally self-defeating. The acts regulate waste disposal in every possible environment (see Table 3); unfortunately, it is not always possible to anticipate the effects of one act in the light of the provisions of the other four, and it is through such interactions that conflicts have evolved. Sewage sludge, for example, could not be dumped at sea after 1981 under the provisions of the amended MPRSA. Its incineration was restricted by the Clean Air Act, and its disposal by composting or lansdspreading was controlled by the Resource Conservation and Recovery Act. Deep-well injection was prohibited by the Safe Drinking Water Act and dumping sludge in or adjacent to "navigable" waters or tidewater was prohibited by the Federal Water Pollution Control Act. So it was that many communities or industries found themselves in a "Catch 22" situation, enmeshed in a tangle of regulations which mandated safe and prompt disposal of waste but excluded all available repositories. An excellent analysis of this emerging dilemma is presented by the National Advisory Committee on Oceans and Atmosphere (NACOA) in the special report, "*The Role of the Ocean in a Waste Management Strategy.*"[41] It would not be useful, here, to repeat that analysis, but it may be helpful to examine some aspects of the waste disposal problem of the coming decades in the light of the special benefits and hazards which ocean dumping affords.

If the 1970's were a decade of environmental awareness and concern, then perhaps the 1980's will prove to be a decade of waste management. During the past ten years, significant progress was made in efforts to control and reduce environmental pollution. It became quite evident, however, that some major environmental problems had resulted directly from improper or illconsidered waste disposal practices of the past. The problems at Love Canal and West Valley, New York, and those at Stringfellow Quarry, California, are familiar examples. In particular, at each of these sites problems resulted from or were the result of the use of conventional disposal techniques in handling unconventional materials. The dangers of radioactive waste, of stable organohalide compounds, 'hard' pesticides, PCB's and similar wastes are not reduced by simple soil filtration, aerobic or anaerobic decay, biodegradation or other processes which have been found sufficient to stabilize or detoxify conventional wastes. Today we face not only a rapidly growing burden of total waste, but a disproportionate increase in the fraction of radioactive and hazardous chemical wastes in that total. While the *per capita* production of domestic waste is expected to increase (over 1975 levels) by 15-25% by 1990, the generation of organic chemical waste (including organochlorine waste) is expected to grow by more than 60% in the same interval.[3] The quantity of radioactive waste is also expected to grow very rapidly as shortages of conventional fuels force greater reliance on nuclear power and as superannuation demands retirement of many existing nuclear power plants and disposal of their highly radioactive components. Ultimate disposition of the high-level radioactive wastes now in 'temporary' storage will substantially increase the waste disposal burden, and ERDA projections indicate that by the close of the century the annual production of high-level waste may be nearly four times as great as the total quantity (in curies) now stored at Hanford, Washington.[42] The problem is compounded by the fact that the decay of radioactive waste may demand tens of thousands or hundreds of thousands of years to reach the level of acceptable risk, and this process cannot be hastened. Many organic chemicals also required extemely long periods to break down under natural conditions. It is imperative, then, that we devise safe and effective means of handling, disposing of and confining these hazardous waste materials for very long periods of time. Under the circumstances, we must proceed with care, and examine every possible repository in the light of the special requirements of these emerging waste types. Like every other potential repository, the marine environment must be evaluated, and utilized in those circumstances and under those conditions where its special characteristics make its use appropriate and advantageous.

The inaccessibility of the deep sea floor, for example, suggests that it might provide a valuable disposal site for high-level radioactive waste.[43] In a region of rapid sediment accumulation, the waste would soon be buried,[44] and in an area of sea floor subduction, the waste and sediment would eventually be drawn into the earth's upper mantel, providing ultimate confinement and security. Unfortunately, the inaccessibility of the sea floor which would protect the waste against casual intrusion also makes it difficult to monitor environmental contamination, and makes corrective action, if needed, virtually impossible. No container now known would resist corrosive failure in the sea floor environment for the 100,000-300,000 years necessary for radioactive decay, and tectonic activity in subduction zones would probably ensure premature rupture of any container which might be used.[24,25] Emplacement in the rock of the oceanic crust offers a possible disposal method which has yet to be fully evaluated.[46]

The deep sea floor is characterized by low temperatures (32-35°F) and very slow rate of microbial activity.[47] These conditions would inhibit organic decay and argue against disposing of biodegradable organic or chemical waste on the sea floor. Nutrient-rich waste, however, might be disposed of in shallow, aerated environments where the nutrients, if free of toxic contaminants, could support or augment the marine food chain. This does not imply, however, that there are no deep ocean environments suitable for the disposal of stable organic wastes, organohalides or 'hard' pesticides. Such materials, whether in the form of high-density liquids or solids might be dumped in anaerobic, density-stratified ocean basins where the natural density gradient would prevent their escape by the extremely slow process of molecular diffusion.

Dredge spoil, which will continue to be a major waste form, should probably be dumped at sea in the future, though greater effort might be made to coordinate dredging and dumping operations to ensure burial of the most contaminated spoil beneath less polluted material. If dumped where there is no risk of generating a navigation hazard, spoil can be allowed to build up a hard-bottom shoal or reef habitat for marine organisms, and can even be used to construct artificial marshlands, islands, harbours or breakwaters as has been done successsfully at Toronto's Leslie Street Extension (Aquatic Park). Even waste heat from industrial cooling systems and nitrate- and phosphate-rich sewage effluents, normally considered pollutants, might be effectively utilized as biostimulators in artificial upwellings and in mariculture.

Finally, sewage sludge, the last major class of waste currently dumped into the ocean, will probably continue to be dumped at sea, though sites farther from shore than those presently in use are likely to be favored in the future.[17] In all probability, however, sewage sludge will soon prove to be too valuable as fertilizer, soil conditioner or energy source to allow its continued disposal at sea.[48,49]

As a departure from past needs and practices, the coming decade will probably see increased at-sea incineration of 'hard' organic chemical wastes and some other materials for which incineration offers the only practicable means of disposal.[33]

CONCLUSION

After centuries of casual dumping of waste in the ocean, episodes of serious marine pollution in the 1960's led to a period of strict government regulation of the practice designed to put an end to all such dumping by the final decade of this century. Research efforts were focused on identifying hazardous components of common wastes, tracing their paths through the marine ecosystem and describing their impact on the ocean environment. Contradictory regulations, enhanced understanding of the ocean and a rapidly growing burden of conventional and exotic waste materials have forced a reevaluation of ocean dumping policies.[16,50] Today increasingly the research emphasis is directed toward an understanding of the entire marine environment and the characterization of various parts of that environment in the light of human needs.

Above all, perhaps, we have come to view the ocean not as one, but as a vast number of diverse resources, not the least of which is the ocean's value as a repository for waste. The exploitation of one resource need not preclude our benefitting from others, and if wisely used, the ocean can continue to provide transportation, food, minerals, oxygen, and recreation, while relieving stresses on the terrestrial environment by accommodating substantial quantities of waste material. In the face of a world-wide problem of waste management, the marine repository represents a resource too valuable to ignore and too vital to abuse.

REFERENCES

1. *Statistical Abstract of the United States, 1982-83* (103rd Ed.) U.S. Department of Commerce, Bureau of the Census, Washington, D.C., 1982.
2. *Environmental Outlook, 1980.* U.S. Environmental Protection Agency, Washington, D.C. 1980. (EPA 600/8-80-003)
3. *Forecasts of the Quantity and Composition of Solid Waste. U.S. Environmental Protection Agency, Office of Research and Development, Washington, D.C. 1980. (EPA 600/5-80-001)*

4. *Ocean Services for the Nation.* National Advisory Committee on Oceans and Atmosphere, Washington, D.C. 1981.

5. *Environmental Assessment: At-Sea and Land-Based Incineration of Organochloride Wastes.* Environmental Protection Technical Series, U.S. Environmental Protection Agency, Washington, D.C. 1978. (EPA 2:600/2-78-087)

6. *Disposal in the Marine Environment, An Oceanographic Assessment.* National Academy of Sciences, Washington, D.C. 1976.

7. Young, D.R., D.J. McDermott, T.C. Heesen, and T.K. Jan (1975). Pollutant input and distributions off Southern California. *in* Church, T.M. (Ed), *Marine Chemistry in the Coastal Environment.* American Chemical Society Symposium Series, #18, American Chemical Society, Washington, D.C., 710 pp.

8. Perkins, E.J. *The Biology of Estuaries and Coastal Waters.* Academic Press, New York, 1974.

9. *Ocean Dumping: A National Policy.* Council on Environmental Quality, (Report), Washington, D.C. 1970.

10. Argo, D.G. & G.L. Culp (1972). Heavy metals in wastewater treatment processes. *Water and Sewageworks,* vol. 119, p. 62 ff.

11. Vernberg, F.J. & W.B. Vernberg (Eds.). *Pollution and Physiology of Marine Organisms.* Academic Press, New York, 1974.

12. Ketchum, B.H. (1970). Biological implications of global marine pollution. *in* Singer, S.F. (Ed.), *Global Effects of Environmental Pollution: A Symposium.* American Assoc. for the Advancement of Science, Dallas, 1968. Kluwer, Boston, 1970, 218 pp.

13. Summers, A.O. & S. Silver (1972). Mercury resistance in a plasmid-bearing strain of *Escherichia coli. Jour. Bacteriology,* vol. 122, pp. 1228-ff.

14. Koditschek, L.K. & P. Guyre (1974). Transmissible multiple drug resistance in Enterobacteriacae. *Science,* vol. 176, pp. 758-768.
Report to Congress on Ocean Dumping Monitoring and Research, January through December, 1978. National Oceanographic and Atmospheric Administration, U.S. Department of Commerce, Washington, D.C. 1980.

15. *Environmental Impact Statement on the Ocean Dumping of Sewage Sludge in the New York Bight.* U.S. Environmental Protection Agency, Washington, D.C. 1978.
Fattal, B., R.J. Vasl, E. Katzenelson & H.I. Shuval (1983). Survival of bacterial indicator organisms and enteric viruses in the Mediterrranean coastal waters off Tel Aviv. *Water Research,* vol. 17, pp. 397-402.
But see also:
Lear, D.W. & G.G. Pesch (Eds.) *Effects of Ocean Disposal Activities on Midcontinental Shelf Environment off Delaware and Maryland.* U.S. Environmental Protection Agency, Washington, D.C., 1975.

16. *The Role of the Ocean in a Waste Management Strategy.* Special Report

to the President and Congress. (U.S.) National Advisory Committee on Oceans and Atmosphere. Washington, D.C. 1981.

17. *Environmental Impact Statement on the Dumping of Sewage Sludge in the New York Bight.* U.S. Environmental Protection Agency, Washington, D.C., 1978.

18. *ibid.*

19. Mahoney, J.B., F.H. Midlige & D.G. Deuel (1973). Fin-rot disease of marine and euryhaline fishes in New York Bight. *Trans. Am. Fish. Soc.,* vol. 102, pp. 596-605.
    Report to Congress on Ocean Dumping, Monitoring and Research, January through December, 1979. (U.S.) National Oceanographic and Atmospheric Administration, U.S. Department of Commerce, Washington, D.C., 1981.

20. Loder, T.C. (1975). Effects of baled solid waste disposal in the marine environment—a descriptive model. *in* Church, T.M. (Ed.) *Marine Chemistry in the Coastal Environment,* Amer. Chem. Soc. Symposium Series, #18, Amer. Chem. Soc., Washington, D.C., 710 pp.

21. Deese, D.A. (1977). Seabed emplacement and political reality. *Oceanus,* vol. 20, pp. 47-63.

22. *Annual Report to the Congress on Administration of the Marine Protection, Research and Sanctuaries Act of 1972 as Amended (P.L. 92-532) and Implementing the International Ocean Dumping Convention, January through December, 1978.* U.S. Environmental Protection Agency, Washington, D.C., 1979.

23. Lettow, C.F. (1974). The control of marine pollution. *in* Dolgin, E.L. and T.G.P. Guilbert (Eds.), *Federal Environmental Law,* West Publishing Company, St. Paul, Minn., pp. 662-667.

24. *Report to Congress on Ocean Dumping, Monitoring and Research, January through December, 1978.* (U.S.) National Oceanographic and Atmospheric Administration, U.S. Department of Commerce, Washington, D.C., 1979.

25. Koppelman, M.H. & J.G. Dillard (1975). An ESCA Study of sorbed metal ions on clay mineral. *in* Church, T.M. (Ed.), *Marine Chemistry in the Coastal Environment.* American Chemical Society Symposium Series, #18, Amer. Chem. Soc., Washington, D.C., 710 pp.

26. Glynis, M., Nau-Ritter & C.F. Wurster (1983). Sorption of polychlorinated biphenyls (PCB) to clay particulates and effects on desorption on phytoplankton. *Water Research,* vol. 4, pp. 383-387.

27. Alden, R.W., III, & R.J. Young, Jr. (1982). Open ocean disposal of material dredged from a highly industrialized estuary: an evaluation of potential lethal effects. *Archs. Environ. Contamin. Toxicology,* vol. ll, pp. 567-576.

28. Martin, J.H. (1979). *Bioaccumulation of heavy metals by littoral and pelagic marine organisms.* U.S. Environmental Protection Agency, Environmental Research Laboratory, Washington, D.C. (EPA 600/3-79-038.

29. *Annual Report to the Congress on Administration of the Marine Protection Research and Sanctuaries Act of 1972 as Amended (P.L. 92-532) and Implementing the International Ocean Dumping Convention, January through December, 1979.* U.S. Environmental Protection Agency, Washington, D.C., 1980.

30. Anderlini, V.C., J.W. Chapman, D.C. Girvin, S.J. McCormick, A.S. Newton & R.W. Risebrough (1975). Heavy metal uptake study. (Appendix H) *in* U.S. Army Corps of Engineers *Dredge Disposal Study, San Francisco Bay and Estuary,* U.S. Army Corps of Engineers, San Francisco District, San Francisco, CA., 1975.

31. *NOAA-MESA Ocean Dumping in the New York Bight. (U.S.) National oceanographic and Atmospheric Administration, U.S. Department of Commerce, Washington, D.C. 1975. (NOAA Technical Report, ERL 321-MESA-2).*

32. *Federal Plan for Ocean Pollution Research, Development, and Monitoring, Fiscal Years 1979-83.* Interagency Committee on Ocean Pollution Research, Development, and Monitoring, Federal Coordinating Council for Science, Engineering and Technology, Washington, D.C., 1979.

33. *Report of the Interagency Ad Hoc Work Group for the Chemical Waste Incinerator Ship Program.* U.S. Environmental Protection Agency, Washington, D.C., 1980.

34. *Environmental Assessment: At-Sea and Land-Based Incineration of Organochloride Wastes.* (U.S.) Environmental Protection Agency, Technical Series, U.S. Environmental Protection Agency, Washington, D.C., 1978.

35. Annual Report to the Congress; Administration of the Marine Protection, Resarch and Sanctuaries Act of 1972 as Amended (P.L. 92-532) and Implementation of the International Ocean Dumping Convention. January-December, 1980. U.S. Environmental Protection Agency, Office of Water and Waste Management, Washington, D.C., 1979.

36. *Federal Plan for Ocean Pollution Research, Development, and Monitoring, Fiscal Years 1979-83.* Interagency Committee on Ocean Pollution Research, Development, and Monitoring, Federal Coordinating Council for Science, Engineering and Technology, Washington, D.C., 1979.

37. Pararas-Carayannis, G. (1973). *Ocean dumping in the New York Bight—an assessment of environmental studies.* U.S. Army Corps of Engineers, Coastal Engineering Research Center Technical Memo No. 39.

38. Lear, D.W. (1973). *Supplemental Report, Environmental Survey of Two Interim Dump Sites, Middle Atlantic Bight.* Project FETCH, NTIS Aquis. #PB-239-257.

39. same as entry #29.

40. Dunstan, W.M. (1975). Problems of measuring and predicting influence of effluents on maine phytoplankton. *Environ. Sci. and Technology.* vol. 9.

41. The Role of the Ocean in a Waste Management Strategy. Special Report to the President and Congress. (U.S.) National Advisory Committee on Oceans and Atmosphere. Washington, D.C., 1981.

42. *Alternatives for managing wastes from reactors and post-fission operations in the LWR fuel cycle.* (U.S.) Energy Research and Development Administration, U.S. Department of Commerce, Washington, D.C., 1976.

43. MacLeish, W.H. (Ed.) *Burying Faust* (Special issue dedicated to radioactive waste disposal), *Oceanus,* vol. 20, ppl-67, Winter, 1977.
    Kelly, J.E. & C.E. Shea (1982). The seabed disposal program for highlevel radioactive waste: public response. *Oceanus,* vol. 25, pp. 42-53.

44. Gross, M.G. (1975). Trends in solid waste disposal in U.S. coastal waters, 1968-1974. *in* Church, T.M. (Ed.), *Marine Chemistry in the Coastal Environment.* American Chemical Society Symposium Series #18, Amer. Chem. Soc., Washington, D.C., 710 pp.

45. *Disposal in the Marine Environment, An Oceanographic Assessment.* (U.S.) National Academy of Sciences, Washington, D.C., 1976.

46. Hollister, C.D. (1977). The seabed option. *Oceanus,* vol. 20, pp. 18-25.

47. Jannasch, H.W. (1978). Experiments in deep-sea microbiology. *Oceanus,* vol. 21, pp. 50-59.

48. Dick, R.I. (1975). Alternatives to marine disposal of sewage sludge. *in* Church, T.M. (Ed.) *Marine Chemistry in the Coastal Environment.* American Chemical Society Symposium Series #18, Amer. Chem. Soc., Washington, D.C., 710 pp.

49. Sittig, M. *Landfill Disposal of Hazardous Wastes and Sludges.* Noyes Data Corp., Park Ridge, N.J. 1979, 365 pp.

50. *Comparative Marine Policy: Perspectives from Europe, Scandanavia, Canada, and the United States.* Center for Ocean Management Studies, Univ. Rhode Island. Praeger Special Studies. J.F. Bergin Publishers, South Hadley, MA, 1981, 260 pp.

51. Ruckelshaus, W.D. (1983). Science, risk and public policy. *Science* volume 221, pp. 1026-1028.

Solid and Liquid Wastes: Management, Methods and Socioeconomic Considerations. Edited by
S. K. Majumdar and E. Willard Miller. © 1984, The Pennsylvania Academy of Science.

Chapter Twenty-One

Bio-Conversion—A Possible Disposal Option

R. P. Albertson

3315 Simmons Street
Oakland, California 94619

There are only three possible options for disposing of waste; in to our atmosphere, our waterways, or on our land. During the past twenty years, all three options have been severely limited by environmental concerns and spiraling cost increases. Water disposal is for all practical purposes no longer an option. Burnng, or atmospheric disposal, is proving a risky alternative because of the increasing mass of industrial and automotive pollution and the inevitable legislative response which can be expected to further tighten emission standards. Present day landfill engineering seems to enjoy almost universal acceptance as being environmentally safe. Nonetheless, growing urban density continues to push landfill sites further from the point of waste generation. Each mile added to the haul brings us closer to prohibitive cost levels.

In an atmosphere of declining options, it may be wise to reexamine some disposal options we have previously rejected. This article addresses one such option; bio-conversion. As used here, bio-conversion means composting, which is the controlled decomposition of organic matter. Bio-conversion does not include all methods of composting. It is a highly mechanized procedure employing sophisticated material handling equipment which produces the ability to closely control the conditions under which the decomposition process takes place. Most bio-conversion programs are conducted entirely inside large industrial structures. Many bio-conversion systems are designed to compost municipal solid waste in combination with raw sewage sludge. This is to be distinguished from windrow composting, which is customarily conducted out of doors with relatively unsophisticatd equipment and lacks the high-tech environmental control offered by bio-conversion systems.

Since your author is not a scientist this will not be a scientific article. The material is offered from the perspective of an attorney and business consultant with ten years of experience in consultation with both municipalities and bio-conversion enterprises. The focus of the article is the economic, political and sociological considerations which appear to have prevented bio-conversion technology from emerging as a viable waste disposal option in the United States. Because of the nature of the subject matter, some technical discussion is unavoidable. In this reagrd, the reader is cautioned that the author's conclusions consist of hearsay and readings from a variety of sources and his admittedly untrained observations. The reader is asked to independently judge the accuracy of these statements.

At the outset, it is the opinion of your author that bio-conversion is not a panacea for American cities. This is not due to the unavailability of proven technology. As is the case in most areas of technological development, there are systems which work extremely well, others which are average in performance and some which barely work at all. There are bio-conversion systems which have been operating successfully, consistent with design, for many years. Most of the successful systems are of European design. There is a facility operating at Madisonville, Kentucky which reports that it is "on line" and performing satisfactorily. It has always seemed strange that this technology, having enjoyed a fairly good record of success world wide, has failed to become a usable disposal option in the United States.

The reason this technology has remained unavailable to large numbers of communities is because of two problems it encounters each time it is considered. First, the existing waste management infrastructure in the United States is not particularly well suited to the needs of a bio-conversion program. Second, the infrastructure necessary to the successful marketing of composted municipal wastes is virtually non existent. Resolving these problems on a national level is a considerable task. Nonetheless, there are probably some communities whose particular circumstances and geographical location are such that a bio-conversion program may be feasible. The purpose of this article is to define the problems and to outline an approach to their resolution.

The undertaking is believed to be worth the effort because this disposal method offers some exceptionally valuable potential benefits, when properly managed: the technology works, it can be performed without insult to the environment and it should prove permanent and highly cost effective once established.

THE MISSING INFRASTRUCTURE

There is a story about Henry Ford that describes the infrastructure problem. No one knows if it's true, but it is a good story. Ford didn't get in to business immediately. The cars worked just fine, but the idea of everyone having one was something for which the country wasn't quite ready. The original Model T was designed to run on alcohol. A group of oilmen proposed that Ford switch to gasoline. Ford responded "why on earth would I want to do that. Anyone can make alcohol, but only oilmen can make gasoline." The oilmen said that was precisely why it was a good idea. They pointed out that the holdup in getting Ford's program moving (for all of its potential to accelerate American economic development) was the absence of highways, street signs, gas stations, repair garages, etc. They added that if Ford could see his way clear to converting to gasoline, they were confident that the government would tax gasoline at the source and use the funds to build the absent infrastructure. We all know what happened.

Bio-conversion equipment and processing are relatively expensive. To establish a successful program, it is essential to be able to sell the composed materials at prices which will serve to reduce the overall cost of disposal to acceptable levels. At this point in time, compost is generally considered to have insufficient dollar value to support an economically viable program. The reason for this situation is the absence of infrastructure. As used in this context, infrastructure consists of a generally held knowledge regarding the optimum utilization of compost in achieving agricultural or land reclamation goals. If the knowledge of how and when to use compost was equal to the knowledge presently held regarding the use of chemical fertilizers, we could say that the infrastructure existed. As will be discussed later, a careful examination of the world wide scientific literature on the subject supports the reasonable belief that compost (once a reliable, continuing supply source is established) will prove to have a considerably higher dollar value than is presently expected. The knowledge of scientists is not equivalent to the knoweldge of the target consumer group. Traditionally, scientific data has served only to induce the potential purchaser of agricultural products to try the material to see if it really works. Farmers and reclamation people believe only that which is growing out of the ground.

Customarily , new agricultural products undergo extensive testing as part of the marketing effort. The purpose of the testing program is to demonstrate the cost effectiveness of the product in achieving the agricultural goals of the intended purchasers. Composted municipal wastes, which is an agricultural product, have not been universally tested in the United States. There is precious little of the material available to test because there are so few bio-conversion programs in existence in this country. Catch-22.

THE OPPOSITE DIRECTION INFRASTRUCTURE

The existing waste management infrastructure is directed at the concentration of waste at specified locations. This makes perfectly good sense when the waste stream is perceived as a dangerous health hazard. Bio-conversion calls for indiscriminate land spreading. The underlying rationale of bio-conversion is that the organic portion of the waste stream is biologically converted from a potentially hazardous substance into a non-hazardous resource. At the moment the waste stream ceases to be hazardous, it is removed from the jurisdiction of the waste management fraternity. At the instant the material passes out of the jurisdiction of waste management regulations it enters the jurisdiction of agricultural regulation. The concerns of these two vertically oriented divisions of government are essentially unrelated. It is analogous to driving the wrong way on a one way street. Before compost can enter the marketplace as a generally usable soil amendment, the conflicting requirements of the various regulatory jurisdictions must be resolved. This task will be complex and require a cooperative attitude, especially when interstate shipment of compost is desired.

Coupled with this jurisdictional problem is the expectable fear that economic displacement will result from a change in the system. This is not intended to attribute an ulterior motive to those that might resist a bio-conversion program on jurisdictional grounds. Waste regulators are trained to believe that the material they supervise is fundamentally hazardous and from that fact it can be assumed that their beliefs are honestly held. Nonetheless, the fear of economic displacement presented by accelerating technological change is a fact of life for most Americans. This fact should not be ignored or fought; it should be acknowledged and managed in a sensitive and concerned fashion.

If one can get past the initial reactions, it becomes apparent that very little displacement is actually required. Collectors still need to collect. It is true that a bio-conversion facility could be located closer to the points of waste generation and this would serve to shorten and/or reduce the collection routes. It is also true that finished compost needs to be trucked to the countryside for use. Transfer trucks, with minimal changes, are ideally suited to perform this work and existing collectors are the ideal choice for the job. Waste management officials need to continue to supervise the waste stream until it ceases to be hazardous. Bio-conversion systems employing aerobic, thermophilic processes are able to manage the decomposition process so that intense heat (172 degrees F) develops in the entire mass of material. This heat must be held long enough to exceed the "death rates" of disease causing organisms. The task of insuring that this procedure is properly conducted would be the mission of waste management officials. The level of supervision required is probably equal to, or greater than, that required to supervise a landfill. People who perform services at landfills will find that many of the tasks to be performed in a bio-conversion facility are quite similar to the work they presently perform. There really isn't any need for an argument. There is a great need for adjustment and accommodation.

These are the principal considerations. If they are not addressed in an appropriate manner, the difficulties which will result may prove insurmountable. The members of the existing waste management infrastructure must be assisted in seeing themselves as an integrated and essential element of a bio-conversion program. The missing infrastructure must be put in place. The citizens of the community must see the project as a community asset and not as an experiment in which they are the guinea pigs. These tasks will most likely require considerable time, expense and sensitivity to accomplish. It is the opinion of the author that these goals can be achieved in a timely and cost effective fashion. The critical ingredient for success will be a realization on the part of the planners of the program that: their waste management problems are not of the slightest concern to anyone but themselves; and, innovative solutions seldom emerge from the use of traditional methods.

THE TECHNOLOGY-HARDWARE

This article is not a technology assessment. There are many different bio-conversion programs throughout the world. There are two or three in the United States. As mentioned before, some are considerably better than others and part of any successful program is a thorough technical review. The elements of a technical review will be discussed later.

Your author is deliberately avoiding a detailed discussion of the available technology. This is not to beg the issue of whether bio-conversion works. A thorough review of operating systems, world wide, appears periodically in Bio-Cycle Magazine published by J. G. Press of Emmaus, Pennsylvania. This would be an excellent starting place for anyone interested in becoming familiar with bio-conversion technology. The article will cover the critical elements of selecting a system which can meet the needs of a successful program.

THE TECHNOLOGY-SOFTWARE

There are technical issues which are appropriate for discussion here. These involve compost rather than composting. Each time the bio-conversion option is considered, two technical problems emerge which are seldom overcome. The first is that there is no marketplace of any significance for composted municipal wastes. The second is that heavy metals and/or toxic substances found in the waste stream will find their way into the food chain if composted municipal wastes are used in general agriculture.

The existence of a market for composted municipal wastes is probably the most significant issue to be resolved before bio-conversion can be a viable disposal option. The true test of any innovative engineering concept is its cost effectiveness. For bio-conversion to be cost effective, the compost must be sold at prices which are considerably higher than it is generally believed to be worth. The method which has been previously employed to determine the value of compost (and which produced the low estimate of market value) appears to be scientifically incorrect.

In almost every previous study of the bio-conversion option, the dollar value of compost has been determined by the weight of Nitrogen, Phosphorous and Potassium in each ton of compost, multiplied by the prevailing price for these chemicals. These three substances are the principal constituents of chemical fertilizers. There are several problems with this approach. The most significant problem is that composted municipal wastes seldom contain more than three percent of each of these chemicals, by weight. Chemical fertilizers customarily contain four to six times more of these chemicals than are found in compost. The result of this computation is an extremely low value for a ton of compost when compared to the value of a ton of chemical fertilizer. This has always been the key point which supported the conclusion that bio-conversion was not cost effective; the expected income from compost sales was considered insufficient to offset higher processing costs.

Compost is not a chemical fertilizer. In addition to Nitrogen, Phosphorus and Potassium, compost contains trace elements and an uncountable variety of living organisms. All of these substances interact amongst themselves and with the soil to increase soil fertility. Chemical fertilizers bypass the soil and provide direct food delivery to the plant. This is not to say that compost is good and chemical fertilizers are bad. In fact, the most successful tests have used a combination of these materials. The entire point is that they are different. Since compost is not a chemical, it is impossible to determine how it will perform in a given situation by simple chemical analysis. The dollar value of compost is measurable only by comparison to the cost of achieving the same agronomic or reclamation objective through alternative means.

The second task which should be undertaken in the development of a bio-conversion program is a serious review of the research which has been done around the world regarding the utilization of composed municipal wastes. The investigator should be satisfied themself that a composting facility can be properly configured and properly operated in a manner consistent with community goals. A critical review of the literature, coupled with the consultation of persons knowledgeable in the fields of agronomy and soil science, should lead to an understanding that compost is able to resolve agronomic and reclamation problems for which the presently available solutions are either unsatisfactory or extremely expensive. Failure to take this step will leave lingering doubt that a bio-conversion program can achieve economic viability. This doubt has often been fatal to the serious consideration of bio-conversion.

A review of the literature will produce an understanding that compost, or other organic matter, works with soil in a manner which produces conditions which no other substance can duplicate. A primary example is the manufacture of fertile topsoil. Topsoil has two principal elements; chemistry and structure. When soils are depleted of their organic matter, soil chemistry changes;

as a consequence soil structure is lost. One small example is helpful. As the organic constituent of a soil is replaced, food is provided for earthworms. The worms drill little tunnels in the soil as they move around. The tunnels allow the infiltration of the atmosphere into the soil. Living organisms in this "refertilized" soil are capable of "fixing" nitrogen from the atmosphere into the soil. Nitrogen is a fertilizer for plants. It is a self replenishing phenomenon which depends entirely on the establishment of an organics based soil chemistry and the subsequent development of soil structure. In certain reclamation problems, no amount of chemical fertilizers will produce a permanent, self sustaining vegetative cover. Organic matter is an indispensable element in these circumstances. In these cases (and many others) compost will have considerable value to those who believe that they must achieve reclamation. Today, there is a great deal of concern regarding the declining inventory of topsoil in our country. The ability to manufacture highly fertile topsoil may prove to be exciting news.

Another factor affecting the dollar value of compost is that the method and materials employed in its production will cause its quality to vary. making compost is analogous to making beer. All brewmasters use essentially the same ingredients, but because their procedures differ, they produce distinctly different products. Each bio-conversion system produces compost of differing quality notwithstanding their use of identical waste streams. Identical bio-conversion systems will produce compost of differing quality if the waste material is varied.

There are experts that can evelute the quality of compost. This is accomplished by an analysis of texture, temperature, particulant size, maturity, microbial population and chemical analysis. This expert is an indispensible member of the technical evaluation team which might be charged with the responsibility of evaluating bio-conversion systems.

The operators of each bio-conversion system take a different approach to marketing their technology. The more enlightened operators have realized that the principal challenge to marketing in the U.S. is the ability to market compost. In fact, some see themselves engaged in the compost manufacturing business and treat tipping fees as an ancillary profit center. These organizations are immediately recognizable by the considerable effort they expend in testing their product's ability to perform in resolving various agricultural and reclamation problems. They also expend considerable effort in seeking markets for the non compostable elements of the waste stream. Look for these signs.

There is second technical issue which any thorough discussion of bio-conversion must address. This is the serious problem of heavy metal and/or toxic chemical concentrations found in many municipal waste streams. Your author has no technical knowledge of the implications of this problem in relation to bio-conversion. However there is more than one position on the subject and it may prove helpful to attempt to balance the dialogue in this article. There is little question that compost containing high levels of these materials should not be used in soils intended for food production. This unfortunate situation has been perceived as an absolute barrier to the establishment of bio-conversion projects. Some knowledgeable persons express the belief that this absolute approach results from oversimplication.

There are many communities, essentially residential, where neither the solid waste stream nor the sewage system contain these dangerous substances. There are major cities where a portion of the community's waste stream does not contain these substances. The source of these substances is almost invariably commercial or industrial. As major urban areas collect solid waste, the tendency is to commingle residential with commercial and industrial materials. This practice renders the entire mass suspect.

It is quite true that even residential waste streams will contain some heavy metals and dangerous chemicals. However, when related to the entire mass of the waste stream, the percentages are normally well below levels considered dangerous. Most bio-conversion methods include a number of grinding, turning, mixing, blending and sifting procedures which tend to homogenize the constituent materials throughout the mass. If the introduction of hazardous materials is occasional and limited in quantity, the overall product is likely to be benign in this respect.

It is highly unlikely that bio-conversion is going to take the nation by storm. The development of this method will be evolutionary rather than revolutionary. It will most likely start in residential communities or in the residential portions of larger communities. Assuming that well structured programs can be successful, their success may serve as an incentive to larger communities to enforce the presently existing laws that make the uncontrolled disposal of hazardous materials a crime. Interestingly, the City of Los Angeles recently tried and convicted the president of a large, nationally recognized water purification company for dumping heavy metals in the sewer system. Instead of the customary nominal fine the defendant was fined $100,000, and sentenced to a jail term. There is no way to be certain, but the reason for this response may have something to do with the fact that the city's sludge is composed by a private concern and sold to the public. The city and county have avoided the need for expansion of their treatment facilities partially because of this relationship.

Heavy metals and toxic waste are serious problems, but not everywhere. Communities which are either free of the problems, or those that wish to address the resolution of the problem, will not find this to be an impediment to the establishment of a bio-conversion program. There are a great many questions which have not been answered exclusively because there have been few samples of the material available for testing. The early developing bio-conversion programs will serve to answer many of these questions.

A BIO-CONVERSION PROGRAM—DESIGN CRITERIA

The Resource Recovery and Conservation Act established that waste management had become a sufficiently serious national problem that it required a statewide perspective. Local communities seldom have the financial resources to secure the services necessary for a thorough evaluation of innovative concepts. Within each state, there are invariably a number of communities which would benefit from the development of more effective methods of addressing their disposal problems. An extremely useful combination would be a community government and its state government cooperating in the development of a bio-conversion facility. The combination would be even more powerful if federal involvement could also be secured.

If state and/or federal involvement is possible, it makes sense to structure the project in a fashion which permits its immediate application in other communities. Assuming there are no other bio-conversion facilities in the area, a joint project should address itself to the need to build the missing agronomic infrastructure and to adjust the existing waste management infrastructure to the point where additional facilities become more feasible.

If the initial facility will be a model, designed for replication in other communities, it is essential that the waste stream selected be representative of that found in most communities throughout the region. If the community selected has a significant concentration of a particular waste substance, this will defeat the universal application of agronomic test results. For instance, if there were a large shoe factory producing leather wastes, the composted material would be very high in nitrogen content. This would produce an extremely high quality compost but it would be a formulation which other facilities could not duplicate. The first facility should produce an average product, capable of replication elsewhre. Super composts are possible using specialized substances, but their development should be a secondary target to the development of a universally applicable system.

The design of the facility should be dictated by the real needs for such systems as they appear in the communities within the region which generate the greatest quantity of compostable wastes. This will often be the major cities. Since space tends to be a problem in dense urban settings, the technology selected should be compact even if not required in the community where the initial facility is established. The procedure should also be capable of being performed entirely inside an industrial type structure, without excessive noise, without odor and in an orderly and sanitary fashion. There are systems which meet this criteria. If universal application is the ultimate goal, these requirements should be imposed even though not necessarily required by the initial site selection.

The collection and landfilling of solid waste is often conducted by a community rather than a private sector enterprise. A technologically complex procedure such as bio-conversion does not lend itself to community operation. Most bio-conversion technologists will be willing to build and operate the facility, providing the contractual arrangement with the community offers the opportunity to make a profit. A private sector response to waste disposal will permit a smoother proliferation of facilities throughout the region because of the private company's ability to standardize its operations. Assuming the program is successful, a private sector group should be able to secure construction and operating capital from the investment community. This would relieve government of the financing burden.

The primary goals of the program are to establish a bio-conversion facility which:

1. Works consistent with design.
2. Is completely benign; environmentally.
3. Produces compost to be employed in testing programs which demonstrate its cost effectiveness in achieving agronomic or reclamation goals, at a price per ton which achieves waste disposal cost stabilization.
4. Stabilizes or reduces the cost of disposal.
5. Provides a permanent waste disposal solution.
6. Makes profits considered attractive by the investment fraternity.
7. Permits the citizens to share in the value of their waste stream as that value develops.
8. Can be replicated within the region at a rate nearly equal to the development of demand for compost at satisfactory prices.

. A BIO-CONVERSION PROGRAM—THE STEPS

STEP ONE—This may be the most important step in the entire program. The planners and the community at large must decide if they are prepared to do a considerable amount of study and work in pursuing the goal. If the traditional method of procuring services, the "Request for Proposals," is used to invite bio-conversion technologists to convince the community that the concept is sound, the effort is doomed to failure. The planners must be initiators rather than responders. The process of negotiating a workable agreement is arduous and seldom produces success if one party is not convinced the goal is worthy. If this attitude does not prevail, it would be wiser to select a disposal option with which the community is more comfortable. A second element of this step is to determine if the community is an appropriate candidate for bio-conversion. A checklist would include an evaluation of:

1. The potential for contamination of the waste stream with heavy metals and/or toxic materials.
2. Existing contractual relationships between the community and others bearing directly or indirectly in the program.

3.Charter limitations of the ability to contract for waste disposal services for extended periods. A bio-conversion facility will need twenty years of operation to justify its expense.

4.The true cost of current disposal methods and other methods presently under consideration.

5.The ability of the community to pass clear title to the waste stream.

6.An alternative method of disposal which could be used if a bio-conversion facility were to fail after it is established.

STEP TWO—The planners should undertake a literature study designed to satisfy themselves that there is reason to believe that bio-conversion technology is developed to the point where it offers a potential solution to the community's disposal problems. The study should also explore the literature which deals with the potential benefits to be derived from compost utilization. If the planners are not attracted to the idea once this step is completed, the effort should be abandoned.

STEP THREE—The planner should consult with agriculturalists and land reclamation experts familiar with prevailing conditions (in their area of interest) within a 400 mile radius of the community. The purpose of this step is to evaluate the market potential for materials which may prove to be cost effective solutions to the problems disclosed. The planners should complete this step believing that there would be a sufficient demand for composted wastes to support the facility if the efficacy of the material could be proven through the means of a testing program.

STEP FOUR—The planners should consult with community officials to determine if there are sites available for a bio-conversion facility which will avoid a negative citizen response to the program. In doing so it should be presumed that the facility will be received as a "garbage factory" until it is better understood. Consideration should also be given to the route which collection vehicles will travel to reach the site. If the available sites are not excellent, in the sense that no one will be disturbed as a result of the program, the planners should anticipate citizen litigation.

STEP FIVE—The planners should investigate existing bio-conversion programs for the purpose of satisfying themselves that the technology works. During this step it should be determined which systems can be configured in a fashion consistent with the design criteria for the program. The systems should be ranked on the basis of overall acceptablity. A sanitary engineer who enjoys the respect of the waste management fraternity at the state and local level should participate in this evaluation. Be prepared for foreign travel if the widest choice of systems is considered desirable.

STEP SIX—A critical evaluation for the compost produced by each of the systems considered satisfactory is necessary. This step will require the assistance of a microbiologist, a land reclamation soil scientist and an agriculturalist with experience in organic agriculture. Each of the persons selected should be known to, and enjoy the respect of, the state regulatory agencies supervising agriculture. The composts should be ranked by quality and the result correlated with the ranking of the bio-conversion system.

STEP SEVEN—Evaluate the operators of the systems considered satisfactory. People with business problems can be expected to create more business problems. It is not essential that the operators have a great deal of money. Having lots of money tends to be inconsistent with being a pioneering technologist. Nonetheless, the operators should display a responsible business attitude toward those resources they control. High marks should be given for those enterprises which demonstrate a preexisting sensitivity to the need for testing and marketing their product. The operator will be primarily responsible for arranging the testing programs which will be essential to building the missing infrastructure.

STEP EIGHT—The planners, should pause and reflect at this point and ask themselves seriously if they still feel that the project is a good idea. This is where the hard work starts.

STEP NINE—The members of the waste management fraternity and the regulatory agencies supervising agriculture must be polled to determine their attitudes toward the project. The presentation made to these groups should be sensitive to the concerns they can be expected to express as described earlier in this article. It would be wise to have the personnel which assisted in the completion of step two and three available at these presentations.

STEP TEN—The planners should pause once again and ask themselves if they emerged from step nine with the belief they were working in a cooperative environment. Enthusiasm is not essential, however, open hostility to the idea portends struggle with entities whose cooperation is essential to the success of the program.

STEP ELEVEN—Assuming the project has survived to this point, this would be an appropriate time to commence negotiations with bio-conversion operators.

STRUCTURING THE PROGRAM

There will always be some special local condition which will require a particularized response. It is impossible to anticipate those concerns here. There are several elements which are generally important to the legitimate needs of the community and the operator of the facility. Foremost is, the contractual arrangements should permit the agreement to work. Driving a hard bargain will serve no purpose here even though it makes sense elsewhere.

The goal of the community is to stabilize its disposal costs and produce a permanent and desirable program. The incentive to the community should be that a highly successful program will reduce its costs. In other words, at the worst, the community should pay at or near the cost of its other alternative, and at the best, a sale of the entire output of compost should reduce those costs.

The incentive to the operator is to operate the facility at a modest profit and to fully exploit the opportunity provided by the first facility to establish other facilities with the region.

The operator should be required to finance the facility without the assistance of the community. The community may wish to participate as required for a bond issue, but the essential element is that the operator is responsible for providing all funds necesary to bring the facility to operational status.

To have any chance of securing financing, the operator must have in hand a binding agreement which will produce sufficient income to operate properly, retire the debt, pay interest and produce a profit. This will be in the form of a contract with the community to provide disposal services and to be paid specified sums for the service. If there is any contingency (other than the satisfactory operation of the facility) which would reduce the operators income below the point where operations can continue, some form of loan guarantee will be necessary. There are a number of state governments and a number of federal programs that could be expected to provide such guarantees for a well conceived program to be executed by competent parties.

A structure which meets the foregoing considerations could be as follows:
1. The total cost of disposal by means other than bio-conversion would be established as the cost per ton of disposal. This fee should be subjected to escalation clauses relating to actual increases in the operator's costs, rather than economic indices.
2. The total cost to the community, on a daily basis, should be determined (price per ton multiplied by number of tons to be disposed). This total sum should be divided by three. The agreement would provide that the community will pay one third of the total as a tipping or disposal fee. It would also provide that the community would pay the other two thirds of the total as the purchase price of the compost produced. All of these fees should be expressed in terms of dollars per ton. An example will be helpful:

   Assume the community will generate 100 tons per day of solid waste and their expected cost by other methods is $50.00 per ton. This produces a total of $5000.00 per day. Dividing this sum by three produces a sum of $1667.00. This amount would be divided by the 100 tons and the disposal fee would be $16.67 per ton. Assume the facility will produce 75 tons per day of compost from the 100 tons of waste (this is the average shrinkage rate for most systems). The 75 tons would be divided into the $3333.00 remaining to be paid, which would produce a price per ton for compost of $44.44.
3. The community would become the consumer of last resort for compost produced by the facility. The operator would be required to sell the compost or use it in testing programs at a nominal cost per ton. Income re-

ceived from this activity, reduced by sales expense, would be employed to reduce the city's obligation.

The foregoing arrangement is designed to counterbalance the community interest in participating in the value of the compost and the operator's need to be assured of sufficient total income to operate the facility. Presumably the compost would ultimately be sold in its entirety, leaving the community with an extremely low disposal cost as their reward for program participation. It is true that the operator would be in a position to do nothing at all and the community would be required to pay for the material anyway. The community may be able to overcome their concern in this regard with the realization that the operator's opportunity for success, beyond the nominal profits offered by the initial facility, is in using the material for tests which will expand the demand for compost. This will lead to the establishment of other facilities.

It is not essential that the arrangement take the form suggested above. It is essential that the agreement be structured in a manner which provides the parties with significant incentives to perform acts which will benefit all of the parties to the agreement.

Some may view this arrangement as a governmental subsidy of private enterprise. In considering the validity of this concern it might be appropriate to point out that there are thousands of private sector enterprises making profits by supplying the needs of the existing waste management programs. Government pays every single penny of these costs. The distinction this program offers is that it would be the only waste management program which held the potential of getting someone else to share some of that expense. Getting socially desirable concepts moving, to the point where they are self sustaining, has always been a legitimate use of taxpayer funds. Solar energy tax credits are a recent example.

CONCLUSION

The human life cycle is perceived by many as starting at conception and concluding at death. This is only half the biological cycle. It might be called the "composition" phase, because it is during this period that elemental substances contained in the food we eat are converted by the human laboratory into the complex substances of which new life is composed. The purpose of "decomposition," which is the other half of the life cycle, is to break down the complex compounds created during composition and return them to their elemental state. Gravity insures that these elemental substances will be stored in the soil where they will be available for the composition of food to support new life. And so it goes.

By continuing to bury and/or burn organic matter, nutrients which were previously caused to be stored in the soil, are misdirected or destroyed. The soil is not self replenishing of these substances. This practice may be an interdiction of our own life cycle.

We often read of the growing mass of "farmed out" lands and frightening losses of precious topsoil. Could it be that the answer to this expanding crisis is as close to home as the garbage can?

Solid and Liquid Wastes: Management, Methods and Socioeconomic Considerations. Edited by
S. K. Majumdar and E. Willard Miller. © 1984, The Pennsylvania Academy of Science.

Chapter Twenty-Two

Toward a Reasonable Solution to Our Waste Utilization and Disposal Problems

John A. Bartone, Sr.

President, Organic Processing Systems, Inc.
409 East 26th Street
Erie, PA 16504

Organic Processing Systems (OPS), Incorporated represents an idea that many Americans share—the idea that our past affects our future. OPS further represents an awareness that our past habits have already jeopardized our ability to continue the comfortable way of life that we Americans have come to expect as one of our rights. Finally, OPS represents a plan, a proposal for action, that can turn our attention back to what we believe are the proper attitudes and values necessary to take us into a future of plenty.

The topic of this article is "Solid Waste Utilization and Disposal Problems," but we will focus on the utilization aspect of the question. OPS believes that part of the solid waste problem stems from the way in which we address the problem itself. We think of solid waste as something that "must be gotten rid of." We look at a pile of organic material, or at a pile of inert material, that has been used once and we call it "waste." Because we look on it as waste we, therefore, try to dump it, throw it away or bury it somewhere. And so we created dumps. And, when the dumps became too smelly and too rat-infested, we created landfills, thinking that we could dig enough holes in the land to bury the *problem* with the waste materials.

The idea of Organic Processing Systems began with a question: "What if we were to reverse our attitude?" What if we look at a pile of organic matter and see, not waste, but raw materials? It is not difficult; we do it all the time. Almost anyone today can look at a forest and tell you that they see the raw materials of paper-making. We expend a large amount of energy in cutting, transporting and *processing* before we see the product—paper. We look at the paper and see the raw material of books, magazines, newspapers, memos and even this article. We expend more energy in writing, typing and printing before we see the "final" product—a letter or a book. But why do we call a letter a "final" product? Does it take that much more vision to look at a letter and to see that by applying a little more energy we can produce yet another product? Apparently that vision *is* hard to come by, or else there would be no need for companies like OPS today.

The idea of OPS is not new but *is* reasonable and timely: recyclable organic waste is removed from the waste stream to create a new product which can then be returned to the earth in an ecologically sound fashion. The new product that OPS processing plants produce is an organically sound soil amendment that can be used in home gardens, on productive and non-productive lands, golf courses, park lands, land reclamation projects and more. Further, by removing organic waste from the waste stream, the bulk of materials going to landfills is reduced by nearly 75%, thus extending the life of current and future landfills by many years.

At the heart of OPS is the Organic Processing Plant (see fig. 1). Unlike the majority of recycling plants that have been tried to date, the Organic Plant is neither complicated nor expensive; it does not even contain a "black box" device that claims to magically transform trash into treasure. The OPS plant consists of multiple hoppers that feed organic materials, both solid and liquid, into a screw auger system that mixes them in organically safe combinations. In many cases the output from the OPS plant can be applied directly to the land without harmful effect. In other cases, the output is fed to on-site reclaimable storage pads (see fig. 2) that allow the soil amendments to compost naturally with neither danger nor offensive odor. Following a composting period the soil amendments can be spread directly on land or reused as inputs for the OPS plant, especially during seasons when one type of organic input is not available to the plant, allowing year-round consistent output.

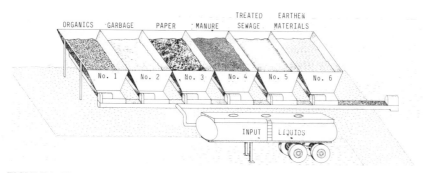

FIGURE 1. The OPS Exclusive Multiple Input Blending & Mixing Pre-Mix or Local Plant.

FIGURE 2. The OPS Exclusive Leaching Controlled Reclaimable Storage Pads, (RSP) Sites where "Mother Nature" Composts them without offensive odors or the need to windrow.

The basis of the Organic Processing System is the OPS plant, but the machines are not enough. Early in the development process, OPS realized that the solution to the problems of solid waste disposal and utilization require a consolidated approach. The problem is complicated by the fact that many individual entities have fragmented the waste question. Cities and municipalities, business and industry have all sought separate answers to the problem in order to meet their own individual needs. It needs to be said that each of these interest groups, working alone, will never solve the problem. We must admit to ourselves that the key to the solution lies in cooperation and effort. Just as there are no individual answers, there are no easy solutions.

Organic Processing Systems thus becomes a plan—a three-point program aimed at achieving the cooperation of all waste generators to aid in one *reasonable* solution. First, Organic Processing Systems proposes *Source Separation* for the resource recovery of our solid waste materials. While many producers of organic materials already practice some form of source separation—for example, grape and vegetable pomace are available in separated form from Northeastern Erie County's fruit and vegetable processing plants—other producers of waste will need to initiate programs to separate waste into various organic and inorganic categories. OPS suggests that the categories and examples found in Table 1 represent a plan that is both useful and achievable.

Notice that the majority of the Table 1 items can be found source separated in many areas today. Of the items, in fact, only household garbage is not currently source separated; and many will, no doubt, object that that is the weak link in the OPS plan. It should be pointed out that OPS is aware of the many problems facing the establishment of a plan that calls for the source separation of household garbage. First, one should remember that business and industrial waste accounts for a larger portion of the waste stream than does household refuse. Second, it is expected that OPS plants, once established, will be operated without household garbage as an input for a period of time with no detriment

TABLE 1

A List of some Solid Waste Materials Available for OPS Soil Amendments

OPS Pre-Mix & Local Recycling Centers will produce OPS Soil Amendments Class I to V from the following solid waste materials:

Organics:	Grass, shrubbery & tree trimmings, bark, leaves, sawdust, wood chips, mash, meal, produce and fruit, grape & vegetable pomace.
Garbage:	Source separated from glass, metals & non-biodegradable plastics.
Paper:	Non-recyclable shredded corrugated and paper, and paper pulp.
Manure:	Dry, semi-dry, and liquid with solids.
Waste Water:	Treated, dry, semi-dry, liquid with solids, and incinerated.
Inert:	Fly ash from coal or wood, and foundry sand.
Earthen:	Sand, silt, and soil.

TABLE 2

OPS Soil Amendments are Classified and Utilized as follows

Class I	Organics and earthen For home gardens
Class II	Organics—manure and earthen For productive lands
Class III	Organics—paper—inert and earthen For landscaping and nurseries
Class IV	Organics—paper-waste water—inert and earthen For the extension of top soil
Class V	Organics—garbage—pape—waste water—inert and earthen For local land reclamation projects

to the plant's output quality and no effect to the financial viability of the venture. OPS does, however, believe that once producers of waste see the valuable output materials, and once demand for those materials grows, it will become fiscally valuable to municipal governments to encourage citizens to source separate their household garbage materials.

The second part of the OPS plan calls for the production of solid amendments using source separated waste materials. The output from an OPS plant depends on the quantity of each type of input materials. Organic Processing Systems has worked closely with the Pennsylvania Department of Environmental Resources to develop five classes of output soil amendments as shown in Table 2. Each of the Table 2 materials has been produced using a prototype plant. Each category has been carefully tested for organic composition and each has been found to be environmentally sound. As noted above, it is expected that OPS production will consist mainly of Classes I through IV during the first year of operation due to the lack of availability of source separated household garbage materials. Table 2 also indicates the expected application for each of the classes of output soil amendments; other, more specific, uses will indeed come to mind as each region of the country researches its own peculiar needs and discovers creative applications for such soil amendments.

While on the subject of the output materials from an OPS plant it is important to clarify the other major difference between the OPS proposal and other recycling schemes of the past. The majority of recycling plants tested to date have produced materials aimed at easing our considerable energy problems in one way or another. In fact, a major facility of this type, costing millions of dollars to build and using large amounts of energy resources to produce burnable briquets, is under planning near OPS' home in Erie County, Pennsylvania. The largest portion of these attempts have been outright failures, others merely inefficient answers. They have failed in the majority of cases to find a market for their output product.

OPS soil amendments are different in two ways. First, our research has shown that, due to the depletion of soil nutrients of many types and due to the consistent application of chemical fertilizers, there is a recognized need among land users for soil amendments, especially for soil amendments that will be very reasonably priced due to the low energy consumption of OPS plants. Second, OPS soil amendments are a more ecologically sound answer to the reuse of waste materials. While there is certainly a social value in the burning of waste matter to solve a portion of our growing energy needs, there is a greater value to society and to our offspring in returning those necessary nutrients to the land to restart the growth process rather than watching them go up in smoke only to return as further pollution in the future. Hopefully, we will be honest with ourselves in admitting our past selfishness in this area.

Returning to the OPS plan for the disposal and utilization of solid waste materials, the third part of the plan deals with local land reclamation projects and landfills. Under the OPS plan, landfills will continue to be necessary for the disposal of those materials that cannot be reused safely; however, the life of those landfills will be greatly expanded because the bulk of the materials deposited in them will be reduced from its current volume. Landfills should, additionally, become less offensive as the materials deposited in them become less so. In the near future, even household garbage will become an input for inoffensive OPS plants.

Part of the third thrust also deals with an important concept in the OPS plan: Cooperation. As noted above, a great deal of cooperation among many special interest groups will be required to initiate the OPS plan for the disposal and utilization of solid waste materials. Cities, municipalities, townships, counties, businesses, industries and ultimately consumers must begin to cooperate in the development of solid waste utilization and disposal plans. Since it is the fervent hope of OPS that soild amendments will ultimately be used cooperatively in land reclamation projects employing Class V output materials, maximum cooperation will, obviously, be required.

In order to clarify some of the points made above, it will be useful to examine the OPS proposal for Erie County, Pennsylvania, which, it is hoped, will be the first completely operational OPS *system*.

OPS research has shown that Erie County, Pennsylvania, with a population of approximately 276,000 and with a mixture of industry and agriculture, generates enough organic waste materials to produce approximately 350,000 tons of soil amendment annually *without* the use of residential and commercial garbage (*with* such use the total rises to over 400,000 tons annually). OPS predicts that 350,000 tons of Class I to IV output will meet only one-tenth of the possible demand for such soil amendments, thus increasing its value and encouraging the source separation of household garbage in order to increase annual output.

OPS recommends a pre-mix recycling center operation within the city limits of Erie with the following input materials from the surrounding area: leaves, grass, shrubbery and tree trimmings, City of Erie treated waste water, shredded corrugated liner board, coal and wood fly ash, foundry sand and others; for details consult OPS brochure, OPS Solid Waste Utilization and Disposal Programs, Erie, Pennsylvania. The processed pre-mix output will then be transferred to local OPS recycling centers operating in cooperative areas in Erie County with each center serving a combined population of twenty to thirty thousand people as shown in Figure 3. The local centers will then use the pre-mix output in combination with other organic waste materials found in that area for the final production of soil amendments that can be directly applied

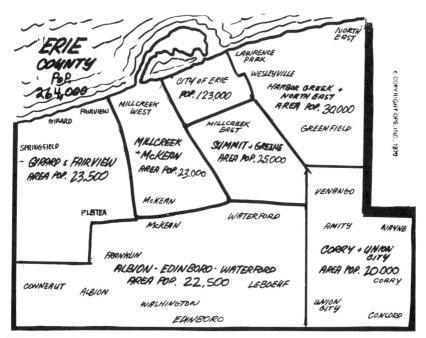

FIGURE 3. Projected OPS Pre-Mix & Local Recycling Centers for the City of Erie and Erie County, PA.

to the land or can be placed in leaching-controlled reclaimable storage pads on-site until composing is complete as noted earlier.

The above proposal, in longer form, has been made to Erie County leadership with positive results. The cooperative efforts so necessary to the success of this project have begun with the reestablishment of the Erie County Solid Waste Authority in order to begin the process of application to the Pennsylvania Department of Environmental Resources for the start of the project. This first step has been taken to making the OPS system a reality.

The pilot program in Erie County is, however, not enough. Before any real progress can be made we must put aside what has become our native selfishness and take three preliminary steps together. We must, first, *educate* ourselves to the value of source separation because most of our solid waste is a renewable resource. This, of course, is where community leaders and educators can be most useful; no progress will be made without education. Second, we must *motivate* ourselves to conserve the resources available to use in the utilization and disposal of our solid waste. Third, we must *cooperate* with each other—municipalities and private enterprise in cooperation—to reduce the high costs of disposal both in real dollars and in environmental damage and to generate new sources of revenue in the utilization process.

Through the solid waste programs of source separation, utilization and cooperation in local land reclamation projects we can take another step toward assuring a reasonable and comfortable future for succeeding generations of Americans. OPS hopes that you agree that solid waste utilization and disposal programs with a little help from "mother nature" make it simple to recycle today's solid waste for tomorrow's growth.

Solid and Liquid Wastes: Management, Methods and Socioeconomic Considerations. Edited by
S. K. Majumdar and E. Willard Miller. © 1984, The Pennsylvania Academy of Science.

Chapter Twenty-Three

Potential Economic Benefits of Recovering and Utilizing Coal Wastes

Teh-wei Hu[1] and James A. Weaver[2]

[1]Professor of Economics
The Pennsylvania State University
University Park, PA 16802
[2]Manager, Economic Studies
Engineering-Science, Inc.

One of the objectives of the 1976 Resource Conservation and Recovery Act (RCRA) is to conserve valuable material and energy resources. The purpose of this paper is to estimate the potential benefits of both recovering chemical constituents from coal wastes and using coal wastes to produce other products (hereinafter referred to simply as "recovery" and "reusing" coal wastes). The benefits of recovering and reusing coal wastes should be considered in terms of improvement in the well-being of the entire society, not just of the industrial or government sector. It can be argued that with the possible increase in the supply of recoverable or reusable materials from coal wastes, consumers or users of these materials will pay a lower price for them.

To estimate the potential benefits of recovering and reusing coal wastes, it is important to first identify the major chemical constituents in the wastes and the technical feasibility of recovering and using these constituents. Section II of this paper will review the physical properties of coal wastes. A second step is to examine the market structure and price elasticities of these materials in order to predict the possible price effect of an increase in their supply. The market structure and price elasticities are presented in Section III. It would be ideal to obtain information on the costs of recovering or reusing coal wastes in order to compare the costs with the benefits. Unfortunately, this cost information is not available. Therefore, the benefits estimated in this paper are gross benefits. Or, if only the consumer benefits are taken into account, the estimated benefits can be considered as long-run benefits. This benefit estimation is included in the Section IV. Section V contains concluding remarks.

II. PHYSICAL PROPERTIES OF COAL WASTES

In predicting the potential recovery and reuse options from coal wastes, it is important to know the chemical composition of coal wastes and the concentration of each chemical constituent.

Coal wastes consist primarily of the inorganic mineral constituents of the coal and, to a minor extent, organic matter not completely burned in the combustion process. The ash residue resulting from coal combustion can range from 3 to 30 percent of the total weight of the coal which was burned (Ray and Parker, USEPA, 1977). In a report by Hart Associates for the Electric Power Research Institute (EPRI), it is suggested that figures of 7 to 15 percent are probably "more representative" (Cooper, 1975, cited in EPRI, p. 12).

The ash residue consists of (1) fly ash, and (2) bottom ash. The proportion of each type of ash produced is a function of the type of coal, the ash content of the coal, the type of boiler (method of firing), and the type of boiler bottom (dry vs. wet).

Coal wastes contain over 60 chemical constituents. A number of these are "enriched" in coal compared to their prevalence elsewhere. In addition, the combustion process further enhances some of these chemical constituents in the coal ash (Ray and Parker, USEPA, 1977). Therefore, some of the heavy metal elements contained in coal are found in greater concentrations in the ash residue formed by coal combustion than in either the earth's soil or in the pre-burned coal itself.

Quite a few constituents are generally found in concentrations greater than 1,000 mg/kg, or 0.1 percent of the solid coal wastes. They include Al, Ca, Fe, K, Mg, Na, S, Si, Ti, and P. The wide variation in content among different coal wastes makes it very difficult to specify an exact chemical composition for coal wastes (See Table 1). In order to estimate the economic benefits of recovery and reuse of resources from coal wastes, it is necessary to narrow the number of constituents and to assume an approximate proportion of constituents that can be reasonably extracted from coal wastes.

Four criteria were used to select the constituents for focus in estimating the benefits of recoverng and reusing coal wastes: (1) the current technological feasibility of recovering or reusing the constituent from the coal wastes, (2) the market value of the recovered or reused constituent, (3) the relative share of the constituent in coal wastes, and (4) the availability of economic data for estimating the benefits of recovering or reusing the particular constituent. Using

TABLE 1

The Composition of Fly Ashes of Typical U.S. Coals

| Constituent | Chemical Composition, Wt. Percent Oxide ± Standard Deviation | | |
	Bituminous	Subbituminous	Lignite
SiO_3	45.6 ± 4.8	46.2 ± 12.3	47.2 ± 11.8
Al_2O_3	21.9 ± 5.4	19.8 ± 3.0	21.6 ± 6.7
FE_2O_3	17.4 ± 4.7	5.5 ± 1.4	5.7 ± 3.8
CaO	3.8 ± 1.8	15.7 ± 9.8	15.8 ± 3.8
MgO	1.0 ± 0.2	3.6 ± 1.8	3.6 ± 2.5
TiO_2	0.9 ± 0.5	1.1 ± 0.4	1.1 ± 0.4
K_2O	2.6 ± 0.7	1.1 ± 0.8	0.7 ± 0.5
Na_2O	0.7 ± 0.6	0.8 ± 0.6	1.2 ± 1.0
SO_3	2.5 ± 3.7	3.7 ± 5.0	1.6 ± 1.5
P_{25}	0.3 ± 0.1	0.8 ± 0.5	0.1 ± 0.1

Source: N. K. Roy et al., "Use of the Magnetic Fraction of Fly Ash as a Heavy Material in Coal Washing." Proceedings: Fifth International Ash Utilization Symposium, February 1979, cited in Engineering—Science and Webster & Associates (1982), Table 7.3.

these four criteria, the following constituents were chosen for this study: cement, iron ore, and alumina. While not an actual constituent of coal wastes, cement can be mixed with fly ash from coal wastes to make concrete. Fly ash concrete can be made either by interbalancing the cement and fly ash, or the cement and fly ash can be added separately in the concrete manufacturing process. Each method has its advantages. Fly ash used for cement or other construction materials is classified as a *reuse* option for coal wastes. Alumina, iron ore, and sulfur are considered as resources to be *recovered* from coal wastes. Substantial proportions of these four elements are contained in coal wastes and technologically it is feasible to extract them. More importantly, these three resources make valuable economic contributions to the U.S. manufacturing industry.

A high-level effort has been made to extract sulfur from coal wastes in recent years. No numerical figure is available for the maximum amount of sulfur which may be extracted from coal wastes.

In summary, based on the economic significance, technological feasibility, and availability of data sources for various constituents, this study will examine the benefits of using fly ash for cement products and the benefits of recovering iron ore and alumina from fly ash. Based on the documented amount of fly ash produced each year in the U.S., it has been estimated that 16 million tons could be derived for use as cement and other construction materials. From that same amount of fly ash one could recover an estimated 5 million tons of iron ore or about 4 million tons of alumina. It should be noted that one cannot extract all of these materials concurrently.

III. THE MARKET STRUCTURE OF THREE MAJOR RESOURCES

In this section, we shall first describe the cement, iron, alumina, and sulfur industries in terms of their supply and demand structures and their respective price elasticities. It is assumed that these estimated elasticities are applicable to the recovered and reutilized resources. In other words, it is assumed that consumers will consider these recovered and reutilized resources to be the same quality and perfect substitutes for the original resources (cement, iron ore, and alumina). These price, quantity, and price elasticity data will be used to estimate the possible economic benefits of recovering or utilizing coal wastes.

The Cement Industry

Cement is the most important binding agent in the concrete and construction industry. Ninety-five percent of all cement is used in concrete and construction. Cement ranks third among all nonfuel minerals in the U.S. in total value of production and in the energy requirement for the production.

Production of cement in the U.S. increased from 52 million tons in 1960 to 75 million tons in 1980, an increase of 44 percent. In fact, 1980 was not the peak production year for the two decades. The highest production was in 1973—85 million tons.

In 1980, about 78 million tons of cement were consumed, an increase of 42 percent compared to 1960. Cement usage in the U.S. is expected to grow at an average annual growth rate of 1.1 percent until the year 2000 (*Minerals Yearbook,* 1980) reaching 115 million tons by the year 2000.

Demand price structure varies with local (regional) markets. The average f.o.b. plant price per ton was $20 in 1960, $18 in 1970 and $51 in 1980. Overcapacity prior to 1970 was responsible for keeping the mill value fairly constant. The price of cement increased faster since 1970 than the Consumer Price Index. In real prices (deflated by the Consumer Price Index), cement increased about 34 percent during the 1970-1980 decade.

To understand the relationship between demand for and price of cement, it is useful to estimate a demand function based on past data so that price elasticity of demand can be established. Price elasticity is defined as percentage changes in the quantity demanded in response to percentage changes in price. No previous empirical estimates are available; thus, it is necessary to rely on estimates obtained in this study as a basis for drawing market implications from compliance with resources recovery or reuse regulations.

A traditional and simple demand function can be considered in which the demand for cement is a function of cement price and national income level. The income variable is considered as a proxy of the national economic condition. The estimated price elasticity of the demand for cement is − .38, which implies that for each 10 percent increase in the price of cement, demand may be reduced by about 4 percent. (For detailed statistical results, see Hu [1983].)

The Alumina Industry

The major source of alumina is the ore bauxite. Alumina can also be recovered from coal fly ash. Alumina or bauxite is an essential raw material, constituting about 82 percent of aluminum production (*Mineral Yearbook,* 1980). Aluminum is used in the transportation and construction industries, in the manufacture of cans and containers, electrical equipment, consumer durables, and mechanical equipment. The first three categories each account for over 20 percent of total aluminum consumption.

Bauxite consumption in the U.S. increased from 8.9 million long tons in 1960 to 14.5 million long tons in 1980, an increase of about 73 percent. However, the peak consumption period was in 1973-74, approximately 16 to 17 million long tons. No systemically published statistical series of bauxite prices is available because of the various forms of the raw material and different import prices. The price of bauxite fluctuates widely over the years. The market value of alumina (calcined aluminum) was $218 per long ton in 1980.

No studies are available on the demand for alumina. A number of studies have estimated the price elasticities of demand for aluminum. These estimates for aluminum range from − .01 (Charles River Associates, 1977) to − 1.15 (Peck, 1961). The Peck study used 1925-40 time-series data, while the Charles River Associates study used 1955-73 data sources. The estimated value of elasticities has a wide range and neither of the data sources is current nor specifically on alumina. Since this study is to estimate the market implications for alumina as a result of compliance with resources recovery, an estimation of the elasticities of demand for alumina would be important. Using the same methodology used in the cement estimation, the estimated price elasticity of demand for alumina is − .85.

The Iron Ore Industry

The U.S. is a major producer and consumer of iron. The major users of iron ore are the blast furnaces industry and the steel industry. Iron ore consumption fluctuated during the period 1960-1980. During the 1973-74 oil shortage, oil and gas exploration stimulated steel consumption in the gas and oil industries, reaching a record level of 219 million tons. On the other hand, the 1980 economic condition resulted in only 89 million tons of iron ore being consumed. The 1980 consumption figure was the lowest during the past two decades.

Iron ore production in 1960 was about 89 million tons while in 1980 it was only 70 million tons. The U.S. relies on imports for about 30 percent of its domestic iron ore consumption. Thus, the U.S. is not self-sufficient in iron ore production.

The price of iron ore has been increasing steadily from $8.70 per ton in 1960 to $10.80 per ton in 1970 to $36.60 per ton in 1980. These price increases were close to the inflationary rate in the economy. Labor costs account for about 50 percent of the operating costs at iron mines and plants. Transportation costs were about 35 percent of iron ore prices (*Mineral Facts and Problems,* 1980). Both these factors are highly related to the inflationary rate in general. The estimated price elasticity of the demand for iron ore is $-.63$.

IV. EMPIRICAL ESTIMATION OF BENEFITS OF RECOVERING OR REUSING COAL WASTES

There are two time dimensions in analyzing economic behaviors: (1) a short-run analysis, and (2) a long-run analysis. The short-run analysis of the firm's behavior focuses on immediate responses to a change in price of costs, without considering expansion or closing of the firm. In the short run, profits or losses may occur. The long-run analysis of a firm's behavior focuses on the ultimate response to a change in price or costs, including expansion or closing of the firm. As long as there are profits, new firms may join the industry until the additional revenue of an additional unit of output is equal to the additional costs of producing an additional unit of output.

In the short-run, a producer (e.g., a utility company) has two options in handling coal wastes: (1) disposing of them as wastes, and (2) reusing or recovering resources from these coal wastes and disposing of the remainder as wastes. In taking the former option, one incurs the cost of waste disposal. In taking the latter option, one has to compare the sum of processing costs of recovering or reusing coal wastes and the price of the recovered or reused resources on the market with the profit or recovering or reusing coal wastes. If a producer finds that the profit is positive, the producer will recover or reuse part of all of the coal wastes rather than disposing of them entirely. In fact, even if a producer finds that the profit of recovering or reusing coal wastes is negative, as long as the loss is less than the difference in the cost of disposing of the entire amount of coal wastes compared to the reduced amount of wastes, the amount of cost savings is a benefit to the producer and the producer will continue to recover or reuse the coal wastes. Since a producer's decision is often made on the basis of incremental changes, that is, calculation of the amount of additional revenue compared to the additional cost of a slight expansion, or the amount of cost savings but loss of revenue associated with a small contraction, a simple mathematical formulation can be outlined.

Let
P_i = price of the ith resource on the market.
MDC_i = incremental (marginal) disposal cost for coal wastes if the ith resource were not reused or recovered.
MPC_i = incremental (marginal) processing costs for the ith resource.

$Q_i =$ quantity of the ith reused or recovered resource.
$PR_i =$ profit or reusing and recovering the ith resource.

The decision to reuse or recover the ith resource is when
$$P_i > MPC_i - MDC_i$$
$$\text{or } P_i + MDC_i - MPC > 0$$
The net profit or benefits for the producer to recover or reuse coal wastes in the short run becomes
$$[(P_i + MDC_i) - MPC_i] Q = PR_i.$$
The increase in the supply of the ith resource may reduce the market price of the ith resource. The producer will continue to produce the ith resource until
$$P_i + MDC_i = MPC_i$$
Based on the above analyses, estimation of the short-run benefits to the producer would require the following information: (1) the expected market price of the recovered or reused resources, (2) the marginal costs for disposing of coal wastes, (3) the marginal costs of recovering or reusing coal wastes, and (4) the quantity of recovered or reused resources. Of these four items, only the cost of recovering or reusing coal wastes is not available, since the technologies for recovering or reusing coal wastes are still in an early stage. Only a few laboratories have conducted such recovery experiments. Therefore, this study cannot provide estimates of the short-run benefits of recovering or reusing coal wastes.

While short-run benefits may be especially useful for estimating producers' benefits, in the long-run producers' benefits from recovering or reusing coal wastes may disappear because of the entry of new producers or an increase in the supply of recovered or reused resources and all benefits will be in the form of consumers' benefits. Consumers' benefits are measured in terms of increased consumer surplus from possible price reduction and the additional resources available on the market. From the governmental policy point of view, the long-run consumers' benefit should be the most important consideration for decision making.

In the long run, it is assumed that the marginal costs of recovering or reusing coal wastes will be equal to the market price of the resources, until all the producers either enter into or withdraw from the industry. In other words, one can assume that if there is a benefit from recovering or reusing coal wastes in the long-run, producers' marginal costs should be at least equal to the market price of the resource. Using this assumption, as long as the market price of the resource is known, one can estimate the long-run benefits of recovering or reusing coal wastes without knowledge of processing costs. Thus, the long-run benefits of recovering and reusing coal wastes wll be examined as the effect of an increase in various resources on the market price and the resultant consumer benefits. The required information for estimating long-run benefits of recovering or reusing coal wastes are (1) the current prices of various resources, (2) price elasticity of demand, and (3) the amount of potential recoverable or reusable resources. The theoretical and conceptual discussions for estimating benefits of recovering or reusing coal wastes are presented in Hu (1983).

In simple terms, one can measure benefits by comparing the price and quantities of resources consumed before and after recovering or reusing the coal wastes. One can measure the change in the difference between what consumers would be willing to pay for a particular amount of resources rather than do without the resources, and the amount they actually have to pay on the market as a result of the additional increase in reused or recovered resources. This estimation approach is often called a consumers' surplus approach. To compute the net gain of consumers' surplus, one requires information on (1) the potential amount of increase in resources, $\triangle Q$, and (2) the possible effects of price reduction of the resources, $\triangle P$.

The information on the potential amount of increase in recoverable or reusable resources, $\triangle Q$, can be obtained from the earlier technical studies of utilization and recovery of coal combusion wastes (Engineering-Science and Webster & Associates, October 1982). It should be noted, however, that these estimates are the most optimistic scenario. Thus, the benefit estimates based on these figures will be the maximum possible benefits of recovering or reusing coal wastes. Furthermore, 50 million tons of coal wastes cannot be used for cement and at the same time for recovering iron ore, and alumina. Thus, the estimated benefit figures for these resources should not be added as cumulative benefits. These estimated benefits are considered as alternative benefits of recovering or reusing coal wastes.

The possible effects of price reduction, P, can be estimated from the current price and quantity consumed, additional increase in quantity, and the price elasticity of demand:

$$\Delta P = (\frac{\Delta Q}{Q} \cdot P) (\frac{1}{E})$$

where $\triangle P$ and $\triangle Q$ are the maximum possible change of price and quantity due to the recovered or reused coal wastes. P and Q are the current price and quantity of the resource and E is the price elasticity of the demand for the resource. The formula for estimating the net gain to the consumer (or the entire society) is $\frac{1}{2}$ x $\triangle P$ x $\triangle Q$.

As assumed earlier, the mineral resources industry operates under a constant cost or close to constant cost condition. The degree of competitiveness among these three resources varies from quite competitive (cement) to noncompetitive (alumina) market structure. The increase in the supply of these mineral resources will reduce the prices of cement, alumina, and iron, depending upon their respective market structure. To estimate their respective effects on prices, two alternative prices are provided for alumina and iron. Under a competitive market, which is the case for cement, in the long run, consumers will receive the entire price reduction as a result of the increased supply. In the case of the other extreme, in the monopoly market, consumers would receive only half of the price reduction under the competitive condition. All these estimates are measured in 1980 prices. The numerical estimates for cement, alumina, and iron follow.

Cement. The average retail price of cement was about $51 per ton in 1980, according to the *Survey of Current Business.* As shown in an earlier section, engineering estimates indicated that 50 million tons of coal wastes produced in the utility industry during 1980 could generate 16 million tons of fly ash that could be used as a substitute for cement. During 1980, about 77.6 million tons of cement were consumed. Given the estimated price elasticity of cement, $-.38$, the increase in the amount of cement resulting from using coal wastes would have reduced the market price of cement by $27.67 per ton ($\frac{16 \cdot \$51}{77.6} \cdot \frac{1}{-.38}$). The resultant net gain to the society as a whole would be $221 million ($\frac{1}{2} \cdot \triangle P \cdot \triangle Q = \frac{1}{2} \cdot \$27.67 \cdot 16$).

Alumina. The average price of alumina was $218 per long ton in 1980, according to the *Mineral Yearbook* (1980). As noted in an earlier section, engineering estimates indicated that 50 million tons of fly ash used in the utility industry could generate 4 million tons of alumina. During 1980, the economy used 14.5 million tons of alumina. Given the price elasticity of alumina, $+.85$, the increase in the amount of alumina resulting from its recovery from coal wastes would have reduced the market price of alumina by $70 per ton ($\frac{4 \cdot \$218}{14.5} \cdot \frac{1}{-.85}$) under a competitive market. Under the monopoly condition, the pricing rule is that where marginal costs are equal to marginal revenue, the price change will be only half of the price change under a competitive market. Thus, the market price would have been reduced by $35 per ton. The resultant net gain to the society as a whole would be $140 million ($\frac{1}{2} \cdot \triangle P \cdot \triangle Q = \frac{1}{2} \cdot \$70 \cdot 4$) under a competitive market and $70 million under the monopoly condition.

Iron Ore. The average price of iron ore was $36.56 per ton in 1980, according to the *Survey of Current Business.* As noted in an earlier section, engineering estimates indicated that 50 million tons of coal wastes used in the utility industry could generate 5 million tons of iron ore. Given the price elasticity of iron ore, $+.65$, the increase in the amount of iron ore due to recovering coal wastes would have reduced the market price of iron ore by $3.14 per ton ($\frac{5 \cdot \$36.56}{89.4} \cdot \frac{1}{-.65}$) under a competitive market and $1.57 per ton under the monopoly condition. The resultant net gain to the society as a whole would be $6.28 million ($\frac{1}{2} \cdot \triangle P \cdot \triangle Q = \frac{1}{2} \cdot \$3.14 \cdot 5$) under a competitive market and $3.14 million under the monopoly condition.

Table 2 summarizes the estimated consumers' surplus to be gained from the

TABLE 2
Summary of Consumer Benefits of Recovering or Reusing Coal Wastes, 1980 Prices

Resource	Retail Price[a] (per ton)	Total Current Consumption[a] (million tons)	Price Elasticity[b]	Increase in Quantity of Coal Wastes Recovered or Reused (million tons)	Amount of Price Reduction (per ton)	Net Gain to Society (millions)
Cement	$51.00	77.6	−.38	16.00	$27.67	$221
Alumina	$218.00	14.5	−.85	4.00	$70-$35[d]	$140-$70[d]
Iron	$36.56	89.4	−.65	5.00	$3.14-$1.76[d]	$6.24-$3.14[d]

Sources: [a]U.S. Bureau of Mines, Minerals Year Book, 1980.
[b]Estimated by this paper.
[c]Estimated according to U.S. DOE report prepared by Engineering Sciences, Inc. (1982).
[d]Ranges of estimates from competitive market to monopoly condition.

recovery or reuse of three major resources from coal wastes. Of these resources, cement ranked the highest, followed by alumina and iron ore. One should not add these net gains to society as total net benefits of recovering or reusing coal wastes since the use of fly ash for cement would preclude the opportunity for recovering alumina or iron ore from the same fly ash. The results indicated that in the long run, using fly ash in cement has the highest benefits, followed by alumina and iron ore. It should also be noted again that these estimates are potential maximum long-run net gains to the entire society, excluding the transfer payments between industries and consumers, such as the cost of processing or profits to producers.

V. CONCLUDING REMARKS

Recovering or reusing coal wastes have both producers' benefits and consumers' benefits. The producers' benefits can be measured by comparing the price of reusing or recovering the material and the cost savings in disposing of coal wastes with the cost of raw materials, processing, and transportation costs. The consumers' benefits are measured in terms of the increase in the quantity of resources and the resultant reduction in market price. In the long run, the societal benefits are composed of consumers' benefits only. Data on processing costs of processing are not currently available; thus, the short-run benefits cannot be estimated. The long-run benefits are estimated assuming that the costs of processing and transportation are at most equal to or less than the sum of the market price of reused or recovered resources and the cost savings in disposing of coal. Otherwise, producers would not recover or reuse the resource. On the other hand, if there is a profit in reusing or recovering the resource, producers would continue to produce additional amounts of the resource until they reached the point where the costs and revenues are equal. Therefore, in the long-run, producers' benefits do not exist.

The long-run estimated net gain to the consumer and to society as a whole has been calculated, given current price and consumption figures, the increase in the supply of these resources which would result from recovering or reusing coal wastes, and the price elasticity of the demand for these resources. The results indicated that of the three major resources examined in this paper, consumers' benefits from the use of fly ash as cement would be about $221 million. While alumina and iron ore rank second and third in contributing additional consumers' benefits, these benefits are much smaller than the benefits of using fly ash as cement. It appears that with the current available technology for recovering resources from fly ash, large-scale recovery of alumina and iron ore is still in its infancy.

Given the relatively reliable estimates of the amount of reusable or recoverable cement from fly ash, the above estimated long-run consumer benefits for cement should be considered reasonable. On the other hand, the figures for alumina and iron ore should be considered rough estimates or perhaps the maximum amount of benefits.

REFERENCES

Charles River Associates and Wharton Econometric Forecasting Associates. *Forecasts and Analysis of the Aluminum Market,* Cambridge, Mass., October 1972.

Cooper, H. B. "The Ultimate Disposal of Ash & Other Solids from Electric Power Generation." In *Water Management by the Electric Power Industry,* edited by E. J. Gloyan. Austin, Texas, 1975.

Engineering-Science and Webster & Associates. *Process Modifications and Waste Management Practices Affecting Disposal, Utilization and Recovery of Coal Combustion Wastes.* U.S. Department of Energy, October 1982.

Hu, Tah-wei. *Estimating Benefits of Recovering the Reusing Coal Wastes.* Prepared for Engineering-Science, Inc., 1983.

Mineral Facts and Problems. U.S. Bureau of Mines, 1980.

Mineral Yearbook. U.S. Bureau of Mines, various issues.

Peck, Merton J. *Competition in the Aluminum Industry, 1945-58.* Cambridge, Mass., Harvard University, 1981.

Ray, S. S., and F. G. Parker. *Characterization of Ash from Coal-Fired Power Plants.* USEPA, January 1977 (EPA 600/7-77-010).

Roy, N. K. et al. "Use of the Magnetic Fraction of Fly Ash as a Heavy Medium Material in Coal Washing." *Proceedings: Fifth Internationl Symposium,* February 1979.

U.S. Department of Commerce. *Survey of Current Business,* 1980.

U.S. Department of Energy. *Impacts of Alternative Combustion/Process Modifications and Waste Management Practices on Disposal of Coal Combustion Wastes.* Assistant Secretary for Fossil Energy, Office of Coal Utilization. Prepared by Engineering-Science, Inc., September 1981.

Solid and Liquid Wastes: Management, Methods and Socioeconomic Considerations. Edited by
S. K. Majumdar and E. Willard Miller. © 1984, The Pennsylvania Academy of Science.

Chapter Twenty-Four

The Economics of Energy Oriented Resource Recovery Systems

John R. McNamara, Ph.D.

Department of Economics
Lehigh University
Bethlehem, PA 18015

A study[5] of the economic feasibility of energy oriented resource recovery in the Lehigh Valley region was recently undertaken with the objectives of (1) reaching a better understanding of the factors which determine the success or failure of regional resource recovery systems, (2) developing a methodology for regional resource recovery planning and feasibility studies, and (3) analyzing the current solid waste management practices in the Lehigh Valley region of Pennsylvania with a view towards the possible integration of one or more regional resource recovery facilities.

The introduction of resource recovery into an existing solid waste management system depending on sanitary landfills requires an entirely new perspective to be successful. Where the objective of solid waste management is the disposal of refuse in an environmentally acceptable manner at least cost, the objective of resource recovery is the production of one or more well defined products which must be successfully marketed. Resource recovery is likely to fail unless the correct objective is employed in all phases of planning and implementation. Resource recovery is a profit-oriented business, as well as a public service.

Well-proven approaches, like waterwall incineration, are preferred over untested technologies or complex systems, especially in areas with moderate population concentrations, because of the cost structure of resource recovery facilities. Such facilities are characterized by large fixed costs and very small variable costs, denoting a high degree of operating leverage. In such cases, losses mount quickly if the system is out of operation for periods of time. The most cost effective and reliable resource recovery approach in use today is waterwall incineration to produce steam and, perhaps, electricity. This approach is also suitable for most regions in terms of waste availability, economies of scale, and existing markets for recovered energy. However, this preferred approach often incurs a loss because revenue from the sale of energy plus tipping charges may not cover total annual costs.

The production of refuse derived fuel (RDF) is a distant second choice for resource recovery because of explosive hazards and reliability problems in the shredding operation, difficulty in separating out undesired materials from the combustible fraction leading to maintenance problems for users, and marketing problems. RDF is a supplement to coal which has the lowest price per million btu of the major fossil fuels. High quality RDF is simply too expensive to produce in view of the price customers are willing to pay, which must be no higher than the coal equivalent price *less* any incremental costs for construction of handling and storage facilities. The cement industry, under appropriate conditions, would be a ready market for RDF.

Non-energy resources like aluminum and glass are best recovered, when market conditions are favorable, by means of source separation programs. Resource conservation initiatives at the state and national levels are also important. Resource conservation is defined as controlling the use of resources at the point of manufacture. Recovery of non-energy resources is not economically feasible in a central facility in an area like the Lehigh Valley. Source separation means segregation of waste materials prior to collection by the hauler.

A central resource facility should be sited so as to satisfy market requirements and so as to be convenient to waste haulers. This implies a site in an industrial area near the center of population. Utilization of the recovery facility by waste haulers must be kept at a high level if the cost per ton is to be kept at an acceptable level. The high fixed cost component of the overall cost per ton means that losses quickly occur if the waste through-put drops off. The tipping charge must, therefore, be kept competitive with landfill charges, considering hauling costs. Where landfill disposal costs are very high, resource recovery system costs can be covered, to a large extent, by avoided landfill site disposal and transportation costs. In the Lehigh Valley, where landfill site disposal costs are relatively low, as evidenced by the wastes flowing into Lehigh and Northampton counties from surrounding areas, including parts of New Jersey, the design of a regional resource recovery system must be well matched to the needs and characteristics of the region.

While regional resource recovery from mixed municipal refuse does not appear to be economically feasible at present, due to the low landfill disposal costs, this situation is rapidly changing and planning must begin now in anticipation of opportunities arising within the next few years.

The "refuse shed" concept is important in resource recovery planning. The refuse shed is defined as a grouping of locations which tend to use a common disposal site for refuse. The common site may be a landfill or resource recovery facility. The limits of the refuse shed are based on transport and disposal cost economics. Haulers select a disposal site so as to minimize the sum of their relevant disposal costs, transport costs, disposal or tipping charges, and maintenance or other costs related to the character of the disposal operation. It should be clear that refuse shed boundaries are functions of the relative economics of hauling and disposal costs and the differences in these costs as location changes. A resource recovery facility, even if located in the center of population, would probably not create a refuse shed unless its tipping charge was less than the combined costs of hauling refuse to a more distant site plus the tipping charge at the site. As a disposal site lowers its tipping charge, its refuse shed expands and vice versa. It should be obvious why some landfill operations choose to levy a range of tipping charges. Nearby customers will always be willing to pay more than distant customers. Of course, there may also be other reasons for variations in tipping charges including whether or not a municipal collector is involved, the volume of wastes discharged, and the nature of the wastes.

As the tipping charge at a resource recovery facility is lowered, its refuse shed expands and larger quantities of refuse flow into the facility until its capacity is reached. Of course, due to cost overruns and poor planning, many resource recovery facilities today operate with a high tipping charge, a smaller than expected refuse shed and higher costs per ton due to lower than planned throughput. The availability of nearby sanitary landfill site capacity at reasonable cost is the primary, and perhaps the only, reason for rejecting the concept of resource recovery in a central facility. A central resource recovery facility must operate at a net cost per ton which competes with the landfill site associated with that refuse shed. Net cost per ton is defined as total operating cost per ton less revenue per input ton.

Recent Experience with Regional Resource Recovery

There are a number of types of mixed waste resource recovery facilities in operation today. The design capacities of these facilities range from 30 to 1600 tons per day. Except for a composting operation in Altoona, Pa., and an EPA-supported wet pulping fiber recovery demonstration plant at Franklin, Ohio, these facilities are primarily or exclusively oriented towards energy recovery. Many are incineration units designed for the mass burning of raw mixed refuse. The incineration units include refractory wall units, waterwall units, and new small-scale modular combustion units of 20 to 50 tpd capacity. In addition, a number of energy oriented facilities are under construction.

The most recent resource recovery facilities to become operational have been weighted towards small scale incinerators with heat recovery modules which had been developed as industrial and institutional boilers. This development is particularly interesting with respect to the Lehigh Valley situation. Pulp and paper manufacturing companies are commonly burning wood wastes in similar systems for steam and electricity generation, and these systems are generally successful.

It should not be concluded that all operational resource recovery plants are fulfilling the expectations of their sponsors. Most facilities have experienced problems and some have been shut down.

A common problem encountered by resource recovery facilities is higher than expected operating costs which necessitate tipping charges of $15 to $20 per ton. These tipping charges, together with the revenues from the sale of energy and recovered materials, must cover operating and amortization costs, and are often higher than local landfill disposal costs. A high tipping charge encourages waste collectors to rely on landfill sites with lower tipping charges.

Market for Recovered Materials

As a part of the feasibility study of resource recovery, the study group gathered and analyzed information on markets for recovered materials in the Lehigh Valley, emphasizing those materials which have met with the greatest success in similar regions. The leading products are clearly (1) direct energy production in the form of steam or electricity, and (2) a fuel supplement or RDF product.

It is critically important to view resource recovery as the production of readily marketable materials which meet required specifications rather than simply as an alternative to landfill disposal. Marketing problems usually occur when the latter perspective is taken by the management or sponsors of a resource recovery facility. Also, the *net* value of the recovered product must be constantly kept in mind. The net value is defined as revenue per input ton less processing and other costs. Even if revenue per ton is attractive, a high processing cost per ton can make the project uneconomic. User specifications for recovered materials are generally tighter than anticipated and consequently yields tend to be lower and processing costs higher than expected. The price per ton for glass, for example, may range from $40 per ton for high purity, color-sorted glass to $3 per ton for a low purity (80-90% glass) product. Transportation costs are also a factor.

The view of the U.S. office of Solid Waste is "Markets First," and a market means a contract. Waste streams and markets are site specific, and a materials recovery project should not be attempted unless the project is suited to local market conditions.

In view of the locally unstable markets for recovered newspaper, glass, aluminum, and ferrous materials and the fact that these materials are more efficiently recovered by specialized scrap dealers or, when conditions are favorable, through source separation programs, this report concentrates on markets for energy products from municipal refuse.

The study group interviewed representatives of the most likely local industrial energy customers, obtained copies of reports prepared by potential industrial users, and studied market surveys prepared elsewhere. The general aversion of investor-owned electric utilities to participating or investing in energy-oriented resource recovery projects is due to technical and economic uncertainties involved in all of the present energy recovery technologies.

Economic Analysis of Solid Waste Management Systems

The use of computer models to assist in the design and analysis of regional solid waste management systems goes back at least 15 years.

The General Electric Company used a very large linear programming model in the development of its plan for the state of Connecticut. The G.E. model was essentially a "transportation" type of linear programming model, and most of its successors have employed this structure.

E.B. Berman[1], an analyst at the Mitre Corporation, has been a leader in the development of regional solid waste management models. Jochen Kuhner[4] used a similar type of model with the feature that the effects of interest groups are included in the analysis of resource recovery strategies. Michael Greenberg[2] applied linear programming to the problem of siting resource recovery facilities in Northern New Jersey. G. Haddix[3], M.J. Skelly[6], and others have also used linear programming for this purpose.

The linear programming models described above attempt to determine whether resource recovery is economically feasible and to determine where resource recovery plants should be sited and what products should be produced, often taking an entire state as the study region.

The RECOVR model[5] considers a smaller region in which at most one or two resource recovery facilities are appropriate, and determines whether or not specific resource recovery options are economically feasible and which sites are superior. The RECOVR model is dynamic in structure, in contrast to many earlier models. The growth of the solid waste management system over time, cost and revenue escalation rates and other changes are simulated in order to determine at what point in the future and under what conditions resource recovery becomes economically feasible.

Social and technical constraints are included in the model equations and/or in the model objective function. The model assumptions are easily varied enabling the user to determine the sensitivity of the system being modeled to alternative data values.

There are two critical economic factors in resource recovery systems which are often overlooked in economic feasibility studies. These factors are:
(1) The extensive economies of scale which exist in resource recovery, es
     pecially when a series of processes is employed to recover different types
     of materials.
(2) The large fixed cost relative to operating costs.

The first factor indicates that the cost per ton of waste processed is greater in smaller systems, and this effect is strongest in those systems employing a larger number of processes. Thus, in two regions with similar landfill tipping charges and market prices for recovered materials, resource recovery may be economically viable in the region with the greater population density and not in the other region.

The second factor indicates that the cost per ton processed is extremely sensitive to the utilization of the resource recovery facility in terms of tonnage processed per unit of time.

These two factors are carefully considered in the economic analysis performed using the RECOVR model. In addition, the standard economic considerations, tradeoffs between economies of scale and transportation cost and the minimization of net system cost through siting and design decisions, are accomplished.

The RECOVR program accepts transportation, processing and disposal cost data and projected escalation rates, both variable and fixed costs, the price of RDF (Refuse Derived Fuel) or other energy produced, the price of recovered ferrous material, the fractions of these materials recovered by the proposed recovery process. The program also accepts the quantities of solid waste generated per unit of time, normally per week, by each of a specified number of sources (communities or geographical areas), and the waste disposal capacities (per week) of each of a specified number of landfill sites and the projected growth rates of these quantities. The maximum capacity of the resource recovery plant may also be specified. The program finds a least system cost solution for each point in time. accommodating all waste inputs while satisfying limitations on disposal capacities. The model runs usually simulate operations over a ten year period.

While an important use of such a computer model is the development of "sketches" of efficient systems, this type of model is also extremely useful in the study of the interactions of various factors in the solid waste management system, the treatment of risk and uncertainty, the handling of non-quantifiable costs such as environmental degradation and, finally, the verification of collected data for consistency. An important problem in solid waste management/resources recovery planning is the lack of firm data on costs, waste stream composition, and market values of recovered materials. To some extent, a model like RECOVR can be used to overcome this lack of precise information because repeated "runs" can be made with alternative sets of data inserted and an understanding of the system and of those data which are really significant to the decision process can be obtained.

However, the basic purpose of the RECOVR model is to test the economic feasibility of a particular resource recovery plan. Resource recovery at a given location and under certain assumptions is included as an option in an economic model of the existing solid waste management system. The solution to the linear programming problem, which is automatically constructed and solved by the RECOVR program, indicates whether or not resource recovery is economically feasible, given a set of data representing the existing system. One may make repeated runs with the RECOVR model using the expected values of future costs, waste loads, and other data and determine how high existing disposal costs must rise before resource recovery becomes feasible.

A very difficult aspect of mathematical programming models for solid waste management/resources planning is the treatment of fixed costs. The usual treatment of fixed costs and economies of scale may substantially increase the size of the problem to be solved to such an extent that the program may only be run on the most sophisticated of computer facilities and only at a high cost per run ($50-$100). The approach used in Program RECOVR is to first assume that every disposal facility is used at design capacity for purposes of computing the fixed cost per unit which is therefore equal to total fixed cost for a facility divided by the design capacity of the facility.

The problem is solved and the utilization levels of the various facilities are noted. If a facility is used at 100% of capacity, then the cost of this facility entered in the objective function is correct. If a facility is not selected at all, then its fixed cost is omitted from the computation of system cost (SCOST). If a facility is used at a low level of utilization and has a relatively high fixed cost per unit, then it is probable that a lower cost solution will result if this facility is removed from the system.

The linear programming problem solved by the RECOVR model consists of an objective function to be minimized and a series of linear inequality constraints. The objective function involves all fixed and variable costs of transporting solid wastes to landfill sites plus the tipping charges to be paid at the landfill sites. The fixed and variable costs of operating a central resource recovery facility are also included. The constraints insure that (1) all wastes generated at various locations are accounted for, (2) all landfill sites capacities are observed, and (3) the operation of the resource recovery facility is represented.

Data for the initial year of the run are read and the linear programming model is constructed and solved. Results are stored and data for next year are read. The process is repeated for the 10 years of the simulation run and these final results are printed.

Conclusion

The economics of current resource recovery options favor simple systems which are reliable and economical. The high fixed cost of a resource recovery facility implies a high level of risk since costs continue even when the system is out of operation and revenues are nonexistent. Complex resource recovery systems with many stages or processes are economical only as large scale facilities. Simpler systems appear to achieve scale economies at much smaller design capacities.

Since the construction and operation of a resource recovery facility involves a large capital obligation with annual fixed costs continuing over many years, it is important to choose a product with similar revenue characteristics. Energy oriented resource recovery seems most likely to yield stable revenues over time since contracts for energy products may be employed. In contrast, the markets for materials like scrap metal and glass are unstable and the revenues from the sale of these materials are bound to fluctuate.

The use of a computer model such as RECOVR, or large scale linear programming model, to simulate the pattern of waste hauling and disposal activities in a region can be very valuable. The use of the RECOVR model to study Lehigh Valley solid waste management activities revealed, among other things, that economic incentives such as taxes or landfills or a subsidy to a resource recovery facility or both are more effective in encouraging resource recovery than closing some of the landfill sites or passing restrictive legislation.

Finally, the resource recovery facility should be located in the area which needs it least, if there are several nearby sites. For example, if cities A and B adjoin one another and A has economical, nearby landfill capacity while B is not so fortunate, locating the facility in A will encourage its use by waste haulers in A while the lack of economical alternatives will encourage the haulers in B to use the facility as well.

REFERENCES

1. Berman, E.B., "WRAP: A Model for Regional Solid Waste Management Planning," Mitre Corporation Report MTR-322, April 1976.
2. Greenberg, M.R., "Application of Linear Programming to the Siting of the Resource Recovery Facilities," *Environment and Planning,* Vol. 8, 1976.
3. Haddix, G., "Application of a Solid Waste Planning Model with Resource Recovery," *Proceedings,* 33rd JSGORAM Conference, Bedford, Mass., 1977.
4. Kylmer, J., "Mathematical Modeling for Regionalization of Resource Recovery," in *Mathematical Models for Environmental Problems*, Pentech Press, London, 1976.
5. McNamara, J.R., et.al., *An Economic, Social and Technical Evaluation of the Feasibility of Resource Recovery in the Lehigh Valley,* Center for Social Research, Lehigh University, Bethlehem, PA, 1980.
6. Skelly, M.J., "Planning for Regional Refuse Systems," Ph.D. Thesis, Cornell University, September 1968.

Solid and Liquid Wastes: Management, Methods and Socioeconomic Considerations. Edited by
S. K. Majumdar and E. Willard Miller. © 1984, The Pennsylvania Academy of Science.

Chapter Twenty-Five

Solid Waste Recycling—
Assessment of Economic Viability In
A Developing Country

Prem Shanker Jha
Senior Assistant Editor
The Times of India
7, Bahadurshah Zafar Marg
New Delhi, 110002, India

The fall in oil prices in nominal terms that began at the end of 1981 and continued till early 1983 has taken the edge off the search for renewable and alternate sources of energy. According to the World Bank, every dollar's fall in the price per barrel has reduced the burden of payments on the oil importing developing countries by $2 billion. This is unfortunate because all that the world has obtained is a respite and quite possibly a brief one from the inexorable upward climb of prices. For the basic fact cannot be denied: oil reserves on the globe are limited and demand continues to climb, although at a reduced rate.

For the advanced countries the problem is not so pressing as it is for the developing ones. Not only do they have greater buying power, they have alternate sources already at hand. These are coal, four-fifths of whose known deposits are to be found in the temperature regions of the northern hemisphere, and nuclear energy to which the opposition, which is based mainly on environmental and safety considerations, is likely to weaken as oil supplies begin to run out.

The quest for renewable energy sources in these countries is therefore a marginal one. Just how marginal was shown by the wholesale abandonment of renewable energy projects in the U.S. and western Europe in the face of the decline in oil prices that occurred in 1982 and 1983. But for the oil-importing developing countries this quest is a matter of life and death. As the studied prepared for the Paris-U.N. conference on the least developed countries in September 1981 showed, in 1980 the 31 least developed countries suffered an outflow of wealth on account of the decline in their terms of trade amounting to 11 per cent of their national product! Country after country now faces economic collapse. And as if this is not enough, all the developing countries are facing a 'second' energy crisis, often termed the 'real' energy crisis, which is the disappearance of the forests that meet their most basic energy need—of fuel to cook their food with.

This is one challenge which the developing countries must meet on their own. Few of them have realised yet that in this area, continued dependence on the rich nations can prove fatal to their aspiration. This is not so only because the rich nations have vastly more money and untapped energy sources at their command, but also because their energy consumption patterns are very different from those of the developing countries.

While more than 97 per cent of the energy consumed in the advanced countries is 'commercial' energy—coal, oil and electricity—the ratio in the poor countries varies from 10 to 60 per cent. The balance is 'noncommercial' energy, obtained by burning wood, agricultural wastes, cowdung and other refuse.

Again, while in the rich nations one-third to two-fifths of the energy is consumed in the home, mainly for heating, in the warm nations of the 'south' where little or no house heating is required, the domestic consumption of commercial energy is negligible. For the same reason, while cooking accounts for only a fraction of the domestic energy consumption in the rich countries, in developing countries it accounts for the bulk of domestic, and up to half of the total energy consumption, both commercial and noncommercial.

Providing commercial energy in the villages or shanty towns, whether coal or electricity, will not solve their problem. Lacking the means to buy their cooking fuel, the poor will continue to use firewood or agricultural residues till such time as none is left, or till these are replaced by other fuels which are available at virtually no cost. This has been vividly demonstrated by the rural electrification schemes in India. In most 'energised' villages there are only a few connections, and these are invariably to the homes or tubewalls of the better-off farmers.

None of the many renewable energy technologies that are being offered today meets this specific problem. Solar cookers, solar heating systems wind driven power generators and small hydro-generators that can be used on canals will all cost money to install and maintain. Even community biogas plants will require the purchase of a simple gas stove, and the installation of a gas delivery system. Unless the government is willing to supply these to the poor at virtually no cost—a task that is beyond its resources—these too will become the preserves of the better-off in the villages. They will therefore at best reduce but not eliminate the pressure on the forests, and on agricultural residues.

The importance of solid waste recycling must be assessed against the background of this dual energy crisis. Municipal Solid Waste—garbage, contains a substantial amount of cellulosic matter, which can, like wood be converted into usuable forms of energy. It can be decomposed to yield compost, it can be burned to yield heat energy, and it can be pyrolised or gasified to yield large quantities of valuable fuel gas.

While composting and incineration are old and well established techniques of disposing of solid waste, pyrolysis and gasification are very new. A large number of processes are in the pilot to semi-commercial stages in the U.S., Germany, Japan, France and elsewhere. The developing countries could wait till these are fully proven, but there are good reasons why they will not serve their interests or their needs. In the industrial countries, interest in waste gasification has been triggered mainly by the growing concern over environmental pollution and the soaring cost of land in and around the major cities.

The problem of space for landfills had become even more acute because the explosive growth of income in the fifties and sixties has caused the volume of garbage to rise at a geometrical rate. While the average city-dweller in India generates 0.4 kg of garbage a day, in Japan he generates one kilogramme, in Europe and the U.K. somewhat more than that, and in the U.S. two kilogrammes a day!

The first solution that the rich nations adopted to this problem was to burn the waste. This decision made sense initially because in Europe and North America the heat that was generated could be fed into the central heating or street de-icing systems of the cities during the long cold winters. But in the late sixties, as concern for the environment grew, the incineration of garbage came, literally, under a growing cloud.

Incineration generates huge volumes of useless waste gas that carries literally hundreds of tonnes of fine ash which has to be orecipitated. What is worse, as the garbage has become more "technological", the proportion of highly noxious and corrosive chloride, nitride and sulphurous gases has increased sharply, contributing to the "acid rain" that has been experienced in more and more western cities in recent years. Cleaning up the huge gas discharge from incineration has therefore become more and more expensive as pollution control standards have been tightened and the proportion of toxic gases in the smoke has increased.

Scientists in the advanced countries therefore began to look for an alternative to incineration as a means of getting rid of urban solid wastes that is pollution free. The gasification of refuse is an obvious alternative, as the only residue—some seven per cent by weight—is a black glassy slag, that is an ideal material for roadbuilding. While the pioneers in this field, like Union Carbide of America, did stress that the fuelgas obtained from gasification was a valuable source of energy in major industrial countries this remained a marginal benefit which did not offset the much higher capital cost of the gasification and gas cleaning plant (estimated variously as 40 to 100 per cent above the capital cost of incineration plants with stringent pollution safeguards).

By contrast in the energy-starved developing countries it is the energy contained by the organic matter in the refuse that is of primary concern. All methods of solid waste disposal have to be evaluated with this in mind.

WASTE RECYCLING: THE INDIAN EXPERIENCE

I. COMPOSTING

Attempts to recycle Municipal solid waste began in India in the early seventies. Because of the high percentage of putrescable matter in it (see below) the first expedient tried out on a fairly large scale was to convert the waste into compost. During the first half of the seventies, no less than 35 composing plants were set up by various city authorities.

By 1972 however, only eight of these were still in operation, and all of these were running at a loss[1] The reasons were partly technical and partly economic.

The technical reasons stemmed from the fact that nearly all of the composting plants were simple open air fermentation plants, with equipment that had been gifted by various donor nations. Nearly all the technical problems stemmed from the fact that the equipment supplied was not suitable for Indian garbage and Indian conditions.[2] To begin with, the garbage had a very high moisture content typically 40 percent and above. This, combined with the large percentage of vegetable matter, made it difficult to shred it with the conventional shredders provided. Secondly, not only was the moisture content high but it varied very widely between the wet and dry seasons. This greatly impeded the efficiency of the composting operation.

But it was the economic non-viability of the composting plants that finally led to their closure. While such compost has found a ready and profitable market in the industrialized countries, notably in Europe,[3] it has proved extremely difficult to sell in India. In fact the various project authorities have been hard put to obtain even the transport cost of the compost from the buyers!

The reasons for this are to be found in the vast differences between farming in India and the advanced countries. While Indian farmers recognize the importance of compost for securing high yields, particular of vegetables, in contrast to the advance countries where dairying is a specialized activity, there are very few Indian farmers who do not own a few head of cattle. They are thus in a position to make their own compost and find little attraction in a compost manufactured in the cities and trucked out to them in the villages.

---

[1]Information obtained from Dr. B. R. Nagar, Adviser, Planning Commission, Government of India.
[2]This was the experience of the Bombay plant. We understand this was not unusual elsewhere.
[3]Discussion with representatives of the French firm SOBEA, which have a plant outside Paris to make compost and Refuse Derived Fuel.

II. INCINERATION

The failure of composting has made city authorities turn to other ways of making use of urban waste. Two such projects are likely to be implemented in 1984. The first is the establishment of a 300 tonne/day incineration plant plus turbogenerator to burn waste and generate electricity in New Delhi, the national capital, and the other is to establish a 200 tonne/day waste pyrolysis or gasification plant to produce methanol in Bombay (the final choice between the two processes has yet to be made).

The incineration plant is of the rotary kiln type, and is to be supplied by the Danish Company Messrs. Volund Miljoteknik A.S. of Denmark.[4] It will consist of two Rotary Kiln type incinerators which, along with associated equipment power generators, civil works, spares and supervision for two years, will cost at March 1983 prices $14.6 million.

The plant requires a total staff of 34 (which in Indian conditions is likely to swell to a hundred) and its operating costs are expected to come to $5.9 million a year. This is based on the assumption that amortization, with depreciation over 25 years and an interest rate of 9 percent, will come to 10.18% of capital costs for machinery, and on slightly different assumptions to 9.78% for buildings; that maintenance will cost 4% of machinery costs and 2% of building costs.

The plant is to use the waste heat from incineration to generate electricity. Calculations of electricity output are based on the calory value of the MSW. There is an element of doubt over this.

Indian government officials gave the composition of the refuse as

        Combustibles   56.34%
        Humidity   14.72%
        Inerts   28.94%

and a net calorific value of 2,853.0 tn/lb = 1,586.3 kcal/kg

But Volund found that this did not tally with the chemical analysis of the waste which is as follows: They pointed out that even if the whole of the combustible matter consisted only of cellulose, (NCV 3860 kcal/kg in dry condition) the calorific value of the garbage would be 2,174.2 kcal/kg. They therefore concluded that the amount of moisture in the garbage had been seriously underestimated.

As a result Volund's calculations are based on the assumption that combustibles make up 40.16% of the garbage, humidity 30% and inert matter 29.84%. On the basis they have assumed a minimum calory value of 1642.5 kcal/kg. After allowing for the heat required to drive out the moisture, they have obtained a net calorific value of 1,462.5 kcal/kg.

---

[4]The details given here are from the project report submitted by Messrs. Volund for the New Delhi Project.

Output: Based on this, they have calculated the incinerator will transfer to the boiler 1249.1 kcal/kg of heat, and that this will gnerate 299.6 kw per tonne of waste. Based on 304 days (6500 hrs) of working per year, this will generate 24,342.5 mwh a year, or about 6 mwh per day. After allowing for in plant use of electricity, this comes to 21,092.5 mwh in the year.

Operating Results
   The project report calls for this power to be fed into the Delhi state power grid. The Delhi electric supply undertaking pays 0.45 (4.5 U.S. cents) per unit. On this basis, its anticipated profit and loss account is as follows:

	Rs. 10.00 = US. $1.00
1. Income from sale of electricity	Rs. 9,491,625
2. Operating Costs	Rs. 5,950,200
3. Operating Surplus	Rs. 3,542,425
4. *Less* Amortization	Rs. 14,729,56
5. Pre-tax profit	= Rs. − 11,188,135

However, the Commission for Additional Sources of Energy, (CASE, now The Department of non-conventional energy sources) which approved of the project pointed out that the real value of the electricity is the current cost of generation at the margin, i.e., by diesel generators that are being used to meet the shortfall in coal and hydel based power by factories in and around Delhi. This was calculated at Rs. 1.40 per unit. On this basis imputed income goes up by Rs. 29,529.500 and pre-tax profit to Rs. 8,850 millions.

Evaluation
   It should be conceded straightaway that on any analysis of social costs and benefits, the above project would be socially viable. The marginal cost of electricity generation given above is the most conservative estimate of this social return. In fact even this is underestimated here, because since the above estimate was made (around March 1982) the cost of diesel has gone up substantially, and the current estimate of generating cost with captive diesel units is Rs. 1.80 to Rs. 2.00 per unit.
   Furthermore, the real returns from power generation come from the additional output and employment they generate further down the production chain. In India it is estimated that one rupee worth of power generates 15 rupees worth of final output.
   But notwithstanding these considerations, the fact remains that the project is commercially non-viable and this makes it most unlikely that it will be taken up on a large scale. To begin with, since the total amount of garbage generated by Delhi is over 2,000 tonnes a day, seven such plants will be needed for the city as a whole. The annual cash loss of the Corporation will therefore be Rs. 78 million ($7.8 million). All city authorities are so short of funds, in the third

world, that it is most unlikely that many of them will accept so great a loss from the very outset.

Secondly, the calcualtions of revenue from electricity generated are actually *potential* and not realized earnings. They assume that all the electricity that the plant can generate will be fully absorbed at all times by the DESU grid. This may be justified when there is one 300 t/d plant generating 6 mwh per day, for this amount is truly marginal to Delhi's needs. But it is not certain that the same assumption will hold true when 42 mwh per day is being generated.

A similar objection applies to using a Rs. 1.40 or Rs. 1.80 per unit social rate of return. Is it justifiable to assume that there will always be a power shortage to meet base loads (and not just peak loads) during the entire 25-30 year life of the plant?

We thus come to the paradoxical conclusion: the plant looks quite attractive when it makes only a marginal contribution to energy supply, but becomes unattractive when it threatens to make a substantial contribution. The Delhi incineration project is thus one more example of the transfer of inappropriate technology from the "north" to the "south."

In the northern countries such plants are more attractive partly because the main purpose from the very beginning is waste disposal. Even on the energy front, such plants are more attractive in the cold countries because the waste heat can be fed into city and street heating systems in winter. But there is no need for such systems in the warmer "southern" countries. In the case of the Delhi plant there was much talk in the approval stage of using the heat to run ice-making plants and cold storages, but nothing has come of this and Volund's project report provides for a cooling tower to vent the waste heat!

III. PYROLYSIS OR GASIFICATION OF URBAN WASTE[5]

The third alternative, to subject the Municipal waste to pyrolysis or gasification, looks the most promising but preparatory work on the pilot/semi commercial project, to treat 200 tonnes/day of refuse in Bombay in this way is not as far advanced as for the incineration plant in New Delhi. The main reason is that the technology itself is relatively new.

While research into the pyrolysis or gasification of urban waste was begun in Denmark, U.K., U.S.A. and elsewhere in the late sixties, till as recently as the end of 1981, there were no proven processes available even to treat the waste of the industrially advanced countries.

However, in September 1981, the first fully commercial plant to gasify municipal waste came on stream. It was set up by Messrs. Showa Denko, of

[5]The information contained in this section is based on the findings of a study team sent out by the Maharashtra government in November 1982, to locate technology for the Bombay plant, of which the author was a member.

Japan and based on Union Carbide's "Purox" process, in which gasification is carried out with the aid of a jet of pure oxygen in preference to air. This was a 150 tonne/day plant at Chichibu some 80 kms from Tokyo.

At about the same time, another Japanese Company, Ebara, perfected a fluidized bed, twin-chamber pyrolysis process to break down waste into char and fuel gases. Their 100 tonne/day plant has been run continuously for as much as 8,000 hours and in 1982 was being used for demonstration runs with U.S. refuse derived fuel.

Both processes are highly attractive because they yield a very large volume of energy rich fuel gases, consisting mainly of methane (CH_4) Carbon monoxide (Co) and Hydrogen (H_2), and around 5 percent by dry weight of input, of blown *char* which can be bricketted and sold as a cheap cooking fuel to the power segments of the urban population.

The fuel gases too can be used in a variety of ways. One would be to separate the hydrogen, and sell the carbon monoxide and methane as town gas. The hydrogen can then be used to manufacture ammonia and then urea. Another is to break down the methane further and synthesize carbon monoxide and hydrogen into methanol, or gasoline (either directly as is being done in the SASOL plant in South Africa, or via methanol by the Mobil process). Lastly, the whole of the gas can be used to make urea, as is done in gas-based fertilizer plants.

The twin attractions of this mode of waste recycling are that firstly, it is the only one that meets the growing and urgent need for cooking fuel in the towns. At present according to a report on the state of the environment in India, fully half this need is being met by firewood obtained illegally from the fast disappearing forests of the country.

Secondly, the end products into which fuel gas can be synthesized are precisely those middle range oil products—transport fuels and gas, a substitute for naphtha for fertilizers—for which demand is growing most rapidly in the developing countries.

Costs and Returns:

An economic evaluation of pyrolysis and gasification plants in a third world country is complicated by the fact that there is as yet no such plant in operation. What is more, at the time of writing the Maharashtra government had not decided which process to opt for, so no detailed offer document similar to that of Volund for incineration had been received.

There is thus an element of uncertainty about the likely costs and returns from the two processes. Notwithstanding this, it can be said without hesitation that provided the pilot plant being set up in Bombay succeeds in establishing the process with the relatively energy poor waste of the developing countries, the processes will prove highly profitable in both social and commerical returns.

(a) Commercial costs and returns: A study team sent abroad by the

Maharashtra government to locate technology for the Bombay plant came across widely divergent capital cost estimates and quotations for various pyrolysis and gasification plants currently under development. The following examples illustrate this clearly:

Pyrolysis

1. C. Otto "Odapyr" Process—6 tonnes/hr.	$ 12 million
2. Foster-Wheeler "Tyrolysis" plant—6 tonnes/hr.	$10.8 million
3. The Ebara "Stardust" 80 process 12.5 tonnes/hr. quotation for wood chips pyrolysis in North America.	$ 18.4 million
4. Showa Denko "Purox" plant—18.75 tonnes/hr. (approximate estimate)	$ 43 million

An informal offer from Ebara to the study team for a 200 tonne/day waste classification coupled with a 60 tonne day gasifier, quoted an equipment cost of $3.5 million.

This wide variation makes it difficult to estimate capital costs. But even with the most pessimistic estimates, namely that to convert Bombay daily output of MSW of 3500 tonnes into methanol the total capital cost will come to $140 million to $220 million.

In the same way, the operating costs will depend very much on whether pyrolysis or gasification is chosen, the degree of automation incorporated into the plant, and the rates of interest and depreciation.

Tentative estimates put forward by the study team appointed by the State of Maharashtra indicate that this could be $20.8 million to $31 million. (Details and assumptions on which calculations are based are given below).

The value of output depends critically on the choice of technology. The Ebara "Stardust 80" pyrolysis system gives an "across-gasifier" energy recovery of 52 percent in the form of a fuel gas. By contrast the "Purox" gasification process adopted by Showa Denko gives an efficiency of 75 per cent or more.

If electricity for the plants' operation is provided from outside the latter will yield almost 50 per cent more methanol than the former. However, based on the very tentative estimates of capital cost available to the team, it seemed that the gasification process would also be correspondingly more capital intensive.

As a result, both processes are likely to give approximately 27 to 29 percent return on capital employed at 300 days working. The details are as follows:

Capital Cost of a plant for Bombay's MSW (a) Pyrolysis

After removing the 40% inorganic matter and reducing the moisture content from 50 to 30 per cent, the total daily input into the gasifiers from 3,500 tonnes of raw MSW will be about 1,680 tonnes a day. If the gasification plants run an average of 300 days a year, this will require the established input processing capacity to be 2050 tonnes a day.

Applying the rule of thumb for process plants that doubling the size of the unit will increase capital costs by 1.66 times, the capital cost of a 2400 tonne/day Ebara pyrolysis plant will be Rs. 86 crores.

Capital cost of methanol plant

This will depend on the total amount of methanol to be obtained, which in turn depends on the heat value of the feedstock and of Methanol, and the energy recovery efficiency of the entire plant.

The heat value of Methanol is 8570 BTU/lb.

The heat value of the above treated feedstock is 4500 BTU/lb.

The energy recovery efficiency of the Ebara gasifier is 52 percent. The energy recovery efficiency of a methanol plant based on natural gas (mainly Methane) is 63 to 64 per cent.

However, since the first stage of such a plant—gas cleaning and methane reformation into CO & H_2 is endothermic, the fact that only 18 per cent by volume of the fuel gas from the Ebara gasifier is methane will reduce both the capital and the energy cost of methane reformation. Thus it is assumed here that a methanol plant working on fuel gas will have an energy recovery efficiency of 70 per cent. The daily output of methanol will be

$$2050 \times \frac{4535}{8570} \times 0.52 \times 0.7 = 435 \text{ tonnes.}$$

The World Bank's report Emerging Energy and Chemical Applications of Methanol: Opportunities for Developing countries (World Bank April 1982) gives the estimated cost of a 1000 tonne/day methanol plant in a developed site in a developing country as $128 million.

Applying once again the engineers thumb rule that the cost of a plant half this size will be 0.6 times, the capital cost comes to $76.8 million.

However, the cost of the methane reformer is 30 per cent of this. Assuming that this is reduced by 2/3rds because less than one-fifth of the fuel gas (composed of Methane) needs to be reformed, the cost of the methanol plant based on fuel gas will be $54.5 million, that is equal to Rs.54 crores.

This total cost of the MSW to methanol plant will be Rs. 140 crores.

Operating Costs:

This is calculated under the following assumptions:

1. Maintenance is assumed to be 3 per cent of capital cost.
2. Depreciation is taken on a straight line basis assuming a plant life of 20 years.
3. A total of 500 workers are employed at the rate of Rs. 20,000 per worker.
4. Managerial costs are 50 per cent of workers' costs.
5. Catalysts for methanol synthesis etc. cost $5.00 per tonne = Rs. 50 per tonne of ouput.
6. Power costs: based on 270 ksh consumption per tonne of Methanol in a modern low pressure methanol plant and assuming that another 50 percent of this is consumed in the upstream plants, a total of 52.85 m kwh is required

valued at about Rs. 2 crores a year.
7. Other overhead costs = 1% of capital costs.
8. Contingencies are 15 percent of above costs less depreciation.
Breakup of operation costs-

1. Maintenance	—	Rs. 4.2 crores
2. Depreciation	—	Rs. 7.0 crores
3. Labour	—	Rs. 1.0 crores
4. Management	—	Rs. 0.5 crores
5. Catalysts	—	Rs. 0.8 crores
6. Power	—	Rs. 2.0 crores
7. Other overheads	—	Rs. 1.4 crores
8. Contingencies	—	Rs. 1.6 crore
Total operational cost		Rs. 18.5 crores

Returns   In Indian Prices:

Ex-factory cost of methanol	—	Rs. 4840 per tonne
Daily output	—	435 tonnes
Annual output (300 days)	—	130,500 tonnes
Value of annual output	—	Rs. 63.2 crores

Gross profit—Rs. 63.2 crores minus Rs. 18.5 crores = Rs. 44.7 crores
Capital employed Rs. 140 + 11.5 = Rs. 151.5 crores
Rate of return on capital employed = 29.5 per cent.
Capital costs: (b) gasification
(a) Gasifier—The estimate of $43 million for a 450 tonne/day plant given to the team by Messrs. Showa Denka, does not include the cost of an SPC-II classifier or a drying unit. It is assumed here that these cost an additional $3 million.
Total capital cost—$46 million.
Therefore cost of 2200 tonne/day plant = Rs. 140 crores.[6]
(b) Methanol plant:
Since the energy recovery efficiency of the Purox gasifier is 75 per cent against 52 per cent for the Ebara Pyrolysis unit, the output of methanol per day on the same 300 day basis as used above will be

$$435 \times \frac{75}{52} = 627 \text{ tonnes.}$$

The cost of a plant of this size using natural gas will be around $100 million. However, in view of the very small fraction of methane in the fuel gas (9 per cent or less) the need for methane reformation will be very small. We therefore can safely assume the plant cost to be $80 million.

	= Rs. 80 crores
Thus total capital cost	= Rs. 220 crores

---

[6]This figure is not comparable with the Ebara figure. The latter is a competitive tender, the former is to the Japanese government.

Operating Cost: (On same assumption as for Ebara plant)

1. Maintenance	—	Rs. 6.6 crores
2. Depreciation	—	Rs. 11.0 crores
3. Labour	—	Rs. 1.0 crores
4. Management	—	Rs. 0.5 crores
5. Power	—	Rs. 3.0 crores
6. Catalysts	—	Rs. 1.1 crores
7. Other overheads	—	Rs. 2.2 crores
8. Contingencies	—	Rs. 2.1 crores
	Total	Rs. 27.5 Crores

Operating Revenue: 627 X 300 X 4840 = 91.04 crores

Gross Profit	—	Rs. 63.5 crores
Capital employed	—	Rs. 220 + 12.3 = 232.3 crs.
Return on capital employed		= 27.4 per cent.

Socio economic benefits

The above studies are based on extremely conservation estimates of the calorific value of Bombay waste (approximately 1200 kcal/kg. mainly because of a higher moisture content) and very high estimates of the cost of the gasifier. The cost of the methanol plant has been based on estimates contained in a World Bank report of April 1982, for a developed site in a developing country. On the gasifiers in particular, offer received from a Japanese company indicates than an actual quotation would be considerably lower. But even under the above unfavorable set of assumptions possible, the two processes are likely to give highly respectable rates of return on capital employed. The main reason is that the price of methanol (about $480 to $500 per tonne) is about twice the c.i.f. price of imported methanol. The margin is admittedly high but at least partly reflects the scarcity of methanol and the high premium attached to the conservation of foreign exchange by the Indian government.

But the social benefits from the process far outweigh the commerical returns. To begin with, the additional revenues will enable the plant to pay the entire cost of garbage collection, now $22.5 million annually, and leave a tidy sum over to finance the expansion of water, transport and other essential services. Secondly, the methanol produced will directly substitute for imports, estimated at 60,000 tonnes in 1981-82, and likely to rise to 100,000 tonnes or more by 1990.

Lastly, either process will yield about 100 tonnes of blown char that can be bricketted and sold as a cooking fuel to the poor. Since such brickettes are four times more efficient as a cooking fuel than firewood, besides being of a higher calorific value, this will substitute for at least 500 tonnes of firewood a day, thereby largely eliminating the inflow of firewood into the city. The reduction of pressure on the remaining forests of Maharashtra will be very considerable.

Part 5
Laws, Regulations, and Socioeconomic Considerations

As the waste materials of an industrial society have grown, there has arisen the needs for regulations and laws to protect, not only the environment, but people as well.

The initial chapter provides an evolution and analysis of Pennsylvania and Federal regulations of solid wastes. Until the passage of the Federal Resource Conservation and Recovery Act of 1976, regulation of solid and hazardous wastes was exclusively a state and local activity. In 1980 Pennsylvania modernized its Solid Waste Management Act (known as Act 97) to qualify the Commonwealth for Federal authorization. Although Federal and state laws are being developed, the local municipal governments play a major role through their zoning power and prosecution of offenders. Issues on health planning, education and politics related to solid and liquid wastes are covered in Chapter twenty-seven.

Becuase the disposal of waste products is usually unsightly, smelly, and possibly noisy, public resistance is developing to having these degrading activities in many communities. Many individuals believe that property values will be lowered by nearby landfills. However, in a study of 10 sanitary landfill sites in Pennsylvania it was found that residential property values were not lower than prices of residential properties in other areas. It was concluded that the fears of many people over substantial losses in property values due to newly established landfills may be unfounded.

This part concludes with a consideration of the human element involved in solid waste disposal. While many of the problems are pragmatic, others are emotional in community resistance to establishing new sanitary landfill sites. While the difficulties are many in gaining community acceptance, it must be recognized that fundamental to this process is communication of the need and dissemination of all pertinent facts to the community so that informed, rational decisions can be reached.

Solid and Liquid Wastes: Management, Methods and Socioeconomic Considerations. Edited by
S. K. Majumdar and E. Willard Miller. © 1984, The Pennsylvania Academy of Science.

Chapter Twenty-Six

Pennsylvania and Federal Regulation of Solid Wastes

Karin W. Carter, Esq. and William R. Sierks, Esq.

Assistant Counsels
Pennsylvania Department of Environmental Resources
Bureau of Regulatory Counsel
505 Executive House,
P. O. Box 2357, 1015 Second Street
Harrisburg, PA 17120

The regulation of nonhazardous solid waste has remained the domain of state and local governments, while the regulation of hazardous waste management activities has become a matter of Federal jurisdiction and significant Federal activity. For those who generate or manage nonhazardous solid wastes in Pennsylvania, or who contemplate doing so in the future, it is therefore essential to become aware of the legal constraints imposed at the state and local level which could alter the course of such activities or prohibit them altogether. For citizen groups, individuals and government agencies who view solid waste management as an activity which poses considerable hazard to the economic or environmental well-being of the host community, it is equally important to have a working knowledge of the ways in which local and state governments, and members of the public to whom these governmental institutions are responsible, can prevent the adverse impacts and excesses which have occurred in the past from happening again. It is the purpose of this paper to summarize for the benefit and assistance of interested readers the respective jurisdictions and powers of the Federal, state and local governments over nonhazardous solid waste management activities in Pennsylvania.

I. FEDERAL REGULATION OF SOLID WASTES IN PENNSYLVANIA

Until Congress passed the Resource Conservation and Recovery Act of 1976[1] (usually referred to as "RCRA"), regulation of solid and hazardous waste management in the nation was exclusively a state and local activity. Because of public pressure for a uniform approach to what was perceived to be an interstate, if not national, problem, Congress asserted Federal regulatory authority over the transportation, treatment, storage and disposal of hazardous wastes by establishing permit and manifest systems for such wastes. Management of wastes which did not meet the statutory definition of, or the regulatory criteria identifying, hazardous wastes was, however, left largely in the hands of state and local government. The only area where Congress insisted upon some Federal jurisdiction over nonhazardous waste management was in the matter of open dumps. Section 4004 of RCRA required EPA to promulgate regulatory criteria for determining which land disposal facilities were open dumps and which facilities were sanitary landfills. Thereafter all open dumps were prohibited unless such facilities were on a schedule to upgrade and achieve compliance within five years. The criteria were published in the Federal Register on September 13, 1979[2] and codified at 40 CFR 257. The criteria provide minimum national standards for landfilling solid waste with respect to protection of floodplains, endangered species, surface water, ground water, food chain crops, air and safety and with respect to disease prevention. Since RCRA required that an annual national inventory of open dumps be published, EPA required the states to use Federal grant funds to develop a plan for closing or upgrading open dumps and compiling lists which EPA could use in assembling its inventory. Pennsylvania has participated in this program since its inception, and the open dump inventory published by EPA in 1983 lists 88 Pennsylvania facilities, many of which were already in litigation. EPA has stated that the listing of such facilities does not constitute a legal determination subjecting any party to Federal sanctions.[3] The list is, however, intended to contain those facilities which have been observed to violate one or more of the criteria. With the exception of the open dump program, regulation of nonhazardous solid waste management activities in Pennsylvania is carried out by state and local government. Moreover, the closing or upgrading of those facilities on the open dump list is being accomplished by the State, using enforcement remedies provided by state statutes and regulations, rather than through RCRA.

---

[1] 42 U.S.C. 6901 *et seq.,* as amended.

[2] Vol. 44, No. 179, p. 53438.

[3] *EPA Guidance Manual for Classifying Solid Waste Disposal Facilities,* SW-828, March 1980.

II. STATE REGULATION OF SOLID WASTES IN PENNSYLVANIA

Regulation of solid waste management is not new in Pennsylvania. Since the effective date of the Pennsylvania Solid Waste Management Act of 1968[4] (usually referred to as "Act 241"), the state has required operators of solid waste processing and disposal facilities to obtain permits for such activities. Enforcement of Act 241 and the regulations adopted thereunder in 1971[5] has been the responsibility of the Pennsylvania Department of Environmental Resources (usually referred to as "DER") since the agency was created in 1970.

In 1980, a new Pennsylvania Solid Waste Management Act[6] was passed which replaced Act 241. The new statute (usually referred to as "Act 97") made vast changes in the state's regulatory approach to hazardous waste management problems in a manner which qualified the Commonwealth for Federal authorization under RCRA. The changes in the state's regulatory approach to nonhazardous waste management, while not so sweeping, were significant.

The most noticeable innovation embodied in Act 97 was a change in the way solid wastes are classified. Instead of broadly referring to "solid wastes," the regulatory scheme now divides the regulated substances into three categories: hazardous wastes, residual wastes, and municipal wastes. The hazardous waste definition is similar to that found in RCRA, as is the regulatory identification scheme. Residual wastes are defined in Act 97 as nonhazardous wastes:

"resulting from industrial, mining and agricultural operations and any sludge from an industrial, mining or agricultural water supply treatment facility, waste water treatment facility or air pollution control facility . . . "

Coal refuse and acid mine drainage treatment residues are excluded from the definition. Municipal wastes are defined as any wastes:

"resulting from operation of residential, municipal, commercial or institutional establishments and from community activities and any sludge not meeting the definition of residual or hazardous waste hereunder from a municipal, commercial, or institutional water supply treatment plant, waste water treatment plant, or air pollution control facility."

The terminology of this classification scheme is now used in permits and applications for permits for municipal or residual waste processing or disposal facilities and will be used in new regulations expected to be promulgated in 1985. Act 97 requires operators of municipal and residual waste processing and disposal facilities to obtain permits from DER, but allows those who merely transport and store such wastes to do so without permits as long as they comply with DER storage and transportation regulations. Hazardous waste transporters, on the other hand, must have licenses; and hazardous waste storage may only be conducted under permit. The determination (usually made by a

---

[4] Act of July 31, 1968 (P.L. 788, No. 241) *as amended,* 35 P.S. 6001 *et seq.*
[5] 25 Pa. Code Chapter 75.
[6] Act of July 7, 1980 (P.L. 380, No. 97), 35 P.S. 6018.101 *et seq.*

generator) of whether a waste is hazardous, residual or municipal thus has significant regulatory consequences. If the waste is residual or municipal, it may be stored and transported without a permit, license or manifest under a regulatory scheme which (except for the open dump criteria) is entirely created and implemented by state and local government. If the wastes are determined to be hazardous, they must be managed in accordance with a more pervasive regulatory structure enforced by both Federal and state government.

A second major change brought about in 1980 by Act 97 was the imposition of bonding and insurance requirements for permitted facilities. Except for municipally-owned landfills used for disposal of a municipality's nonhazardous wastes, all permitted facilities must be bonded with a surety or collateral bond of a type authorized in Section 505 of Act 97, payable to the Commonwealth and filed with DER prior to permit issuance. The amount of the bond must be at least $10,000 and is set at an amount sufficient to perform final closure of the facility and such monitoring, post-closure care and remedial measures as are necessary to prevent adverse effects upon the environment. The operator of a permitted facility must also carry an ordinary public liability insurance policy in an amount to be specified by regulations expected to be promulgated in 1984.

A third change brought about by Act 97 was an increase in the types and ranges of enforcement remedies available against violators of Act 97, DER solid waste regulations, DER solid waste permits or DER solid waste orders. Civil penalties of up to $25,000 per day per violation may now be imposed upon violators, as well as criminal fines of up to $25,000 per day and imprisonment of up to one year. The Commonwealth may continue to bring actions in equity to restrain violations, and DER continues to have available the administrative remedies of revoking or suspending permits and issuing orders. Violations of Act 97, DER solid waste regulations, DER permits and DER orders are declared to be public nuisances.

One other change embodied in Act 97 which is of particular significance to municipalities is the increased ambit of the municipal solid waste planning responsibility. Under Act 241, only those municipalities with a population density of 300 or more inhabitants per square mile were required to submit for Department approval a plan committing the municipality to a means of providing for its municipal waste disposal needs for a ten-year period. Act 97 has now extended that requirement to municipalities with a population density of less than 300 inhabitants per square mile where DER has identified a waste problem or a potential waste problem.

III. MUNICIPAL REGULATION OF SOLID WASTE MANAGEMENT

The impact of a solid waste management facility on the community which

332 Solid and Liquid Wastes: Management, Methods and Socioeconomic Considerations

surrounds it is a concern and a responsibility of local government as well as state governments. Even though Pennsylvania's regulatory system designates a state agency as the primary regulator of solid waste management in the Commonwealth, local governments retain or have been given four important sources of authority to assist them in assuring both that the solid waste needs of their citizens are taken care of and that the health and safety of their residents are not jeopardized by solid waste management activities.

The first source of municipal authority which affects the location of solid waste management facilities in Pennsylvania is the zoning power. There is now a considerable body of case law affirming the principle that a zoning ordinance is an independently enforceable set of requirements which applies to solid waste management facilities independently of DER requirements. In other words, it is not sufficient for a facility operator to obtain a DER permit; in order to operate legally, he must also obtain whatever zoning variances or approvals are required by the applicable local zoning ordinance. If he cannot obtain such approvals, he cannot legally operate. *Greene Twp. v. Kuhl, et al.* 32 Pa. Commonwealth 592, 379 A.2d 1383 (1977). However, municipalities may not use their zoning powers to zone out solid waste facilities. Exclusionary ordinances which prohibit waste disposal altogether, or ordinances which only allow such activities in clearly infeasible locations have been invalidated by courts.[7] However, an ordinance which banned privately operated (as opposed to municipally operated) facilities has been upheld.[8] Even in industrially zoned areas, the operator of a solid waste facility must comply with setback and other restrictions in order to operate legally.

With respect to other local ordinances not of a zoning nature, the case law is incomplete. In *Greater Greensburg Sewage Authority v. Hempfield Twp.*, 5 Pa. Commonwealth 495, 291 A.2d 318 (1972), Commonwealth Court invalidated an ordinance which established a local government permit program, partly because of language contained in Section 10(b) of Act 241 which preempted local ordinances conflicting with the state regulatory program. Similar language is contained in Section 202(b) of Act 97, but no cases interpreting that section have yet reached Commonwealth Court for decision. The best indication of Commonwealth Court's opinion on the proper roles of local and state government in regulating solid waste management facilities is that contained in *Green Twp. v. Kuhl, supra*. In that case, Commonwealth Court quoted with approval a lower court opinion which stated:

"[A]local municipality cannot set geological or engineering standards stricter than those established by DER for issuance of its permit. However, factors other than geological ones, such as those involving aesthetics,

---

[7]*General Battery Corp. v. Zoning Hearing Board,* 29 Pa. Commonwealth 498, 371 A.2d 1030 (1977); *Kefo, Inc. v. Greenwood Twp.,* Crawford County Court of Common Pleas, Civil No. 160, February Term 1978 (May 29, 1979).
[8]*Kavanaugh v. London Grove Twp.,* 486 Pa. 132, 404 A.2d 393 (1979).

population density and accessibility govern the selection of a landfill site, and these factors are the appropriate subject of local land use planning." 32 Pa. Commonwealth at 595.

More extensive discussion and guidance from the Court will undoubtedly be forthcoming in future litigation, as public pressures cause municipal officials to assert as much regulatory authority over solid waste management facilities as the courts will allow.

Another power possessed by local governments which is applicable to solid waste management facilities is the power of prosecution. Section 604 of Act 97 explicitly empowers county district attorneys and municipal solicitors to bring suits in equity to enjoin public nuisances and violations of the statute or DER regulations. The statutory declaration of nuisance in Section 601 of Act 97 for violations of DER regulations, orders or permits not only implements the equity action provision in Section 604 but also invokes the nuisance abatement powers which local governments have traditionally exercised. Moreover, the criminal prosecution power is exercised in Pennsylvania by county district attorneys as well as by the Attorney General, and criminal sanctions authorized by Act 97 may thus be pursued on the initiative of a local government which has the evidence to prove that crimes defined in Act 97 have been committed.

Act 97 has also given local governments the opportunity to participate in the DER solid waste facility permit process. Section 504 of the statute requires that solid waste facility permit applications be reviewed by the host county, county planning agency, county health department (where one exists), and host municipality, for the purpose of recommending to the Department approval of, disapproval of, conditions upon, or revision to the permit application. Local governments are given sixty (60) days within which to perform this task. If no comments are submitted within that time period, the local governments are deemed to have waived their right to review and comment. If the comments are submitted and the Department disagrees with them, the Department must publish in the Pennsylvania Bulletin its justification for overriding the local government's recommendations. Within thirty (30) days after the permit has been granted or denied by DER, a host municipality or county which believes DER's decision to be wrong may appeal the decision to the Environmental Hearing Board.[9] This three-member board, which is independent of DER, functions much like a trial court in that it holds adjudicatory hearings (presided over by an individual Board member) which are adversary proceedings conducted according to the Pennsylvania Rules of Civil Procedure, and it issues adjudications containing findings of fact and conclusions of law which are binding upon DER and the parties. These adjudications may then be appealed to Commonwealth Court, whose opinions may in turn be appealed to the Pennsylvania Supreme Court.

---

[9]This right has recently been affirmed in *Franklin Twp. v. Commonwealth, 499 Pa. 162,* 452 A.2d 718 (1982).

Finally, Pennsylvania local governments have the power to control and direct the flow of municipal waste to specific facilities. This power has in recent years come under attack in Ohio and in Iowa[10] because municipalities which have invested in resource recovery facilities or other expensive municipal waste disposal technology have insisted on committing their wastes to these facilities, thus depriving the private competing waste disposal facilities of business and the haulers which patronize them, of profit. The basis for the attack has been the claim that the municipalities involved have engaged in anticompetitive practices in violation of Federal antitrust laws. Because Pennsylvania municipalities may well face the same challenge, it is worth examining how the Commonwealth's laws have been drafted to give municipalities the legal basis for defending themselves if this antitrust claim were ever brought against them.

Act 97 establishes a broad and comprehensive program for the planning and regulation of solid waste storage, collection, transportation, processing, and disposal. Section 202(c), which is the key provision for antitrust purposes, states:

" . . . In cases where the planning agency determines and the governing body approves that it is in the public interest for municipal wastes management and disposal to be a public function, the plan shall provide for the mechanisms. *Municipalities are authorized to require by ordinance that all municipal wastes generated within their jurisdiction shall be disposed at a designated facility."* (35 P.S. §6018.202(c), emphasis added)

Several other sections of the Act are also relevant to the antitrust issue. In Section 202(a), the General Assembly makes each municipality responsible for the collection, transportation, processing, and disposal of municipal waste which is generated or present within its boundaries (35 P.S. §6018.202(a)). The legislature further prohibits municipalities from delegating their responsibility for collection, transporting, processing and disposal of waste. (35 P.S. §6018.202(c)). Finally, many municipalities are required under Section 201 of the Act to prepare and implement plans for municipal waste management systems; such plans must address procedures for the collection, transportation, treatment and disposal of their municipal wastes.

Section 306(A) of the Municipal Authorities Act of 1945, 53 P.S. §301 *et seq.,* also indicates that the legislature contemplated anti-competitive activity in the area of waste disposal. The Municipal Authorities Act restricts municipal authorities from construction, improvement, maintenance or operation of any project which duplicates or competes with existing enterprises serving the same purposes. However, collection and disposal of waste are expressly excepted from this general limitation.[11]

It is thus apparent that the Pennsylvania legislature has expressly authorized its municipalities to restrict disposal of municipal waste to designated land-

---

[10]*Glenwillow Landfill, Inc. v. City of Akron,* 485 F. Supp. 811, 654 F. 2d 1187, vacated, ____U.S.____, 102 S. Ct. 1416, 71 L. Ed. 2d 640. *Central Iowa Refuse Systems, Inc. v. Des Moines Metropolitan Area Solid Waste Agency, et al.,* Civil No. 79-32-1 (S.D. Iowa, Dec. 10, 1982).

fills. The next question is whether this type of authorization is sufficient to exempt municipalities from Federal antitrust laws. This question can only be answered by examining Federal case law on the subject.

Under the Sherman Antitrust Act, an agreement in restraint of trade is unlawful. (15 U.S.C.A. §§1,2.) However, the courts have long recognized that actions of the State itself are not covered by the Sherman Act. This "state action" exemption from antitrust laws was first articulated by the U.S. Supreme Court in *Parker v. Brown*, 317 U.S. 341, 63 S.Ct. 307, 87 L.Ed. 315 (1943). In *Parker*, a California raisin grower sued state officials who had established a proration system authorized by state law which allowed the state to impose marketing programs for raisins. It was unlawful for growers not to follow the program. The Supreme Court assumed that such a program would violate the Sherman Act if it were carried out by private parties. However, the Supreme Court held:

"We find nothing in the language of the Sherman Act or in its history which suggests that its purpose was to restrain a state or its officers or agents from activities directed by its legislature . . . The Sherman Act makes no mention of the state as such, and gives no hint that it was intended to restrain state action or official action directed by a state." (317 U.S. at 350-51, 63 S.Ct. at 313)

The Court in *Parker* concluded that because the Sherman Act is directed against the state as sovereign, state regulator programs which impose restraints as an act of government could not violate it. *Id.*, 317 U.S. at 351-52, 63 S.Ct. at 314.

The Supreme Court has reviewed the "state action" exemption announced in *Parker* in several recent cases. In *California Retail Liquor Dealers Assn v. Midcal Aluminum, Inc.,* 445 U.S. 97, 100 S.Ct. 937, 63 L.Ed.2d 233 (1980)), the Supreme Court indicated that *Parker* and subsequent cases establish a two-prong test for antitrust immunity under the "state action" exemption:

"First, the challenged restraints must be 'one clearly articulated and affirmatively expressed as state policy'; second, the policy must be 'actively supervised' by the State itself." (445 U.S. at 105, 100 S.Ct. at 943, quoting *City of Lafayette v. Louisiana Power and Light Co.*, 435 U.S. 389, 410 (1978).

In *Midcal,* the court held that a California system for wine pricing met the first part of the test, since the legislature had clearly stated its intent to permit

---

[11]Section 306(A) states: "This limitation shall not apply to the exercise of the powers granted hereunder for facilities and equipment for the collection, removal or disposal of ashes, garbage, rubbish and other refuse materials by incineration, landfill or other methods, if each municipality organizing or intending to use the facilities of an Authority having such powers shall declare by resolution or ordinance that it is desirable for the health and safety of the people of such municipality that it use the facilities of the Authority, and if any contract between such municipality and other person, firm or corporation for the collection, removal or disposal of ashes, garbage, rubbish and other refuse material has by its terms expired or is terminable at the option of the municipality or will expire within six months from the date such ordinance becomes effective."

resale price maintenance for wine. However, the second standard was not satisfied, since the state simply authorized price setting and enforced prices established by private parties. Such conduct did not rise to the level of "active supervision." In *Midcal,* Justice Powell nonetheless specifically noted that a *comprehensive* regulatory program would be exempt under *Parker.* (445 U.S. at 106, 100 S.Ct. at 943, ftnt. 9).

Under the "state action" exemption, the State itself is exempt from antitrust scrutiny if the challenged activity meets the two-part "state action" exemption test set forth in the *Midcal* case. Municipalities and other political subdivisions of the State may also qualify for the "state action" exemption.

The two leading cases concerning the "state action" exemption as applied to municipalities are *City of Lafayette v. Louisiana Power and Light Co.,* 435 U.S. 389, 98 S.Ct. 1123, 55 L.Ed.2d 364 (1978) and *Community Communications Company, Inc. v. City of Boulder, Colorado,* _____ U.S. _____, 102 S.Ct. 835, 70 L.Ed.2d 810 (1982).

In *Lafayette*, the petitioners were municipalities which operated electrical companies for profit to provide power both within and outside their city limits. Louisiana Power and Light Company, a competing electrical utility, argued that the cities' activities violated antitrust laws. The Supreme Court, in a plurality opinion by Mr. Justice Brennan, explicitly recognized that the "state action" exemption may apply to "anticompetitive conduct engaged in as an act of government by the State as sovereign, or, *by its subdivisions, pursuant to state policy to displace competition with regulation or monopoly public service." Id.,* 435 U.S. at 413, 98 S.Ct. at 1137, emphasis added. Unlike private parties who must be compelled by the state to take anticompetitive activities in order to enjoy the protection of the "state action" exemption, municipalities need only have "affirmative state authorization", as Justice Brennan explained:

> "This does not mean, however, that a political subdivision necessarily must be able to point to a specific, detailed legislative authorization before it properly may assert a Parker defense to an antitrust suit . . . *(A)n adequate state mandate for anticompetitive activities of cities and other subordinate governmental units exists when it is found 'from the authority given a governmental entity to operate in a particular area, that the legislature contemplated the kind of action complained of.'" (Id.,* 435 U.S. at 415, 98 S.Ct. at 1138, emphasis added)

While the municipality does not need specific, detailed legislative authorization to successfully assert the "state action" exemption, the Supreme Court held in its 1982 *Boulder* decision that the state legislature must have *affirmatively addressed or contemplated* that the municipality perform the anticompetitive activity which has been challenged. A general grant of authority from the state legislature, such as broad "home rule" powers, is not sufficient to shelter a municipality from antitrust scrutiny.

In *Boulder,* a cable television company operating in the city challenged an "emergency" ordinance enacted by Boulder which prohibited the company from expanding its business for a three month period. During that time, the City planned to draft a new cable television ordinance, under its home rule powers, which would encourage increased competition for *Boulder's* cable business. The Court in *Boulder*, applying the two-part test set forth in *Lafayette* and *Midcal*, noted: "A State that allows its municipalities to do as they please can hardly be said to have 'contemplated' the specific anticompetitive actions for which municipal liability is sought." 102 S.Ct. at 843. The fatal weakness for the City of Boulder was that it could point to no State authorization or legislative intent that contemplated the *particular* kind of cable television regulation which it had undertaken. Each home rule city was free to choose a different method of regulation. The Court in *Boulder* concluded:

"... (W)hen the State itself has not directed or authorized an anticompetitive practice, the State's subdivisions in exercising their delegated power must obey the antitrust laws." (102 S.Ct. at 843-44, quoting *Lafayette, supra,* 435 U.S. at 416, 98 S.Ct. at 1138).

It should be noted that the court in *Boulder* clearly affirmed its holding in *Lafayette* that the State as sovereign could sanction anticompetitive municipal activities if it clearly expressed its intention to do so. 102 S.Ct. at 840. Such municipal activities were exempt from antitrust laws, since the State had chosen to act through its cities and towns.

There have been several cases in which lower courts have applied the two-part "state action" exemption test in light of the *Boulder* decision, allowing the exemption if the test is satisfied, denying the exemption if either part of the test is not met. The Courts have held, without exception, that an anticompetitive activity established under clear statutory authority as part of a comprehensive state regulatory program is exempt from the antitrust laws. For example, a Kansas City ordinance allowing only one ambulance service to operate in the city, which was adopted under a comprehensive state program which clearly authorized the ordinance and which was supervised by both the state and the municipality, satisfied the test. *Gold Cross Ambulance v. City of Kansas City,* 538 F.Supp. 956 (W.D.Mo 1982). Similarly, an Arizona program regulating dentists, established under a detailed statutory scheme and supervised by a State Board of Dental Examiners, also satisfied the test. *Benson v. Arizona State Board of Dental Examiners,* 673 F.2d 272(9th Cir. 1982). The regulation of jockey fees by the Pennsylvania Horse Racing Commission pursuant to a comprehensive statutory program regulating the horse racing industry also met the "state action" exemption test. *Euster v. Eagle Downs Racing Association,* 677 F.2d 992 (3rd Cir. 1982). See also, *Limeco, Inc. v. Division of Limestone of the Mississippi Dept. of Agriculture and Commerce,* 546 F.Supp.868 (N. Miss. 1982); *Kartell v. Blue Shield of Massachusetts,* 542 F. Supp. 782, 790-91 (D. Mass. 1982); and *United States v. Southern Motor Carriers Rate Conference, Inc.,* 672 F.2d 469,472 (5th Cir. 1982). Finally, in *Central Iowa Refuse*

Systems, Inc. v. Des Moines Metropolitan Area Solid Waste Agency, et al., Civil No. 79-32-1 (S.D. Iowa, Dec. 10, 1952), the federal district court upheld the municipal defendants' requirement that all solid waste generated within the Metro area be deposited at the Metro facility, finding that municipal regulation of waste pursuant to state law was protected by the state action exemption from the antitrust laws. The Ohio law, unlike Pennsylvania's statute, did not expressly condone anticompetitive municipal activity or authorize municipalities to dispose of solid waste generated within their boundaries. The law did authorize construction of municipal solid waste facilities and provided for their financing through the issuance of revenue bonds.

The Court concluded that the Iowa legislature must have intended for municipalities to do what was necessary to assure revenues and protect bondholders, including the enactment of anticompetitive ordinances and regulations. Therefore, the Court found that the Defendants satisfied the "state action" exemption test by inference because their anticompetitive activity was "a necessary and reasonable consequence of engaging in the authorized activity." (Slip Opinion at 8.)

It is readily apparent that a Pennsylvania municipality which adopts an ordinance requiring disposal of its municipal waste at a designated disposal site satisfies both parts of the "state action" exemption test and consequently is not subject to Federal antitrust laws. First, there is a clearly articulated and affirmatively expressed state policy authorizing municipal anticompetitive activities of this type, since Section 202(c) of Act 97 expressly authorizes municipalities to "require by ordinance that all municipal wastes generated within their jurisdiction shall be disposed at a designated facility. In the *Gold Cross* case, Kansas City had established its ambulance service monopoly under similarly clear authorization by the state legislature. The Court in that case highlighted the significance of this fact:

> "To satisfy the *Lafayette* requirement such specific detailed legislative authorization is not required, but its existence points clearly to the 'authorization or direction' contemplated in *Lafayette*. Sections 67.300 and 190.100-195 contain the clear articulation and affirmative expression of the state policy that cities may operate ambulance services in their respective localities." (538 F.Supp. at 966)

The language in Section 202(c) of the Act authorizes municipal anticompetitive activity more clearly than the statutory language in the *Gold Cross* and the *Central Iowa* cases. Therefore, a court should even more rapidly conclude that a Pennsylvania municipality acting under Section 202(c) meets the first part of the "state action" exemption test.

The presence of a comprehensive regulatory program of solid waste management in Pennsylvania further strengthens the conclusion that municipal waste facility designation ordinances satisfy the first part of the "state action" exemption test. The Courts have stressed the significance of such detailed regulatory programs in demonstrating the existence of a "clearly articulated and affirmatively expressed" state policy, distinguishing these cases from the *Boulder* "home rule" situation. *See, for example, Gold Cross Ambulance v. City of Kansas City*, 538 F.Supp at 965: "This is totally unlike the grant of blanket 'home rule' power at issue in *Community Communications*. In sections 67.300 and 190.100 Missouri has announced a state policy to regulate the ambulance industry;" *Central Iowa Refuse Systems, Inc. v. Des Moines Metropolitan Area Solid Waste Agency, supra* at 8; *Benson v. Arizona State Board of Dental Examiners*, 673 F.2d at 275; *Euster v. Eagle Downs Racing Association*, 677 F.2d at 994-95; *California Retail Liquor Dealers Assn. v. Midcal Aluminum*, 445 U.S. at 106, ftnt. 9.

Another factor which has been examined by the courts in these cases is whether there is an important state interest in regulating the challenged activity. For example, in the *Gold Cross* case, the Court noted the clear public need for regulation of ambulance service, stressing that, unlike cases in which the state merely acquiesced in essentially private anticompetitive activities, there was a clear state interest in assuring quick and efficient ambulance service for the protection of the public. 538 F.Supp. at 965, 967. Similarly, in the *Eagle Downs* case, the Third Circuit noted the strong state interest in preventing corruption and maintaining high standards and public confidence in the horse racing industry. 677 F.2d at 995-95, quoting *Gilligan v. Pennsylvania Horse Racing Commission*, 492 Pa. 90, 422 A.2d 487, 489-91.

In Act 97 the Pennsylvania General Assembly has emphasized the critical need for proper management of solid waste activities, stating: "Improper and inadequate solid waste practices create public health hazards, environmental pollution, and economic loss, and cause irreparable harm to the public health, safety and welfare . . . " (35 P.S.§6018.102) This strong public interest in controlling solid waste management practices further strengthens the conclusion that the "state action" exemption from antitrust laws is appropriate and necessary in order to carry out the legislature's intent.

The second half of the "state action" exemption test requires "active supervision" by the Commonwealth. Municipal waste facility designation ordinances could meet this requirement in two ways. First, the Commonwealth, through DER, has extensive regulatory authority over any municipal waste disposal facility. In *Gold Cross*, the court noted that a State agency licensed ambulance vehicles and personnel, inspected ambulances, issued, reviewed, suspended, and revoked licenses, and adopted and enforced state rules, regulations, and standards. The Court in that case held that such activities constituted active state supervision over Kansas City's ambulance program. 538 F.Supp. at 966. Similarly, in *Arizona State Board of Dental Examiners,* the court found sufficient state supervision over restaints on the practice of dentistry because the challenged regulatory system "is supervised by a State agency." 673 F2d at 275.

Similarly, DER's extensive permitting, licensing, investigation and enforcement powers under Act 97 satisfy the active state supervision requirement. Every municipal waste disposal facility requires one or more DER permits. Likewise, DER must review and approve any municipal waste management plan revision and any waste flow control ordinances contained in those plans. DER can order a municipality to implement its municipal plan if necessary. (35 P.S.§6018.201(1)). DER also enforces rules and regulations concerning the project's waste management activities. Such extensive State involvement is more than sufficient to satisfy the "active state supervision" test. *See also, Bates v. State Bar of Arizona,* 433 U.S. 350, 362, 97 S.Ct. 2691, 2698; *Euster v. Eagle Downs Racing Association,* 677 F.2d at 995-96.

In addition to supervision by DER, municipalities which required disposal of their wastes at designated facilities exercise regulatory authority granted to them by the legislature in Sections 201 and 202 of Act 97. Through the adoption and enforcement of plans and ordinances requiring the disposal of wastes at a specified facility, municipalities carry out their duty to ensure the proper disposal of municipal wastes generated within their boundaries. The *Gold Cross* court concluded that similar enforcement of its ambulance service ordinance by Kansas City itself could satisfy the "state supervision" requirement:

"We are satisfied that the action of Kansas City in enforcing its ordinance through its Director of Health, Physicians Advisory Baord, and Medical Advisor constittutes active state supervision *since its regulation of ambulance service is exercised under the authorization and direction of state policy."* (538 F.Supp 966-67, emphasis added)

The same conclusion should apply to Pennsylvania municipalities which direct the disposal of their municipal wastes under the express authorization of the Pennsylvania legislature.

Solid and Liquid Wastes: Management, Methods and Socioeconomic Considerations. Edited by
S. K. Majumdar and E. Willard Miller. © 1984, The Pennsylvania Academy of Science.

Chapter Twenty-Seven

Health Planning, Education, and Politics Related to Solid and Liquid Wastes

William C. Livingood

Professor and Chair
Health Department
East Stroudsburg University
East Stroudsburg, PA 18301

Issues involved with solid and liquid waste disposal are acutely interwoven with our general health status. There is a common belief that our general health status has improved in the last few centuries due to improved medical techniques, but in reality most of the improvements can be attributed to prevention in areas such as sanitation, housing, nutrition, and immunization. In particular diseases such as tuberculosis, typhoid fever, plague and cholera have been controlled and in some cases eradicated in this country due to sanitation.

At one time our environmental health concerns focused on sanitation related to organic waste and communicable disease. Today our environmental concerns are much broader, but they are as interrelated with health as they ever were. This is reflected in the highly acclaimed Surgeon General's Report, *Healthy People,* which refers to studies that attribute environmental factors to 20% of the leading causes of death in this country. While environmental factors are not limited to the physical environment, the Report does indicate that "Influences in the physical environment that increase risk include contamination of air, water, and food; work place hazards; radiation exposure; excessive noise; dangerous consumer products; and unsafe highway design." The Surgeon General's Report further addresses concerns related to solid and liquid waste with the statement, "The environment has become host to many thousands of synthetic chemicals, with new ones being introduced at an annual rate of about 1,000—and to byproducts of transportaiton, manufacturing, agricultural and energy production processes."

The Surgeon General's Report is significant because it is the foundation for the federal health planning efforts. It serves as a basis for other federal health planning documents such as *Promoting Health/Preventing Disease: Objectives for the Nation* and the "Public Health Services Implementation Plan for Attaining the Objectives for the Nation." It also serves as a major reference document for many of the state and local planning efforts.

Objectives for the Nation specifically addresses issues of solid and liquid waste. The following statements are a reflection of some of the issues included in *Objectives for the Nation*. "From over 2,200 contaminants of all kinds identified in water, 765 were identified in drinking water. Of these, 12 chemical pollutants were recognized carcinogens, 31 were suspected carcinogens, 18 were carcinognic promoters and 59 were mutagens. It is not known what the additive effects of these chemicals will be on the total cancer burden." "As water resources become in shorter supply, more and more surface water, used for drinking water, will be recycled or reprocessed, continuing the recycling of pollutants unless adequate water treatment measures are taken."

An example of a recommended education and information measure included in *Objectives for the Nation* is a program of "Informing the public that exposure to hazardous agents is serious, but manageable, and the government control measures are essential; through providing information on the control of environmental and occupational health hazards to teachers and students in elementary and secondary schools within the context of comprehensive mandatory classroom health education."

Several examples of legislative and regulatory recommendations are: "Establishing priorities and developing more standards for hazardous substances in both air and water," and "Withholding from introduction into commerce new chemicals that pose a significant public health threat unless the manufacturer can demonstrate that there are safe and practical methods for their manufacture, intended uses and disposal."

The recommended economic measures contained in *Objectives for the Nation* include taxation and legal redress, such as "effluent/emission taxes (using effluent/emission taxes as supplements to, and not replacements for, regulation to create additional incentives for hazard abatement): favorable tax treatment of investment in pollution control; legal redress for harm resulting from exposure to toxic agents; and tax policies encouraging capital investment in redesigning process technology to emphasize process improvement over add-on technology."

Objectives for the Nation also includes specific objectives for 1990. An example of an objective related to reduced risk factors is, "By 1990, at least 95 percent of the population should be served by community water systems that meet Federal and State standards for safe drinking water, (In 1979, the level was 85 to 90 percent for the National Interim Primary Drinking Water Standards.)" Another objective related to reduced risk factors is "By 1990, there should be virtually no preventable contamination of ground water, surface water, or the soil from industrial toxins associated with wastewater management systems established after 1980."

The concerns of "Increased Public/Professional Awareness" within *Objectives for the Nation* include objectives such as: "By 1990, at least 75 percent of all city council members in urban communities should be able to report accurately whether or not the quality of their air and water has improved or worsened over the decade and to identify the principal substances of concern." Another objective states that "By 1990, at least half of all people ages 15 years and older should identify the major categories of environmental threats to health and note some of the health consequences of those threats." Related to toxic wastes, an objective proposes that, "By 1990, the Toxic Substances Control Act and the Resource Conservation and Recovery Act should be fully implemented to protect the U.S. population against hazards resulting from production, use, and disposal of toxic chemicals."

Objectives for the Nation identifies many of the major issues related to solid and liquid waste and it includes measures that should significantly improve environmental quality with emphasis on programs such as public education, regulation and economic incentives. Unfortunately the "Implementation Plans for Attaining the Objectives for the Nation" are less ambitious. In fact, it is difficult to find concerns about liquid and solid waste in the "Implementation Plans" other than some very indirect reference in the Research and Education sections of the Chapter on Toxic Agent and Radiation Control. This is somewhat understandable since the U.S. Public Health Service does not have primary responsiblity for the environment within the federal structure, but it does not change the fact that the "Implementation Plans" are the only federal plans for implementing the "Objectives." "It contains the plans for the action steps the Federal Government plans to take in its efforts to achieve better health for Americans through attainment of the five major goals established by the Public Health Service and published in *Healthy People: The Surgeon General's Report on Health Promotion and Disease Prevention.*" In other words *Objectives for the Nation* identifies what needs to be done related to waste, but the "Implementation Plans" indicate that there isn't any planned implementation of waste control programs that have been identified in the federal health planning efforts.

Health Planning also occurs at the state and regional levels within states. *The State Health Plan 1982-1987* is the general health planning document for Pennsylvania. The Goal, cited in the State Health Plan related to solid, and liquid waste, states that, "THE COMMONWEALTH SHOULD CONTINUE TO PROTECT HUMAN HEALTH AND ENVIRONMENTAL QUALITY BY CLEANING UP POLLUTION, RESTORING DEGRADED NATURAL RESOURCES, MINIMIZING RADIATION EXPOSURES TO CITIZENS, AND ENSURING THAT ENVIRONMENTAL PROTECTION STANDARDS ARE ACHIEVED STATEWIDE." The need for the government's role in protection and restoration of contaminated areas is further recognized with a specific objective. The objective states that, "The Department of Environmental Resources should provide assistance for needed environmental restoration and capital improvement projects." The State Health Plan also identifies two related actions that should be pursued to implement that objective. They are RECOMMENDED ACTION 9,1,1,1 which states "The Department of Health and the Department of Environmental Resources should within existing resources, work to improve programs for protecting human health and environmental quality," and RECOMMENDED ACTION 9,1,1,2 which states, "The Commonwealth should seek available federal funding to rehabilitate water supply systems."

Another objective included under the previously stated goal is "The Department of Health and Department of Environmental Resources should direct their existing programs to apply existing knowledge on environmental health activities." RECOMMENDED ACTION 9,1,2,1 more specifically addresses the issue. It states that, "The Department of Health and the Department of Environmental Resources should improve public education, economic incentives, and compliance/monitoring activities for non-point source pollution control programs."

A third objective of the State Health Plan is, "The Commonwealth should assess the viability of conducting research on human health effects of long-term to low level dosages of toxic substances." The objective is accompanied by RECOMMENDED ACTION 9,1,2,1 which states, "The Department of Health should upgrade environmental data collection so that it is useful to environmental epidemiologists." RECOMMENDED ACTION 9,1,3,2 is also included under this third objective related to research. It states, "The Department of Health should provide medical backup and support to the state agencies, engaged in regulation and control of hazardous and toxic substances."

A fourth objective related to toxic waste is identified in the *State Health Plan*. The objective is, "The Commonwealth should continue to adopt and implement regulations and procedures for the transportation, disposal, and storage of toxic and hazardous wastes." This objective includes the recommended action that, "The Department of Environmental Resources should establish management procedures to correct past unsafe practices."

Some of the Regional Health Plans are even more ambitious than the *State Health Plan*. For example, the Health Systems Plan for Carbon, Berks, Lycoming, Monroe, and Northampton Counties has very strong recommendations for control of waste. The Goal or general statement of direction is, "To prevent unreasonable risk to the Public Health and Welfare due to exposure to Chemical, Physical and Biological Agents." An objective under the goal states that "By 1985 an adequate supply of safe water should be available to all residents." To accomplish this objective the Regional Health Plan recommends that:

a. "The HSC (Health Systems Council) shall encourage the appropriate agencies to monitor known areas of contaminated drinking water and take whatever action necessary to purify the present water source.
b. The HSC shall encourage the development and implementation of proper educational programs for farmers concerning proper land care, and the treatment of hazardous sustances (pesticides, fertilizers, etc.)."

Another objective under the same goal states, "By 1983 there should exist disposal areas that are environmentally safe, and will protect the environment against any type of hazard: or the production of those wastes should be prohibited." This restriction of the production of wastes that cannot be safety disposed is supported by the following recommended actions:

a. "The State Department of Environmental Resources in cooperation with industry, should prepare information campaigns to inform all users of their obligation to reduce wastes at their sources, and to dispose of these in a proper manner.
b. "The HSC shall be diligent in urging the Boards of Health and all relevant bodies in municipal governments to prevent the production of hazardous wastes (and substances) for which there is no safe means of disposal."

The need to determine nature and extent of contamination is addressed with the objective, "By 1982 the Pennsylvania Department of Environmental Resources should determine the extent and containment characteristics of hazardous waste disposal sites." It is also supported by the recommended action, "In cooperation with local municipalities and industry, Pennsylvania Department-ment of Environmental Resources shall develop recommendations encouraging a program of reducing hazards and, if necessary, setting required standards to control waste disposal." Another recommended action concerning identification of health risks associated with waste products is, "The HSC in cooperation with local groups should work together with Pennsylvania Department of Environmental Resources to identify those unreported risks which could lead to public health problems."

It should be clear that Health Planning Efforts recognize many of the environmental health needs related to solid and liquid waste. This is official recognition in that the plans are legislatively mandated documents and in the case of the State Health Plan, it is signed by the Governor. However, it is still little more than recognition. Although the Health Plans have major regulatory authority for the construction of health care institutions such as hospitals and nursing homes, the health plans have little authority for public health in general. The Department of Health does not even appear to be bound by the State Health Plan related to most health issues. Other government agencies such as the Department of Education appear to completely ignore the document. The wording of the recommended actions in the Regional Health Plan is a good indication that the Regional Planning Agency has little authority in this area.

The lack of influence that health planning has on liquid and solid waste disposal stems from a lack of commitment to health planning in general. Health planning efforts can be a forum for addressing the issues of liquid and solid waste disposal. Health planning may even provide some official recognition of what needs to be done. Health Planning can not be expected to actually resolve any of our environmental problems because it has no public health regulatory authority and there is no evidence that government structures are even sensitive to health planning.

The lack of commitment to health planning is tied to the lack of federal commitment to prevention. The Surgeon General's Report states that "only 4% of the Federal Health dollar is specifically identified for prevention related items." While the federal expenditures for treatment of illness continue to increase, the tight budget, particularly for domestic spending, indicates that the emphasis on prevention can only get worse.

It is true that legislation related to the environment has been introduced and passed, but the legislation could hardly be characterized as the result of national health promotion and illness prevention programs. The new environmental laws have come into existence in the same manner as many of the federal health statutes in the last century. They are reactions to a great deal of public attention surrounding major health catastrophes. An example is the passage of the 1938 Food, Drug and Cosmetic Act following a number of deaths. The Comprehensive Environmental Response, Compensation and Liability Act of 1980 is similar in that it can be linked to considerable public attention generated by the Love Canal incident.

There is no doubt that legislation will continue to be introduced in the same manner, but reliance on catastrophes to stimulate legislation is obviously less than a sound prevention program. Reliance on legislation for prevention, regardless of soundness, is dangerous. Laws are not the final solution. Sound legislation still requires enforcement for it to be effective. Allegations that the 1980's has been an era of reductions in the enforcement of many health laws are common. The 1983 EPA Superfund Scandal was a case where the lax enforcement was found unacceptable because of the public outrage. Citing public concern about wrong doing, President Reagan withdrew the claim to executive privilege in blocking the congressional investigations of Superfund Administration. Of course this was followed with the firing of Rita Lavelle and the eventual resignation of EPA Administrator Anne Burford. Unfortunately the public outrage which stimulated the President's actions was based on the public being informed through the media attention surrounding Love Canal, Times Beach and similar environmental catastrophes. Reliance on catastrophes to stimulate public awareness is obviously an unacceptable approach to public health protection.

The dynamics of politics makes public awareness of liquid and solid waste issues essential. There are many special interests that gain from inadequate laws or lax enforcement of adequate laws. The laws at least are designed to protect the public interest from special interests that either take advantage of the public or disregard the public welfare. There can be great financial gain in either case. If the public is apathetic, the national legislators and regulators receive pressure in only one direction. National legislators, in particular, depend on vested interests to finance costly election campaigns.

Although regulators are not elected they are also sensitive to pressure from special interests. Regulators who do not bow to the pressure can be subject to budget cuts and layoffs by the legislators who are likely to be very sensitive to pressure. Non-enforcement does not require the recision of laws which could draw public attention. In the case of the regulators, the vested interests can gain from simple neglect. Those dedicated to the public interest are not likely to be any match for the financial and other pressure which might be associated with lax enforcement unless there is support from an aware general public.

The pressure which is brought to bear on the regulators can be seen in other areas of health regulation. In 1982, the Pennsylvania legislature failed to appropriate half of the available federal funds for health planning. The Bureau that would have received the federal money was forced to furlough a substantial number of the agency's staff. The unappropriated money of approximately 500 thousand dollars went unused. The Bureau which experienced the furloughs was the regulatory agency for nursing home construction. The fact that the chairman of the House Appropriations Committee had a vested interest in a nursing home did not go unnoticed. One can only speculate about the effect of the funding cuts on the regulatory agency, but the staff should have been intimidated even if they weren't. It should be pointed out that public support or even awareness of the agency's cut backs was almost non-existent. Some of the legislators on the Appropriations Committee even claimed to be unaware that federal funds were being denied to the agency.

The above example illustrates how an agency can be at the mercy of small but well positioned vested interests if the public is unaware or unconcerned. Yet it is too disadvantageous to have a catastrophe on a regular basis to stimulate media attention and public concern. One solution is to depend more heavily on the education system for an informed and concerned public. Some of the previously mentioned recommendations from *Objectives for the Nation* support this solution.

Another rationale for having a well informed and concerned public is that a lack of public awareness is not the only danger. The public's interests are frequently more concerned with immediate rather than long term needs. Parents with starving children are seldom concerned with financing their children's college education while their children are starving. Obviously the parent's attention would be primarily directed to the immediate problem. Similarly, workers who are unemployed are going to be more concerned with making mortgage payments and putting food on the table than with long term health risks which may not show up for decades. It has not been uncommon for workers to protest enforcement of environmental standards when employers threaten to close a factory as a result of the enforcement. The immediate concern of workers in such a threatening situation will take precedent over the long term concern. If there isn't a larger citizenry committed to protecting long term health, it is extremely unlikely that the long term needs will be given priority. Again, education is critical if the long term interests are to prevail.

Public education is also important because the public is directly responsible for some of the hazardous substances that are placed in the water systems. Some of these substances are actually designed for the septic and sewage systems such as drain cleaners. There are many other hazardous substances which are inappropriately disposed of in the sewage systems such as petroleum and paint products. Limiting the availability of some of these products may be a partial solution, but an aware and concerned public appears to be the only practical solution for controlling the overall problem.

Regardless of legislation and regulations, public support for environmental health is essential if the public's interests are to be protected. This is as true in Pennsylvania as it is in other states. Some of the other states such as Colorado and California have enacted legislation which requires environmental concerns as part of the health education curriculum. Pennsylvania does not include environmental health as a part of required health education subjects although the Pennsylvania curriculum regulations do require environmental education in the secondary schools. Unfortunately, the environmental studies can be integrated into other courses or taught as a separate course of only 30 hours, which does not provide much assurance of an adequate program. Furthermore, it is not a required subject at all in the elementary schools. The fact that the new regulations have also been criticized for the lack of enforcement provisions should have little impact on environmental education because there really isn't much to enforce.

Health planning and health education should play a major role in public health protection related to the environmental problems of liquid and solid waste. Although the various health planning documents provide considerable recognition of the role of health planning and health education in the control of solid and liquid waste, the impact of these doucments appears to be minimal. Other recognition of the importance of education to these environmental issues also appears to have little impact in Pennsylvania, particularly as evidenced by Pennsylvania's new Curriculum Regulations.

Many professional organizations have requested that the country have a national health policy. This would enable health promotion and disease prevention at least to be given a priority within the government and hopefully have priority in the governments relationships with the private sector. Without a national health policy, meaningful health planning, or even environmental education, it is unlikely that our society will find long term solutions to our solid and liquid waste problems or most of our other environmental problems.

REFERENCES

1982, *Health Systems Plan,* Health Systems Council of Eastern Pennsylvania Allentown.

1979, *Healthy People: The Surgeon General's Report on Health Promotion and Disease Prevention 1979,* U.S. Department of Health, Education, and Welfare, Washington.

1980, *Promoting Health/Preventing Disease: Objective for the Nation,* Department of Health and Human Services, Washington.

1983, "Public Health Service Implementation Plans for Attaining the Objectives for the Nation:" *Public Health Reports,* Sept.-Oct., Suppl.

1983, "Reagan Orders Investigation of EPA Superfund Allegations:" *Congressional Quarterly,* Feb. 17, 363.

1983, *State Health Plan 1982-1987.* Statewide Health Coordinating Council and Pennsylvania Department of Health, Harrisburg, Feb., 1983.

Solid and Liquid Wastes: Management, Methods and Socioeconomic Considerations. Edited by
S. K. Majumdar and E. Willard Miller. © 1984, The Pennsylvania Academy of Science.

Chapter Twenty-Eight

Effects of Sanitary Landfills on Property Values and Residential Development

Hays B. Gamble, Ph.D.[1] and Roger H. Downing[2]

[1]Associate Director
[2]Research Assistant
Institute for Research on Land and Water Resources
The Pennsylvania State University
University Park, PA 16802

Most people, given a choice, prefer to live in neighborhoods or communities free from obnoxious or annoying activities. If they already live in such a community, they tend to strongly oppose any efforts to bring in unsightly, noisy, smelly, or otherwise degrading activities. Popper has called these LULU's— locally unwanted land uses. They are pervasive in our modern society: power plants, airports, expressways, low income housing, sewage treatment plants, junk yards, shopping malls, factories, hazardous waste deposit sites, sanitary landfills, etc.

A LULU always threatens its surroundings in one way or another, and this is why few people want to live near one and why many people opppose attempts to site new ones. The adverse effects—the threats—from LULUs can take many different forms: scenic degradation, traffic congestion, higher noise levels, undesirable odors, water and air pollution, more crime and vandalism, and possible health effects as from nuclear power plants or toxic wastes. Many homeowners believe that in turn these effects will lower residential property values, probably the main reason why there is such strong opposition when a LULU is proposed in a community.

People's perceptions vary, and what may be a LULU to one person may be a wanted activity to another. The factory may be unwanted by nearby homeowners, but wanted by the workers who depend on it for their livelihood. What may not be considered a LULU today may become one tomorrow: residents who just moved into new homes in a recently created subdivision next to a dairy farm may object to the noise, flies, dust and odors, while many environmentalists would argue for the protection and preservation of that farmland.

Most people know, of course, that almost all kinds of LULUs are beneficial and vital to the smooth functioning of our society. One doesn't need much imagination to realize how harsh our life style would be in the closure of power plants, interstate highways, or sites for the proper disposal of all the wastes generated as a result of our affluence. The asymmetry between the incidence of the cost and benefit of LULUs is very difficult to rectify. The costs, called negative externalities by economists, generally fall on only a relatively few people—the neighbors of the LULUs. The benefits, on the other hand, generally accrue to a much broader segment of society spread over a wide area. We do not have the institutional arrangements whereby the losers can readily be compensated for their loses by those who gain or benefit. Moreover, the beneficient majority is generally quite apathetic, in part because the benefits are not nearly as intense on an individual bases as are the costs, and also because beneficiaries rarely see or come in close contact with the LULU: out of sight, out of mind. I want the energy, but don't put the power plant or power line in my back yard.

Increasingly, our nation is faced with the problem of siting facilities sorely needed at the regional and even national level but objected to by people who must live near them. As a society we need these projects, but as individuals we do not want them in our communities; we want them miles away. Figure 1 shows the results of a national public opinion survey in which people were asked how far away they would choose to live from various kinds of activities.

FOCUS OF THE CHAPTER

This chapter focuses on two likely effects arising from one specific kind of LULU—sanitary landfills. The two effects examined are: changes in the amount or rate of residential development, and changes in residential property values. The material in the chapter is based on a study of 10 sanitary landfills in Pennsylvania sponsored by the Pennsylvania Bureau of Solid Waste Management. For a more detailed account of the study, see Gamble (1982b).

The 10 sites used in this study were selected from a list of 33 sanitary landfills, all operating under the permits from the Department of Environmental Resources. Site selection was based primarily on those sites which would most likely reveal the kinds of effects we sought; there had to be residential development without other commercial or industrial activities. Not too many approved landfills in Pennsylvania meet these criteria. The 10 sites finally chosen are located in nine counties scattered throughout Pennsylvania. All of the land-fills except one are located either in rural agricultural areas or in the developing fringes of major urban areas. Because these sites were not randomly selected from the total population of sanitary landflls in the state, the findings from this study cannot be used to project likely impacts on other existing or potential sites.

To the best of our knowledge we know of no rigorous study that has examined the effects of sanitary landfills or hazardous waste disposal sites on property values or residential development.

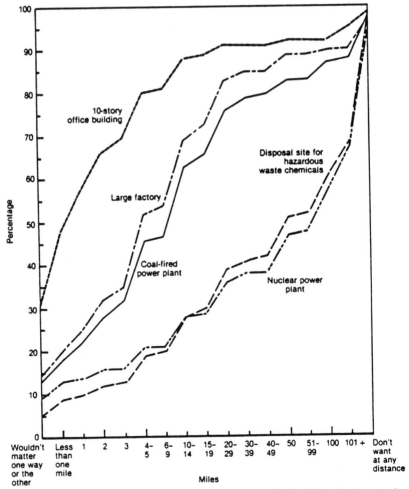

FIGURE 1. Cumulative percentage of people willing to accept new industrial installations at various distances from their homes. *Source:* Council on Environmental Quality, Department of Agriculture, and Environmental Protection Agency, *Public Opinion on Environmental Issues: Results of a National Public Opinion Survey* (Washington, D.C., GPO, 1980), p. 31.

Effects on Residential Development

Our hypothesis is that there are people in sufficient numbers choosing not to live closely to sanitary landfills such that the rate of residential development will be adversely affected. We defined "close to a landfill" as the area (impact zone) within one mile of the landfill center, containing about 2,011 acres. It is reasonable to expect that if there are adverse effects on residential development they will occur within this area. Because the landfills occupy from 50 to 260 acres, ample space remains within the 1-mile zone for development.

Our purpose is to compare the rates of residential development (new house construction) over a six-year time span (1975-1981) in the impact zones to the rates in surrounding control areas. Four control areas around each landfill site were randomly selected, each control area being 3 miles from the landfill and containing about 500 acres in an area circumscribed by a circle of one-half mile radius. Thus the total area in the control and impact zones were approximately the same.

All residential properties within the impact and control areas comprise the data base. Individual property characteristics and data, such as the nature of property improvements and when made, data on property sales, sales prices, etc. were obtained from the property record cards in County Tax Assessment Offices. Two distance zones within the impact zones were identified—0 to ½ and ½ to 1 mile. Six size classes of properties were specified and the data kept accordingly. Comparing the rate of new residential development in the impact zones to the rates in the control areas provides an objective measure of the effects of landfills on development.

Findings

In this portion of the study one of the landfill sites is omitted because of its location near Pittsburgh, which precluded the selection of suitable control areas. Tables 1 and 2 show the distribution of properties by size classes and those with and without improvements, for 1981.

It is apparent that there is no particular distinction between the distribution of parcels by size categories among the impact zones and the control areas (Table 1). There is virtually the same proportion of small parcels, upon which one finds

TABLE 1

Percent distribution of residential properties by parcel size class and location, nine landfill sites, 1981.

Parcel Size	0 - ½ mile	½-1 mile	Control Areas
Acres	%	%	%
>100	1.8	1.1	1.3
26-100	2.2	3.3	3.6
11-25	2.2	2.0	2.6
6-10	3.9	2.6	2.8
2-5	9.7	7.2	9.1
0.1	80.0	83.8	80.5

TABLE 2

Percent of parcels with residential development and vacant properties, by parcel size class, nine landfill sites, 1981.

Parcel Size	0-½ mile		½-1 mile		Control Areas	
	Residences	Vacant	Residences	Vacant	Residences	Vacant
Acres	%	%	%	%	%	%
> 100	69	15	100	0	97	3
26-100	42	42	80	13	68	24
11-25	58	26	50	42	64	29
6-10	68	29	57	29	69	23
2-5	71	24	68	20	71	23
0-1	80	18	77	20	81	17
Total	76	20	75	20	79	18

TABLE 3

Residential constructoin by years, 1971-1981, nine landfill sites.

Year Constructed	0-½ mile		½-1 mile		Control	
	No.	% all prop.	No.	% all prop.	No.	% all prop.
		%		%		%
1971	23	4.72	21	0.93	51	1.98
1972	9	1.81	59	2.54	50	1.91
1973	9	1.78	54	2.27	95	3.49
1974	22	4.17	68	2.78	83	2.96
1975	18	3.30	45	1.81	81	2.81
1976	15	2.68	81	3.15	88	2.96
1977	21	3.61	89	3.35	77	2.53
1978	15	2.52	51	1.88	101	3.21
1979	41	6.46	122	4.31	82	2.54
1980	5	0.78	30	1.05	46	1.40
1981	4	0.62	14	0.49	35	1.06

residential development, near the landfills as in the control areas three miles distant. The proportions of vacant land and parcels with residential development in the three smallest size class (less than 11 acres) are about the same for the impact and control areas (Table 2). The landfills do not seem to have influenced the nature of land subdivision or the proportion of parcels developed.

Table 3 shows the rate of new residential development by years from 1971-1981 around the landfill sites and their control areas.

There is no conclusive evidence from the data in Table 3 to indicate that landfills exerted much influence on the rate of residential development. In 5 of the 11 years, the rates of development in the nearest impact areas (0½ mile) exceeded those in the control areas. In 1979 development around landfills was much more extensive than in the control areas, although it should be noted that much of the development that year occurred on only one site, the Boyertown landfill. Development declined significantly in 1980 and 1981, probably due to high morgage interest rates and the scarcity of funds. The rates of new development in 1980 and 1981 are considerably lower in the impact zones than in the control areas. This may be starting to reflect the growing concern among people over the handling and disposal of wastes, particularly hazardous wastes. Perhaps there is a growing environmental awareness, fostered by such incidents as TMI and Love Canal, that we are just beginning to pick up in our data. Only time will confirm this.

Table 4 arranges the data somewhat differently. Residential development between 1976-1981 is shown for the individual landfill sites at the top fo the table, with the data in the lower part of the table grouped by tons of waste handled on a daily basis. The first four landfills listed had lower rates of development in the 0½ mile impact zone than did their respective control areas. Rates of residential development around the last 5 sites exceeded rates in their respective

TABLE 4

Increase in number of parcels wtih residential development, 1976-81.

Landfills	Tons/day waste	0 - ½ mile No.	%	½-mile No.	%	Control No.	%
York Authority	300	1	6	4	7	14	25
Hanover	60	1	4	7	27	16	23
Lara	500	0	0	4	4	9	13
Lakeview	850	1	5	13	8	52	9
Community	150	11	137	7	32	5	14
Strasburg	1000	13	72	78	101	116	47
Lower Paxton	60	24	45	26	3	82	12
R & A Bender	200	3	21	2	3	2	2
Boyertown	200	32	15	165	32	47	11
Total Number		86		306		343	
Avg. growth (percent)			21		16		15
Grouped by Tons/Day:	Average area occupied by landfill (acres)						
> 300 Tons/day (3)	221	14		95		177	
Ave. growth (%)			19		28		20
≤ 300 Tons/day (6)	110	72		211		166	
Ave. growth (%)			22		14		12
Avg. No. homes per site							
> 300 Tons/day		29		175		352	
≤ Tons/day		68		357		258	
Housing Density: No. homes/100 acres							
> 300 tons/day		10.3		11.6		17.5	
≤ 300 tons/day		17.3		23.7		12.8	

control areas. For all sites, there was a 21 percent rate of residential development from 1976-1981 in the 0½ mile impact zone as compared to a 15 percent rate of development in the control areas.

Is there a relationship between the size of landfill (in terms of tons of waste handled per day) and rate of residential development? When the data in Table 4 are grouped into landfills handling more than 300 tons per day and those handling 300 or less tons per day, there is no strong relationship to suggest this. The rate of development within one-half mile of the three large landfills is almost the same as in their control areas (19 and 20 percent, respectively). Around the 6 smaller landfills, the rate of development between 1976 and 1981 is significantly higher close to the landfill as compared to their control areas (22 and 12 percent, respectively). However, it should be pointed out that most of the development close to the 3 large landfills occurred around only one site—Strasburg. There was virtually no development within one-half mile around the other two sites.

When we compare the average number of houses per site in 1981 grouped by large and small landfill, we find that there are about twice as many homes in both 0½ mile and ½-1 mile zones around the smaller landfills than around the larger landfills, although there were more homes in the large landfill control areas. This does not necessarily mean the large landills inhibited residential development at some point in time. The larger landfills could have originally been sited in more remote areas where little residential development had occurred or was likely to occur. This seems to be substantiated by the bottom two rows of data which account for the area occupied by the landfill and subsequently not available for development. Potentially developable areas divided by the 1981 housing stock were used to derive these figures. Housing density is lower around the large landfills than around their control areas or the impact zones of the smaller landfills. From the data in Table 4 there is no evidence that landfills have adversely affected residential development from 1976 to 1981. In fact, the data show both the rate of development and density of housing to be decidedly greater within ½ mile of the small landfills than in their control areas. Perhaps lower lot values in the vicinity of the landfills explain in part this latter phenomenon. The next section examines this possibility.

EFFECTS ON PROPERTY VALUES

Theory of property value effects.

If many people are concerned over the health and environmental aspects they perceive to be associated with solid waste disposal facilities, logic suggest that when choosing a residential location they will not select a property near a sanitary landfill or other disposal facility unless they obtain some form of offsetting compensation. Compensation could arise when long-term locational or site disamenities, such as those arising because of proximity to a solid waste disposal site, result in lower prices for the property.

The various attributes or characteristics of a house and its location provide a "flow of services" for the occupants and owners. When people choose a house and its location, they reveal preferences by their willingness to pay for certain housing and locational characteristics. If people value large lots, family rooms with a fireplace, quiet, nearness to employment, or remoteness from an undesirable or hazardous activity, the real estate market should reflect those preferences.

An economic relationship is therefore expected to exist between market price and the quality and quantity of services that any given dwelling provides the occupant. Location is one attribute that can provide a number of such services: nearness (accessibility) to employment, schools, and shopping, as well as distance or remoteness from undesirable environmental variables such as noise, congestion, odors, or perceived hazards from a solid waste disposal site. This relationship implies that at equilibrium in the housing market, price differentials will arise among various locations which compensate consumers for the differences in housing services associated with specific locations. Otherwise, consumers would not remain at particular locations. Equilibrium requires that for two properties identical in all respects except that location 1 is near a solid waste disposal facility and location 2 is well removed, the price of the first property must be less than that of the second by an amount which will just compensate buyers for the undesirable effects they perceive at location 1. Otherwise, the consumer would be better off at location 2.

Valuing locational disamenities by examining their negatively capitalized effects on property values has been widely applied in recent years to environmental intrusions such as air and noise pollutants. Freeman (1979) reviews the extensive literature on this subject. Gamble (1974, 1982) has studied the effects of highways and nuclear power plants, including the Three Mile Island accident, on residential property values. These studies are based on the hedonic price model developed and refined by Freeman (1974), Griliches (1971), Nelson (1975), Rosen (1974) and others, whereby sale prices are expressed as a function of property characteristics. Hedonic prices represent compensating price differentials, since individuals are assumed to choose locations such that price differences among different housing and site characteristics are equalized at the margin (equilibrium willingness to pay).

Residential sales values.

Recording all valid sales of single family houses and vacant lots between 1977 and 1981 within the impact zones and control areas at all 10 landfilling sites, and correcting these actual sales values for inflationary effects by using quarterly price deflators, we were able to compute mean sales values as shown in Table 5. As in the previous analysis, we grouped the landfills into large and small sites

TABLE 5

Mean residential sales values, 10 landfill sites, 1977-81, converted for inflation (n observations) (average year built). Value in $1,000.

	0- ½ mile	½-1 mile	0-1 mile	On landfill access road	Control area
Residential Properties					
All sites (10)	$45.4 (54)	$49.1 (159)	$48.4 (313)	$35.5 (14)	$44.1 (182)
	(1958)	(1965)	(1964)	(1936)	(1958)
Lge sites (3)*	52.9 (3)	49.4 (65)	49.6 (68)	42.1 (7)	45.6 (84)
	(1962)	(1964)	(1964)	(1952)	(1963)
Sm sites (7)	45.0 (51)	48.9 (194)	48.1 (245)	28.9 (7)	42.9 (98)
	(1957)	(1965)	(1964)	(1922)	(1954)
Lots					
All sites	7.1 (15)	9.2 (50)	8.7 (65)	5.4 (4)	10.3 (30)
Lge sites	—	3.7 (3)	3.7 (3)	—	8.1 (10)
Sm sites	7.1 (15)	9.5 (47)	8.9 (62)	5.4 (4)	11.4 (20)

*Sites handling more than 300 tons of waste per day.

based on waste volume handled on a daily basis. Developed residential properties near the landfill sites were not lower in price than in their control areas, regardless of whether they were near large or small landfills. The average age of the houses (in terms of year built) has little significance when comparing properties near landfills to properties in control areas.

However, properties on the main road leading to the landfill (access road) are significantly lower in value than other properties near the landfill and properties in the control areas. Apparently the garbage trucks, daily going past these properties to get to and return from the landfill, are nuisance enough to depreciate residential property values. Also, these houses were considerably older than other houses in the area. At the landfill near Pittsburgh, over the noon hour, we observed more than one truck per minute delivering wastes to or returning from the landfill. Such a situation obviously is harmful to the character of a residential neighborhood.

We were able to compute the average value of lot sales, although there were too few sales around the large landfills to be able to make meaningful interpretations. Around the seven small sites, average lot prices were significantly lower in the two zones near the landfill as compared to the control areas. As one would expect, prices were lower in the 0½ mile zone than in the ½-1 mile zone. Average prices of lots along the access road (there were only 4 sales, not enough on which to place much reliance) were only about half the prices of lots in control areas. It appears that some people may be attracted to sites near landfills because of the lower land prices, and with the savings thus realized construct homes of higher than average values, thus accounting for overall slightly higher property values near landfils as compared to control areas. A long run effect may also be present, in that buyers may feel that in time the landfill will close and the site will become a park. Portions of the Lakeview and Lower Paxton sites are already in use as parks. The next section, by estimating the hedonic price function by means of multiple regression, will attempt to statistically determine if there is a relationship between landfills and residential property values.

Estimating the hedonic price function.
The procedure used in this study to draw inferences about the relationship between residential property values and the presence of landfills is based upon the theory of hedonic prices. The standard approach to estimate hedonic price functions, multiple regression analysis, was used in this study.
The first step is to specify both the correct list of variables and the functional form. Both of these aspects of specification are problematic. The correct list of variables is all utility-bearing characteristics of residential properties in the market area. To identify this list, however, is economically and probably also technically unfeasible. With regard to the choice of correct functional form, hedonic theory offers little guidance. The correct functional form is determined in general by features of the commodity and the underlying market equilibrium and can be expected to be complex, nonlinear, and variable across markets.
Butler (1982), however, agrues that the biases due both to omitted variable and improper functional forms may not be serious enough to severely qualify the results obtained from misspecified models. Specifically, Butler suggest that investigators have considerable flexibility in the choice of functional forms and that significantly fewer characteristics need be included than is generally supposed.
The linear form of the multiple regression model was used to explain variation in the selling price of houses, the dependent variable, expressed as follows:

$$V_i = b_o + \sum_{j=1}^{n} b_j X_{ij} T_u$$

where
V_i = the deflated selling price of the ith residential property,
b_o = constant term
X_{ij} = independent variables from 1 to n associated with the ith property, and
u = an error term, assumed to be randomly distributed, reflecting all other unexplained variations.

In this analysis the independent variable we are most interested in is a binary variable: close to the landfill. If the landfill is perceived to adversely affect the attributes of residential properties, then the prices of properties close to the landfill (within one-half mile) should be lower, as indicated by a coefficient with a negative sign.

The results of the linear regression for all landfill sites are presented in Table 6. The magnitudes of the coefficients are close to what one would expect and the signs of the coefficients are as predicted. We hypothesized variable 24, close to landfill, to be negative, and it is, but it is far from being significant in explaining variation in housing prices. Variable 23, on landfill road, signifies a house on the main access road to the landfill, and is negative and weakly significant (at the 5-10 percent level). The results of the regression, as shown in Table 6, conform to our earlier findings: that the landfills in this study apparently have little effect on housing price, except possibly for houses along the main access road serving the landfill.

The hedonic price model assumes uniformity in real estate markets. Given the fact that we are dealing with ten sites over a period of 5 years, model misspecification may explain in part the failure of "close to landfill" variable to be significant. To "standardize" the markets, we ran a multiple regression on the data for one site and for three separate years, the Boyertown landfill, the only site having a sufficient number of sales. Rather than "close to landfill," the variable relevant to measuring landfill effects was "distance to landfill," which should have a positive coefficient since price is hypothesized to be directly related to distance. The results for both the linear and logarithmic functional forms of the model were much the same as those just presented. For one year the distance variable was weakly significant (at the 5-10 percent level) and positive; the other two years it was not at all significant, and was negative one of those years. Given our data set, we have been unable to find a significant relationship between housing prices and sanitary landfills.

RESULTS AND CONCLUSIONS

There is no evidence in our results to suggest that the well managed landfills included in this study had any adverse influences on the rate of residential development near them between 1976 and 1981. In fact, the rate of residential development within one-half mile of landfills handling 300 or less tons of waste per day was considerably higher than the rate of development in control areas three miles from the landfills. The average number of residential properties around the smaller landfills was almost twice the average number around the large landfills. This may be because the larger landfills originally selected more remote, less developed areas in which to locate or occupied more space; we did not examine this feature. There is some weak evidence that people may be more environmentally conscious today perhaps because of the TMI and Love Canal publicity, and that a study similar to this for the current years might show different results.

The average price of residential properties near landfill sites, whether they

TABLE 6

Regression results: 10 landfill sites in Pennsylvania.

Variable	Regression Coefficient	(t values)
Constant	16,125	(4.48)***
1. Built before 1915 (b)[a]	− 7,826	(− 3.70)***
2. Built 1915-1933 (b)	− 6,151	(− 2.74)***
3. Built 1934-1945 (b)	− 6,810	(− 3.33)***
4. Built 1946-1967 (b)	− 3,506	(− 2.73)***
5. Built after 1972 (b)	212	(0.16)
6. Poor grade (b)	− 3,676	(− 2.02)**
7. Good grade (b)	9,933	(5.03)***
8. Poor condition (b)	− 5,167	(− 2.78)***
9. Good condition (b)	151	(0.15)
10. Distance to big employer	− 18	(− 0.23)
11. No. floors	4,833	(3.26)***
12. No. of bathrooms	2,034	(2.04)***
13. First floor area	8.44	(5.08)***
14. Second floor area	1.33	(5.08)***
15. Attached garage (b)	4,090	(7.27)***
16. Garage detached (b)	842	(1.27)
17. Garage, integral (b)	3,206	(4.66)***
18. No. fireplaces	2,497	(3.35)***
19. Central air conditioning (b)	8,580	(5.86)***
20. Modern kitchen (b)	6,879	(4.61)***
21. No. bedrooms	307	(0.46)
22. Control area (b)	− 91	(.012)
23. On landfill road (b)	− 4,121	(− 1.90)*
24. Close to landfill (0-$\frac{1}{2}$) mile (b)	− 351	(0.30)

R²	.718	
F	49.97	
Residual degree of freedom	470	
Mean values	$46,856	

Notes: * significant at the 5-10 percent level
        ** significant at the 1-5 percent level
        *** significant at the 1 percent or better level
        [a] (b) refers to a binary variable

were big or small landfills, were not significantly lower than prices of residential properties in control areas. There was little significant variation in ages of houses (year built) in impact zones as compared to control areas. However, residential properties on the main access road to the landfill site did sell for considerably less value than other properties, either within one mile of the landfill or within a control area. We conjecture that the frequent passing of garbage trucks going to and from the landfill accounts for this. Also, houses along the access road were older than other houses in the vicinity of the landfill.

The average price of lots around seven small landfills (there were not enough sales around the large landfills to make meaningful interpretation) were significantly lower than the average price of lots in control areas. As we expected, lot prices within one-half mile were lower than lot prices within one-half to one mile. People may be attracted initially to areas near landfills because of lower land prices, and then construct more expensive homes from the savings realized. Analyzing residential property sales values using a hedonic price model gave no evidence that the landfills exerted any strong effects on property values.

Our conclusions are quite straightforward: nothing in this study suggests that the well-managed landfills included in this study had significant adverse effects on the rate of residential development within one mile of the landfill or on residential property values. There was evidence to show that the prices of building lots were lower near landfills and that the values for residential properties located on the main access road serving the landfills were lower than other properties in the area.

The results of this study are somewhat surprising; intuition (bias?) tells us we should have found lower preperty values near the landfills. Perhaps the particular landfills used in this study are not representative of most landfills. Because landfills, for the most part, locate in rather remote rural areas there is not much residential development around them, which makes meaningful measurement of property values difficult. People today may be more environmentally sensitive to locally unwanted land uses (LULU's), such as landfills, than they were between 1976 and 1981, the years of the study, as a result of the media attention to such events as TMI and Love Canal.

The reasons homeowners choose certain locations are many and varied and include accessibility to various services and needs, availability of land, and so forth. Apparently, the presence of a well-managed landfill does not seem to be as strong a factor as perhaps we thought it to be; other characteristics of houses, the site, and its location, appear more significant than a landfill in explaining differences in sales prices. This does not imply that all people are indifferent to living near a landfill; some people hold strong negative feelings about such locations, but apparently, at least during the years of this study, there was sufficient depth in the real estate market to ameliorate such attitudes.

The latter point needs to be emphasized, as we feel it is the key to understanding the results of this study. In Niagara Falls, New York, according to Popper (1983), there was strong demand for the homes abandoned by the residents of Love Canal. Over the remainder of the year following the March, 1979, accident at TMI there was no measurable negative effect on property values (Gamble 1982a and Nelson 1981). Although many people may have strong negative feelings about living near LULU's, there are enough people who do not hold those attitudes such that the real estate market is little affected. In other words, given the nature of the demand curve for residential property, there is not the effect one would suppose from the attitudes of many people about certain kinds of land uses. However, it may take longer for properties near LULU's to bring the prices anticipated, and this can impose costs on the sellers.

If our findings are correct, they mean that the fears of many people over substantial losses in property values due to a newly established sanitary landfill in the area may be unfounded. The findings, if substantiated by additional studies, imply that future sitings of new landfills may be made somewhat easier by the offer of monetary compensation to property owners who believe they will suffer property value losses. If such losses are small or non-existent, as implied in this study, public agencies may be more willing and able to adopt a compensatory policy for the siting of new landfills, thus improving the likelihood of public acceptance.

LITERATURE CITED

Butler, Richard V. 1982. "The specification of Hedonic Indexes for Urban Housing." *Land Economics* 58 (February):96-108.

Freeman, A. M. III. 1979. "The Hedonic Price Approach to Measuring Demand for Neighborhood Characteristics" in *The Economics of Neighborhoods* (O. Segal, Ed.). Academic Press, NY.

Freeman, A. M. III. 1974. "On Estimating Air Pollution Control Benefits from Land Values Studies." *Journal of Environmental Economics and Management* 1:74-83.

Gamble, H. B. and R. H. Downing. 1982a. Effects of Nuclear Power Plants on Residential Property Values." *Jn. Regional Science* Vol 22:4, pp. 457-477.

Gamble, H. B., R. H. Downing, J. S. Shortle, and D. J. Epp. 1982b. *Effects of Solid Waste Disposal Sites on Community Development and Residential Property Values.* Final report to Bureau of Solid Waste Management, Dept. of Env. Resources, Commonwealth of PA. Inst. for Research on Land and Water Resources, PA State Univ., University Park, PA, 60 pp.

Griliches, Z. 1971. "Introduction: Hedonic Prices Revisited," in *Price Indexes and Quality Change* (Griliches, Ed.), Harvard University Press, Cambridge, Mass.

Nelson, J. P. 1981. "Three Mile Island and Residential Property Values: Empirical Analysis and Policy Inplications." *Land Economics* Vol. 57, No. 3, pp. 363-372.

Nelson, J. P. 1975. *The Effects of Mobile-Source Air and Noise Pollution on Residential Property Values.* DOT-TST-75-76, U.S. Department of Transportation, Washington, DC.

Popper, F. J. 1983. 'LULUs", *Resources*, No. 73 Resources for the Future, Washington, DC, pp. 2-4.

Popper, F. J. "The LULU: Coping with Locally Unwanted Land Uses." Manuscript in preparation, Resources for the Future, Washington, DC.

Rosen, S. 1974. "Hedonic Prices and Implicit Markets: Product Differentiation in Pure Competition," *Journal Political Economy*, 82:34-55.

Solid and Liquid Wastes: Management, Methods and Socioeconomic Considerations. Edited by
S. K. Majumdar and E. Willard Miller. © 1984, The Pennsylvania Academy of Science.

Chapter Twenty-Nine

Sociological Considerations of Siting Facilities for Solid Waste Disposal

Robert C. Bealer and Donald M. Crider

Professors of Rural Sociology
Weaver Building
The Pennsylvania State University
University Park, Pennsylvania 16802

The task of this chapter is to move beyond the physiographic, technological, managerial and related types of considerations in waste disposal problems and probe some of the often overlooked human-side elements involved in the issue. In a word, we shall examine the kinds of attitudinal and behavioral phenomena focal to social science research.

We had the opportunity to explore this important area of solid waste problems in a recent monograph that is readily available to the public[1] The exposition which follows will use the cited publication as a major resource. We shall précis it, laying out what we, as social scientists, see to be especially salient in that report. Further, we will summarize the relevant literature which has come to our attention in the interim between the earlier publication and the writing of this chapter. The literature of our interest is that which is empirical, reporting research on the sociological/psychological dimensions of solid waste problems in the United States and similar societies. Our intent here is to sort out what can stand the test of scientific validity and, therefore, may be a sound basis for trying to meet the manifold human puzzles involved in solid wastes. Finally, we shall examine also certain sociocultural aspects of the general problem that practical experience suggests to us as worthy of note.

LIMITS TO DISCUSSION

The sociological dimensions of solid wastes are varied. We shall limit our concern, however, to landfills and, more specifically, the problems of siting such facilities. This emphasis results partly from the restriction of what can be handled in a brief chapter, and from recognizing the realities of our present society.

The United States *is* adept at producing material wastes, which need to be disposed of somehow. Were we as a people less encouraged than we are to a "throwaway" mentality, and culturally less organized to facilitate the use and quick discard of things,[2] the weight of waste disposal would be greatly reduced. Avoiding material wastes in the first place is the surest way not to have their disposal later as a problem. But what might be possible, and what in fact is reality, need to be clearly distinguished. A reasoned perspective on waste of any kind—as with other societal ills—must start with people where they are and not with where we might wish them to be.[3] Landfills are the dominant mode of solid waste disposal today in America,[4] and are likely to remain so.

Recycling is only a partial alternative to using landfills. It is possible to reprocess many throwaway products (such as aluminum or steel cans, glass, and newsprint) into useful ingredients for further manufacture, reducing dramatically the total amounts of new materials necessary for production processes. Recycling, as a form of waste reduction at the outset, enjoys a rather strong verbal show of support[5] but actions to follow works are often less evident.[6] People talk a stronger line of support for conserving resources than they are able to show in relevant behavior. Thus it has been noted[7] that: "In total volume, more is being recycled today than 10 years ago. But our per capita generation of solid waste continues to rise. We are recycling more, but we're making garbage even faster."

Landfills are ubiquitous. Increases in recycling activities, including incineration, where that disposal form is used to produce heat and/or energy, help to reduce the volumes of solid waste interred in landfills and can greatly prolong the use life of a given site. However, the residue from incinerators ends up in landfills as well as anything that cannot—or will not—be recycled. Hence, a landfill of some sort is likely to be the final step in the disposal of solid waste for a long time. Moreover, landfills still remain the cheapest acceptable disposal method. Solid waste disposal in the United States today means largely landfill utilization.

Therein lies a major snag: "The greatest single obstacle to proper land disposal is citizen resistance to sanitary landfill sites," according to the National Science Foundation.[8] The National Center for Resource Recovery's state-of-the-art report on landfills[9] is of the same voice: "The most serious problem in acquiring landfill sites is invariably obtaining the acceptance and support of the community. All too often, sites ideal from engineering, economic, and regulatory points of view have been lost because of adverse public pressure." Without a site, landfills are impossible. Siting is a crucial juncture where sociological aspects create bottlenecks.

Whether the concern be solid wastes or the much more explosive issue of toxic/hazardous/radioactive wastes, the location of handling facilities is public issue laden. Our intent, however, is to limit the range of observation; we assume that the typical community refuse facility would *not* be receiving radioactive wastes or other forms of hazardous (toxic) materials. This assumption may be inaccurate, of course, for there is already a wide spectrum of substances diagnosed as nonbenign, and the upgrading of information continuously makes suspect more and more materials once thought to be nontoxic. Combined with the culpability, deviousness, criminal guile, and sheer stupidity of the human creature, this means, undoubtedly, that toxic wastes often do get dumped at ordinary landfills when they should—for the commonweal—receive special handling as cautioned by the label "hazardous."[10]

Hazardous waste problems are an exceedingly important issue in their own right. For the most part, we will set them aside in our exposition, not because they lack import, but because there are more than enough difficulties just trying to understand problems of solid waste disposal with relatively benign materials, and because there seems to be a critical order of difference between the two cases. The response of citizens to siting a sanitary landfill in their neighborhood or community is often negative. In part, we suspect, this refusal is due to misperceived estimates of assumed dangers from the facility, of seeing adversity and potential harm that dispassionate observers would not be able to substantiate.[11] By definition, however, hazardous wastes present very real dangers under even the best of circumstances. Opposition to siting of such operations becomes a somewhat different order of fact. At this point of limited knowledge it seems judicious to defer hazardous wastes in our discussions until we can gain a clearer perspective on, perhaps, the less prickly case of nontoxic materials.

THE RESEARCH LEGACY

Despite the wide acknowledgement that sociological elements are a thorn in the side of those seeking to site solid waste facilities, comparatively little empirical research about the matter has occurred. The void becomes particularly vivid in the context of reported research conducted on the more malleable aspects of solid waste disposal—for example, things such as cost modeling to lay out the best routes of refuse collection, evaluation of efficiencies to alternative technologies, hydrogeological assays for identifying sites with the best geography, engineering matters.[12] The number of such studies is enormous and only underlines the irony of the void. All of the types of research cited, while entirely commendable, can easily go awry over the less tractable human side of the situation. The best laid plans of engineers and other scientists for a sanitary, well-run solid waste facility can be killed or crippled by delays through citizen opposition to the siting of a landfill. Nonetheless, the amount of research concerned with trying to find out how siting may be effectively done or, conversely, how it stumbles about or fails is miniscule.

Perhaps the most extensive source of data bearing directly on siting experience is a report to EPA by a private consutling firm.[13] That study contains 21 case histories from across the United States involving hazardous waste facilities, most of which (N = 18) involved landfills. The cased studied comprised five types of situations: a new facility was created with no public opposition (N = 4); despite public oppostiion, a fresh facility was brought into operation (N = 5); citizens' displeasure stymied the attempt to site a new operation (N = 7); an already operating facility, though challenged by public opposition, continued to operate (N = 3); a going concern was forced to cease operation because of the public's disapproval (N = 2). The relative distribution of cases among the five types of situations is not necessarily representative of the overall frequency of occurrence of each category[14] but reflects a desire to get at least some instances of each type. Also, the cases obtained mirror problems of getting relevant information. The often inflammatory nature of the siting situation led the researchers to reduce their originally proposed sampling frame and take what cases they could.

While we specified earlier our exclusion of hazardous wastes, the data set just noted is by far the most extensive—and pertinent—encountered. Moreover, we think that the general conclusion in the report transcends hazardous wastes and fits our case equally well. The researchers[15] concluded:

There is no . . . certain or simple solution to the problem of public opposition to facilities, either to avoid public opposition or to mitigate it once it has arisen. . . . No single action or set of actions could be recommended for all siting situations. . . . [No] factors are necessarily important for all sites; in each situation the sponsor [of a landfill or other operation] will have to face a different set of problems.

What appears to work in getting a facility sited successfully in one location and at a given time may not be effective at another time/place.

This type of conclusion can be illustrated by some material drawn from our Pennsylvania research.[16] We looked at two instances of new landfills coming into existence. In one case (Warren County) an interarea facility came into being without great upheaval. In a second (Lycoming County) the same type of operation was finally realized, but only after a decade of protracted struggle. The two cases were similar in many ways. But, whereas, in the Warren situation part of the success could be attributed to the landfill being sited on federal forest lands, where it was relatively out-of-sight (and mind), and where land acquisition was neither advantageous nor disadvantageous to anyone, in the Lycoming case the same attempt to maintain a low-profile using federal lands was a significant part of the original and continued opposition to the facility by citizen groups.[17]

The overall conclusion to be drawn here is not that social science research is inherently unable to provide definitive, defensible answers for use in much the same way that chemistry or agronomy can provide proscriptive information on landfill leachates and ways to curb their possibly detrimental effects. Rather, it is to note where we are presently and to remind ourselves that the maddening spectrum of human-side problems makes it easy to hope for and expect more solutions from the social scientist than he/she is really able to bring to public issue impasses. Solid research for understanding siting snarls is but embryonic. Scattered and relatively unsystematic observations of landfill siting situations, however, provide some helpful hints to alert would-be problem solvers.[18]

The social science research record also provides, sometimes, unexpected pictures. For example, in a Pennsylvania study[19] 844 nearby residents to eight operating landfills in the Philadelphia area were, among other things, queried thus: "In *your experience,* to what extent are the following [items] important problems because of the disposal [i.e., landfill's] operation?" Specific complaints inquired about included: truck traffic and noise, litter dropped from trucks, paper blown out of the disposal site, rodents, smells, smoke, insects, noise from the site operation, dust, and surface or ground water pollution coming from the landfill. The level of displeasure about such matters varied widely among sites studied with no apparent pattern of answers standing out except that there was a surprisingly lower level of expressed dissatisfaction than might have been expected. All of the items elicited an overwhelming majority of responses that the particular, potential problem was "not important"! In this regard the Pennsylvania study findings parallel those of a four-state (North Dakota, South Dakota, Nebraska, and Texas) survey of 1,167 households done in 1975-76 by members of the Northcentral Technical Research Commitee of the participating Land-Grant universities. The Committee's report noted that: "When asked to identify solid waste disposal problems [like smells, rodents, etc.] experienced in the three years prior to the survey, a total of 905 (78 percent) responded 'None' ".[20]

Opposition to siting can be clamorous, but landfills are not inevitably a source of overwhelming citizen discontent. Indeed, one can wonder further about the degree of displeasure with siting to begin with. The typical impression, widely held, is one of an aroused citizenry, time and again—as a "normal" course of events—overturning the carefully crafted plans of this or that community, solid waste authority, or other relevant body trying to site a new facility. If this be true, however, it means that solid waste disposal planning is an exception to the modality of behavior in our society. Despite the stereotype, the fact is that in most sectors of American life, the citizenry is relatively passive.[21] Citizen participation, overall, does not greatly influence the ongoing operations of governmental/private sector programs. Is solid waste atypical?

As is true with many aspects of this discussion, the kinds of data which speak to the question are sparse. However, one study[22] has published pertinent statistics—a nationwide survey of county governments. A total of 539 counties were surveyed, including all of those in the nation with a population over 100,000 and where the county had some formal responsibilities for solid waste disposal. Asked whether they had attempted, within the past 10 years, to get additional land for landfill sites, 68 percent said, yes. Of that figure, more than four-fifths (80.8 percent) got the land they sought, though 32 percent of those counties being successful had to use more than one attempt to reach the goal. Overall, 52 percent of the reporting counties were unsuccessful in their first try at getting more or new landfill territory.

These figures carry a certain contradictory note. On the one hand, that 80 percent of the counties who tried reported ultimately getting land for solid waste disposal tends to question the inevitability of crisis which characterizes much of the writing about siting landfills. Apparently, most counties most of the time in the last decade have been able to succeed in their efforts. This is not to argue that siting problems for solid waste facilities are simply a tempest in a teapot or wholly imagined. There is also, the other part of the data to interpret. At least on their first attempts, 45 percent—a near majority of the counties surveyed—reported an inability to get the land they wanted for a landfill. Moreover, citizen opposition was cited as the leading obstacle to a successful landfill location. When the respect usually given to citizen clamoring is not especially great a "rejection" rate of 20 percent begs attention. Even where some new facility comes into operation after repeated attempts to thwart it have been turned aside, the likely cost is much waste of human resources which, with greater understanding, might have been avoided. The need for continued research remains.

RECENT RESEARCH ON SOLID WASTE

The interim between the completion of our earlier review and the writing of this chapter has shown no great surge in applicable social science research. Although there are no indications that the rates at which we as a nation generate solid wastes have abated, attention to the problems of siting landfills for benign refuse disposal appears to have waned. We were unable to locate a single instance of relevant research reported.

The primary reason for the deficiency is probably the overriding public concern focused on the cleanup and disposal of hazardous wastes. In the light of citizens'—and politicians'—near panic reactions to these matters, solid waste disposal of nonhazardous materials has slipped easily into the shadows of neglect. More recently, the prolonged political brouhaha surrounding EPA, including the ultimate firing of some top eschelon staff, has only intensified the neglect.

The dismal record for the period at issue did not improve greatly by shifting attention from solid wastes to the more inflammatory case of hazardous refuse. While we did not engage in the same level of intense, systematic search here as we reported for solid wastes, empirical research on the problems of facility siting were not conspicuous. To be sure, the amount of writing about hazardous waste questions is enormous. That is, legion are the essays, books, papers and the like which treat hazardous waste issues and which readily opine about social problems, such as siting. Lacking in these sources, however, is the demonstrably sound research grounding giving the scientific bases for opinions and/or advice. Nonetheless, such literature often makes pronouncements about sociological elements, frequently with an authority and conviction that is not scientifically warranted. There may well be a law-like function at work here in human affairs, which may be stated as follows: the less that is known empirically as fact, the more cocksure are writers' assertions about them! We need to avoid this kind of media poisoning of the public mind like the plague.

SOCIOCULTURAL ELEMENTS IN SITING PROBLEMS

What makes people act and think as they do? Why do they behave as observed by those involved in waste disposal planning and decision making? While no sure answers can be given, the insights of social scientists, gathered from broad experience with human responses to varied life situations can be helpful. As sociologists, we see three lines of thought as important bases for better understanding human behavior in regard to solid waste siting. We shall briefly explicate each.

Language as a Communicator of Values
   Central to many of the problems associated with solid wastes siting are words and their uses.[23] Behind all of what is thought and said and written are meanings which have their roots in the historical derivations and current connotations of such expressions. Whether by sight, sound, or sense of meaning, words are symbols. Just as the alphabet spells out words for us to see and read, or in the same way we hear messages with our ears, there are conjured in our minds the ideas—or sense—that these sights or sounds, as symbols, convey. Sense of meaning is developed in experience. "Waste," and its various synonyms, have negative connotations in everyday language.[24] This simple fact makes it a real bugbear. As people in the vanguard of fighting for a landfill site know, one of the more effective techniques for turning a public meeting (or a public's response) toward opposition is to allow the topic of discussion to shift from the referent "landfill," to "gargabe dump," to just plain "dump," to more vulgar, i.e., four-letter vernacular, terms. With movement through such language, the tendency to dispassionate discourse decreases. One may wish that words were neither so important to human commerce nor so readily debased in meanings.

But so it is—and, often with great impact. For example, rural communities, desperately needing the revenues involved, have emphatically turned down proposals for transportation of solid waste from urban centers because they could not handle the prospects of being labeled "the junk capital of the state" or "king of the garbage heap."[25]

These kinds of reactions, while not inevitable, dramatically underscore a lesson worth pondering. "Solid waste," by whatever name, is often not an entity of neutral interest which science is free to appropriate for its own purposes. Handling the subject best requires high sensitivity to the feelings and disposition of others.

Localization as a Currency of Power

The distribution of America's citizenry is such that cities, given their large populations and economic functions, are the prime sources of solid wastes, while rural areas, with their typically low-density occupancy, are ideally suited to provide landfill sites. Smooth interarea cooperation on jointly managing solid wastes, however, is not assured. Quite the contrary. Sharp conflict may ensue when cities look to the countryside for help. Why? Among other things, with all that is common or shared in our society, there are yet differences within parts of the whole. The generation and disposal of refuse in the nation historically, and the attitudes developed about it, are not of *a* piece.[26] The legacy of these can infect interarea deliberation and create strain, especially among those who see rural places as inherently good, waste as bad, and the city as rapacious in forcing its will. Solid waste facility siting becomes, then, another chapter in an evolving story of urban "rot," in which the city is seen as trying to despoil all it dominates. The circumstances of a shift over time in political (and other forms of) power can be critical. Power is rarely given up readily. It is defended when challenged. In this context, the need for landfill siting offers an unusual reversal in opportunity made more acute and incisive by the symbolic disvalue of the subject at issue.

History is not something that can be easily undone in a community; it can—and should—be carefully noted. Inquiry into past events and local idiosyncracies may be as critical to getting a good landfill site as is awareness about underlying geological strata.

Even without endemic rural-urban conflict, siting a landfill often encounters difficulties because of the nature of such facilities to exceed localization. Economies of scale have been recognized in laws like RCRA.[27] Getting cooperation from more rather than a few civic parcels—towns, cities, suburbs or other political entities—can cut per unit costs and make for a technologically better facility. But the regional thrust bumps into the jurisdictional disjunction of diverse (and often jealous) decision-making bodies. Power here is also not automatically foregone. Communities, as aggregates of individuals, are not necessarily given over to more rationality than the parts of which they are composed. To suppose or hope for this condition is to let desire get ahead of fact. In planning, it is well to contemplate that, as put by a panel on public acceptance of waste facilities:[28] "Society has no [inherent] obligation to be [scientifically] logical about risks." Assessments of costs and benefits to people is not simply a matter toted up on a scientific abacus.

Because of the frustrating nature of sociological elements in siting problems, one of the potential solutions widely discussed in some form of governmental preemption.[29] Whatever its form or name—"rights of eminent domain," "nonrecourse licensing," and so on—the basic idea is the same: some physical areas are to be selected as solid waste sites notwithstanding local complaints. If the citizenry is declared by some manner of fiat to have no (or no simple) legal recourse to a decision made by government agencies about siting, the phenomenon of popular opposition presumptively pales. It would be naive to assume, however, that discontent will become necessarily inconsequential. If opposition to siting is deeply felt, if citizens' values have been transgressed, then the *form* of discontent and its *mode for expression* may be altered, but not its fact. This general outcome has been nicely documented regarding hazardous wastes,[30] so much so that the authors of the cited work could appropriately subtitle their book "Local Opposition and the Myth of Preemption." Our present society certainly deviates considerably from the ideals of a literal democracy, but it is not so far evolved away from that form to render ineffective widespread and deeply rooted citizen resistance. The folly of trying to legislate "right" thinking and behavior about alcohol through Prohibition laws and the ultimate negotiated (and to some, shameful) shutdown of the unpopular hostilities of Vietnam are historic cases bearing out the lesson that an aroused citizenry can be a formidable force; legality is not necessarily an effective ban on its presence.

Institutionalization as a Process of Organization

Throughout this chapter, we have suggested that it is impossible to separate the sociological elements from the physiographic/technological aspects of solid waste disposal siting problems. So intertwined are these two sets of factors that ignorance of, or indifference to, either courts disaster for a proposed project. At the least, there may be undue delay in its implementation. Greater awareness of potential issues, therefore, and better understanding of the social-psychological processes which generate them can be useful to all concerned.

A typical scenario in establishing a solid waste disposal facility goes something like this. Public officials become alerted to the impending fill-up or otherwise unsatisfactory nature of a traditional dump. Alternatives are considered. Information is gathered on new technologies in handling garbage, and fresh locations adaptable for presumed needs are considered. At this point, discussion is usually casual and decisions are tentative; no need is seen for extending the deliberations beyond the halls of elected or appointed officials. But a "leak" to news media occurs. The public becomes involved; issues emerge; opinions polarize. Not infrequently, a champion comes forward to defend the citizenry against officialdom. A group is formed and there may ensue: a war of words, public hearings, lawsuits and countersuits, injunctions to delay action, and, sometimes, vandalism and violence. Eventually, agreement of sorts is reached and a new facility is opened.

Repeated observations of the phenomena just sketched have led to their identification as social movements. Relatively united, citizens are caught up in a kind of "bandwagon," sharing a distinctive perspective. A strong sense of solidarity is fostered with an emphasis on ideals and an orientation to action. Often, what begins with a champion and a handful of causetakers, continues across time as a formal organization. In such cases, the opposition has become institutionalized.[31]

At such a juncture the maintenance and continuity of the citizens' group *per se* has gained significance *as an end in itself.* Formalization of the associational structure usually occurs complete with officers, rules of governance, membership criteria, and stated objectives. Free expression of opinions, or individual actions thought to be in the interests of opposition, come under increasing control. Heretofore acceptable spontaneity, controlled only by tradition or the give-and-take of personal and group interaction, becomes subject to explicit rules and regulations. Failure to conform to group expectations is cause for censure or expulsion. Ideals, perspectives, and objectives particular to the group's attempts at opposition are infused with values: some traceable to the larger society, but others generated from the interaction of members in the enthusiasm for their cause. New recruits are selected for their ability to contribute to stated objectives or for their acquiescence to group ideals. Far from being representative of the general public, such citizens' groups tend to entertain and espouse only those ideas and actions which promote their segment of society. It is not strange, therefore, that otherwise anti-social or illegal behavior becomes acceptable for the good of the cause. Defeat and disbanding come about only after every avenue of thwarting has been exhausted, rarely because of failure to win on any single issue.

Meeting the challenge of institutionalized opposition has no easy answer. The specifics of the group's dissatisfaction may change often as the agenda of issues are adjudged one after the other, unfounded, irrational, or untenable. At each turn of events, a new *cause celebre* is ready so that continuity of the group is assured. Admittedly, some questions and objections posed in the adversary stance may be sincere and insightful, leading to noteworthy improvements in the overall design of original proposals. All too often, however, activities seem to be deliberately obstructionist, serving no purpose other than interference in the orderly pursuit of public works. Inordinate postponement and delay in implementation of critically needed services for the commonweal may result.

While the chain of events outlined above has become the stereotypic "horror story" in attempts at solid waste disposal siting, such an outcome is not inevitable. Many municipalities and their leaders are alerted to the potential for deep and long-term dissension associated with mishandling citizen participation in public-service-related decisions. Moreover, a variety of mechanisms for involving relevant publics in the decision-making process have been used.[32] Still, no unassailable principles of action, universally applicable, can be suggested to avoid all problems. Experience, however, does commend certain approaches.

The assumption that secrecy may be maintained in official deliberations is ill-advised. It is inevitable that the citizenry will be involved, sooner or later. Democratic processes are such that representative government can be challenged on virtually any point, and at almost any time, by an aroused citizenry. Reluctant acceptance of the "Sunshine Law" after "the cat is out of the bag" may often be too little, too late. Hence, whatever can be done to involve the public in decisions relative to waste disposal, from the outset of such deliberations, seems advisable; broad citizen participations, initiated as early as possible, is a good bet as a correlate of successful siting.

Though it may go without saying that citizen participation must be real rather than simply "window dressing," the ends and means of that involvement should not be left to chance. Whatever mechanisms for information dissemination are used, they should assure: knowledge of the facts which pertain to the local case including the technical and managerial data, awareness of the risks involved together with steps being taken to minimize their effects, understanding of tradeoffs among positive and negative aspects of alternative proposals, and an appreciation of the total context in which a single issue must be considered.

Each of these ends can not be elaborated here, but it should be obvious that what is intended is the development among the citizenry of the same kind of information base upon which duly elected or appointed officials would be expected to make objective decisions for the commonweal. That this ideal is seldom realized will be readily admitted. At the same time, it might be suggested that herein lies both the explanation for failure, and the primary secret of success in siting solid waste facilities.

REFERENCES AND NOTES

1. R. Bealer, K. Martin, D. Crider, 1982. *Sociological Aspects of Siting Facilities for Solid Wastes Disposal,* Penn State U., University Park, Pa. Copies may be obtained by adressing requests to the authors at 205 Weaver Bldg. Zip code 16802. Ask for AE&RS pub. no. 158. Self-addressed, stamped ($1.56 for first-class handling) envelopes submitted with requests will expedite orders. The report is 22x28 centimeters in size.
2. *Ibid.,* pp. 9-18, 35-37.
3. R. Bealer, 1980. *Compost Sci./Land Utilitzation* 20, 8.
4. R. Bartolotta in *Municipal Year Book,* 1975 (Wash., D.C.: Int. City Mgt.

Assoc., 1975), pp. 232-241; T. Bulger in *County Yearbook,* 1978 (Wash., D.C.: Nat. Assoc. of Counties, 1978), pp. 88-112; *Waste Age* 13, 25 (1982).

5. R. Anthony, 1982. *Environ. 24,* 14.
6. R. Bealer, K. Martin, D. Crider, *op. cit.,* pp. 47-49; O. Albrecht, E. Manuel, Jr., F. Efaw in *Municipal Solid Waste* (Cincinnati, Oho: EPA, 1981), pp. 238-250.
7. J. Marinelli, 1982. *Environ. Action 13,* 16.
8. *Program Solicitation: Decision-Related Research in the Field of Urban Technology,* 1976. Wash., D.C.: NSF.
9. *Sanitary Landfill,* 1974. Lexington, Mass.: D.C., Heath.
10. J. Marinelli, G. Robinson, 1981. *Progressive 45,* 23; A. Purcell, 1980, *The Waste Watchers,* Garden City, N.Y.: Anchor Books.
11. R. Bealer, K. Martin, D. Crider, *op. cit.,* pp. 41-47.
12. *Ibid.,* p. 40.
13. Centaur Associates, 1979. *Siting of Hazardous Waste Mgt. Facilities - Public Opposition,* Wash., D.C.: EPA.
14. Accurate date on the relative frequency of occurrence of these types of situations with public access to it are lacking or, at least, did not turn up in our search. The literature on siting is not a readily available pool, see R. Bealer, K. Martin, D. Crider, *op. cit.,* pp. 40-41.
15. Centaur Associates, *op. cit.,* pp. 22 and 9.
16. Penn State Ag. Exp. Station Project #2337.
17. R. Bealer, K. Martin, D. Crider, *op. cit.,* pp. 67-68.
18. A major confounding circumstance for a definitive sociology is the tendency in human systems for complex interactive effects to occur among relevant variables and, on the other hand, the general moral proscription to wait for naturally arising cases to emerge. Manipulating the latter toward the ideal of science - investigator control on all factors except the one at issue - is nearly impossible to realize. A sensitizing function may be the most sociology can provide to problem solutions for the time being.
19. R. Coughlin, H. Newburger, C. Seigner, 1973. *Perceptions of Landfill Operations Held By Nearby Residents* (Philadelphia: Regional Sci. Res. Instit.).
20. P. Gessaman, et. al., 1978. *Consumer Perceptions of Selected Community Services in the Great Plains* Lincoln: Instit. of Ag. & Nat. Resources, U. of Neb.
21. R. Bealer, K. Martin, D. Crider, *op. cit.,* pp. 28-35.
22. Bulger, *op. cit.* Elaboration of the data and caveats about its limitations occurs in R. Bealer, K. Martin, D. Crider *ibid.,* pp. 59-64.
23. R. Bealer, K. Martin, D. Crider *ibid.,* pp. 9-14, 41-49.
24. *Ibid.,* pp. 14-18.
25. E. Pollock, 1974. *Solid Wastes Mgt. 16,* 80; P. Philipovich, 1968. *Compost Sci. 9,* 29.
26. R. Bealer, K. Martin, D. Crider, *op. cit.,* pp. 18-24, 51-57.

27. EPA, *The Resource Conservation & Recovery Act of 1976,* 1977, (Wash., D.C.: Govt. Printing Office).

28. *Proc. Symp. on Waste Mgt.,* R. Post, ed., 1981, Tucson: U. of Arizona, p. 446.

29. J. Manko and B. Katcher, 1978, *Compost Sci. 19,* 10.

30. D. Morell and C. Magorian, 1982. *Siting Hazardous Waste Facilities* (Cambridge, Mass.: Ballinger).

31. L. Schneider in *Dictionary of the Soc. Sci.,* 1964, (New York: Free Press, pp. 338-339).

32. M. L. Hendrickson, and S. Romano, 1982. *Public Works 113,* 76.

Appendix

Acid Rain Research—A Special Report*

A. Clarifying the Scientific Unknowns
B. Tracing the Pathways of Atmospheric Conversion
C. Discerning the Change in Waters and Woodlands

Acid Rain: An Overview

A. Clarifying the Scientific Unknowns

Research is shedding light on the critical scientific issues of acid rain and opening up new control alternatives.

The term *acid rain* first issued from the pen of British chemist Robert Angus Smith in 1872. But it was not until the 1972 United Nations Conference on the Human Environment, exactly 100 years later, that the acid rain concept crossed the threshold of public awareness.

Today acid rain has become an issue of international import. It is a source of tension between Great Britain and Scandinavia in Europe, between the United States and Canada, and even between different regions of the United States. The charge is that emissions from burning fossil fuels, sometimes carried over long distances before returning to Earth, are contributing to harmful acidity in the waters, forests, and fields of near and not-so-near neighbors.

Such transboundary air quality problems have no precedent. Further, the issue contains many elements that cannot be proved or disproved, given our current state of knowledge. Nevertheless, there has been increasing pressure to curb emissions immediately on the theory that irreversible damage could occur during the time it will take to establish solid proof.

Now that momentum exists for establishing new controls, the challenge will be to keep our knowledge evolving apace. Without such knowledge, there will be no way to evaluate the many alternative strategies proposed, now and in the future.

Here we look first at acid rain as a physical/chemical phenomenon. Then we scan some of the major scientific questions surrounding it, the options for controlling or mitigating its effects, and the research now planned or under way to shed light on its scientific unknowns. Finally, we note the complexities of trying to resolve a scientific problem that has inherent economic and political dimensions.

*Reprinted with permission from EPRI Journal, Volume 8, Number 9, November 1983, Brent Barker, Editor in Chief. For complete information refer to Nov. EPRI Journal.

What is acid rain?

Acid rain refers to a mixture of wet and dry acidic deposition from the atmosphere. The wet part may be rain, snow, hail, sleet, fog, dew, or frost. The dry part, estimated to be about half of the total in the northeastern United States, consists of gases and solid particles that settle to Earth. Dry deposition can go into solution and begin behaving like acid rainfall as soon as it contacts surface moisture.

A solution's acidity is measured on the pH scale by its concentration of hydrogen ions. Because the pH value is a negative logarithm of this concentration, pH falls as acidity rises, and each full unit of change on the pH scale represents a tenfold increase or decrease in acidity. The range goes up to 14: a pH value of 1 is highly acidic (like battery acid), a value of 7 is neutral, and a value of 13 is highly alkaline (like lye).

If there were no contaminants except carbon dioxide in the air, rain and snow would be somewhat acid because they are tinged with carbonic acid that is formed continuously in the atmosphere when carbon dioxide reacts with airborne moisture. That influence alone can lower precipitation pH to a mildly acidic 5.6. Other naturally occurring contaminants also have an effect. Dust, for example, can bring the pH up to 7.0 or 8.0. And nitrogen oxide from lightning, chlorine from sea salt, and sulfur oxide from geysers, decaying vegetation, or the ocean can bring it down again to the 4.5-5.0 range. Measurements of rainfall pH in such remote areas as the Indian Ocean, the mid-Pacific Ocean, and the Amazon Jungle have been recorded at these low pH levels.

Man-made emissions of sulfur and nitrogen oxides from burning fossil fuels can add significantly to this atmospheric loading of acid precursors—substances that precede and are the source of the acids that form. The heavily industrialized northeastern quadrant of the United States, embracing the Midwest through New England, produces man-made sulfur and nitrogen emissions that far outweigh those from natural sources. The region's 4.0-4.5 precipitation pH reflects this human contribution and makes it the geographic center of the nation's acid rain controversy.

Man-made emissions of sulfur and nitrogen oxides are now believed to be roughly equal in that region, although the relative role of nitrogen emissions is expected to grow. The sulfur emissions issue mostly from burning coal and oil in industrial plants, with electric utilities accounting for some two-thirds of the total. Utility combustion also accounts for about one-third of the nitrogen emissions, and almost half flow from the exhaust pipes of automobiles, trucks, and buses.

Sulfuric and nitric acids in the air can come from the atmospheric conversion of these precursor sulfur and nitrogen oxide emissions (SO_x and NO_x). Even though the quantitative side of the conversion reactions remains unclear, the sequence is partially understood: the oxide emissions can react with atmospheric oxidants, such as ozone (O_3), to form sulfates (SO_4) and nitrates (NO_3), which can then combine with water vapor (H_2O) to form sulfuric acid (H_2SO_4) and nitric acid (HNO_3).

The routes of long-range transport remain in doubt, but we know that SO_x emissions converted to airborne sulfates can travel hundreds of miles under the right mix of weather conditions. The consensus that such long-range transport can occur does not, however, rule out a major role for local emissions sources in acidic deposition. Local sources are thought to account for the bulk of the sulfur dioxide gas and sulfate-bearing particles that are dry-deposited on exposed surfaces.

U.S. sulfur emissions, except for the southeastern region, have actually declined in recent years, due largely to the retirement of older coal-burning power plants. But it is hard to tell what effect, if any, this emissions decline has had on rainfall acidity. One of the few sites in North America that has monitored rainfall pH for more than a decade, a site in New England, shows no substantial change in acidity but a slight decline in sulfate deposition. Nine sites in and around the state of New York, on the other hand, show mixed trends for both pH and sulfate—some slightly up, some slightly down.

Assessing the problem

The main concern about acid rain is cumulative ecosystem damage, especially to lakes and forests. In the United States the focus is on midwestern power plant emissions and their effect on the waters and woodlands farther east in New York and New England. The big questions are whether power plant emissions are in fact damaging these natural systems; how the potential effects can be controlled or mitigated; and who will bear the very substantial direct and indirect costs of implementing such procedures.

Pinning down exactly what power plant emissions do after they leave the stack is the first step in coming to grips with acid rain's scientific unknowns. The challenge is to describe the connection between emissions and deposition and, in particular, the geographic link. To clarify what is called the source-receptor relationship, we need a clearer picture of the connection between individual source areas (where emissions go up) and individual receptor areas (where they come down). We have to know how much and where acidic deposition will fall to Earth, given a known emissions profile.

Linearity is a relationship of direct proportion between two quantities. If one varies by a certain percentage, the other will vary by that same percentage. Nonlinearity is the norm for many atmospheric processes. In Europe the relationship of industrial emissions to rainfall acidity appears to be decidedly nonlinear.

In the eastern United States, however, with its quite different topography and climate, the situation may be different. A recent report issued by the National

Academy of Sciences (NAS) found "no evidence for strong nonlinearity" in the relationship between sulfur emissions and deposition. In other words, its findings imply that a 50% reduction in sulfur oxide emissions in states east of the Mississippi can yield close to a 50% reduction in deposition of wet and dry sulfate in this same broad region.

Some link clearly exists between what goes up and what comes down when viewed from the perspective of large regions and extended periods of time. But even if this link is nearly linear in character, as the NAS report suggests, we still do not have the information we need to protect those specific areas where acid rain is suspected of causing environmental change. Can a 50% reduction in emissions from midwestern power plants guarantee a reduction in rainfall acidity in the Adirondack Mountains of upper New York state? Can it guarantee reduced deposition in the New England states farther east? No one can say.

Clarifying the geographic source-receptor relationship is one of the most important challenges in acid rain research because optimal control strategies require this knowledge. Emissions controls are very expensive, so we must know where they will do the most good in order to apply them in a selective, cost-effective manner. Without better information on the source-receptor link, we cannot guarantee that reducing emissions will pay off by reducing deposition in the specific areas targeted. We cannot even tell whether control of local or distant sources will be most cost-effective in protecting those areas considered to be ecologically sensitive.

A parallel puzzle is the relationship of deposited acidity to ecosystem change. The potential for ecosystem change in response to acidic deposition has been demonstrated in the laboratory. And in the real world, geographic correlations between acid rainfall and acid lakes furnish circumstantial evidence that such changes are actually occurring. But we cannot assume that a correlation is a causal connection, nor can we attach numbers to these changes. We need comprehensive, long-term field data to quantify the magnitude and the rate of change—both major factors in understanding the extent of the problem.

Lake acidification is a highly charged environmental concern among residents of the northeastern United States and the adjacent provinces of Canada. Sportfishing and related activities are not only a part of local lifestyles, they are also part of the economic underpinning in such areas as the Adirondacks, where tourism and recreation are basic industries. What we know for sure is that some lakes in the Adirondacks are now too acid to support sport fish. What we do not know is how they got that way or whether the apparent trend toward increasing acidity in certain lakes is a strong or a weak one.

Core samples of lake sediments, which reveal a lake's accumulated chemical history, show that some of these lakes have been acidic for hundreds to thousands of years, long before exposure to man-made emissions. And comparative studies of Adirondack lakes show that they can differ markedly in pH even when the acidity of their rainfall is the same. Such findings do not rule out a contribution from acid rain, but they do make it clear that acid rain cannot be the only factor contributing to lake acidity.

Overall, the lake watershed is a complex geologic-biologic-chemical laboratory that can both produce and consume acids. Bedrock, soil depth and composition, live vegetation, organic debris, and the habits of local wildlife all play roles in the fluctuating pH balance. These are some of the factors that make one lake better equipped than another to cope with man-made acidic input. In some cases, acid rain falling within the watershed will follow a terrestial pathway through deep mineral soils, particularly soils rich in calcium, which can neutralize the moisture before it reaches the lake.

Forests also draw anxiety about acid rain effects, especially following recent reports of forest damage in Vermont and in Germany. The main fear is soil impact. Acid-induced mobilization of aluminum ions in forest soil has been linked with reported damage because it is known that such ions can be toxic to the root systems of certain trees. On the other hand, some scientists find drought, pathogens, or heavy metal deposition more likely to be the source of the problem. In no case has actual forest damage been proved to result from acid rain.

The effects of acid rain on crops, materials, and human health are also being explored, but the connections here ar more tenuous and the concern less urgent. What remains the biggest worry about acid rain is ecosystem impact. And the important thing to remember here is the context, the theater of operations. The Earth itself and all its subsystems are in constant flux between acidifying and alkalinizing influences. The amount of man-made acidity that a particular site can tolerate without damage—the safe threshold—depends heavily on the dynamic balance of other factors that are already operating at that site.

Meanwhile, because the potential for ecosystem change in sensitive areas is recognized to exist, vigorous efforts are under way to improve control of the precursor emissions that lead to acidic deposition.

Control/mitigation options

Many options are now available or are being developed to control the emission of acid precursors from power plants to the atmosphere. No single choice is best for all situations. Plant design, fuel type, and location all help determine what is most appropriate. In addition, there are many nontechnologic considerations in what may appear to be a simple technologic choice because the billions of dollars at stack in such decisions can have profound impact on the economic health of various industries and regions of the country.

One approach, already used, is to retire older coal-burning power plants. Such units have been replaced by new coal-burning systems with far more efficient emissions controls or by hydro or nuclear units that burn no fossil fuels. A variation on this theme is the practice of environmental dispatch, which restricts the use of older coal-fired plants when the weather makes pollutant buildup and conversion likely.

Flue gas desulfurization systems, commonly known as scrubbers, are the most efficient means of removing sulfur from stack gases of coal-burning power plants. They are also the most expensive, particularly if the attempt is made to retrofit older plants. Other retrofit options that are already available and require less capital investment include switching from high-sulfur to low-sulfur coal and physically cleaning coal before it is burned. On the frontier of emissions control, injecting limestone into the furnace where coal burning occurs to capture most of the emissions by chemical reaction may be commercially feasible for some plants in a few years.

But the most fundamental advance will be to generate power from coal in ways that minimize or eliminate airborne emissions and the consequent need for controls. New power generation technologies now being developed, such as fluidized-bed combustion and gasification—combined cycles, promise to produce electricity from coal in a manner that is not only intrinsically cleaner but also intrinsically more efficient than conventional coal-burning power generation systems. Investing in these new technologies, in addition to work on advanced retrofit emissions control, is important for the future of coal combustion, allowing the nation to move forward to the next generation of coal plants rather than tying us to continued use of older, less-efficient units that have been modified for environmental compliance.

Supplementing control measures are mitigation measures aimed at helping to reverse whatever environmental changes might c ccur. For example, lake acidity can be neutralized by adding lime or limestone to the water. This measure may allow fish to repopulate and sportfishing to be restored. Because every lake is different, individual treatment programs must be developed that are responsive to the needs of each lake. But the costs are still quite low compared with retrofit emissions control costs.

What all mitigation measures have in common is that they focus directly on the problem, be it acid lakes or corroding monuments or contaminated drinking water, rather than on the emissions that so far we can only suspect or assume to be a source of that problem.

Reversibility, like the magnitude and rate of ecosystem change, is another critical issue in acid rain research. What we really must find out is whether measurable and irreversible change is occurring, say, every year. Such a situation would clearly justify greater urgency than one in which change is minimal, occurs on a timeframe of decades, and is reversible at any point.

Besides NAS, the Environmental Protection Agency, the Office of Science and Technology Policy, and the Government Accounting Office (GAO) have all addressed the acid rain problem during the past year and have all stressed the need for better information. What is being done to provide this information?

International research

The United States federal acid rain research budget will total about $27 million in 1984. Canada is spending over $10 million a year. The United Kingdom is spending about $1-$1.5 million annually, and Scandinavia has a similar budget, but the yearly expenditure is less than $5 million for all Europe. Most of the worldwide research effort on acidic deposition is concentrated in the western hemisphere, with the governments of the United States and Canada and the electric utilities, through EPRI, taking the lead.

EPRI's environmental work, the world's largest privately funded research program on acid rain, centers in the Environmental Assessment Department of the Energy Analysis and Environment Division, where studies are under way on both atmospheric proceses and ecosystem change. In addition, a special Economic and Environmental Integration Group within the same division is looking at a number of areas where economics and the physical sciences intersect. And the Coal Combustion Systems Division is exploring a spectrum of possibilities for emissions control.

In the realm of atmospheric processes, a major research target is the elusive source-receptor relationship. The NAS panel has already credited EPRI's Sulfate Regional Experiment (SURE) with providing the best data now available on air quality in the northeastern United States, and upcoming efforts focus on narrowing down the uncertainties that remain. For example, a massive tracer experiment is being developed to complement ongoing chemical and physical studies. This experiment will track the emissions transported from midwestern power plants to their final destinations—whether in the Adirondacks, in New England, or closer to their source.

In probing ecological effects, EPRI and governmental research agendas are complementary. Federal studies are surveying the actual status of individual receptor systems, particularly those that are deemed most susceptible. EPRI is exploring the general processes by which ecosystem preservation, change, and restoration may operate. At what point, for example, can we expect lake acidity to begin affecting fish? And how long will a liming treatment be effective?

EPRI's funding to answer such questions has taken a recent jump. The 1983 budget was about $10 million, up from an average of about $3 million in the preceding five years. In 1984 that figure will rise to $13 milion and again to about $15 million in 1985. For the period 1984-1988, in the Energy Analysis and Environmental Division alone, a total of $75 million is slated for the study of the scientific phenomena associated with acidic deposition.

The Coal Combustion Systems Division has another $100 million budgeted to study emissions control over the next five years. Much of it will apply to the acid rain problem. For example, how can we bring down the cost of scrubbers and lessen their adverse impact on power plant performance? Other research efforts continue to explore alternative retrofit options and to develop entirely new generating systems that can sidestep the need for postcombusion emissions control.

Weighing alternatives

We know far more about acid rain now than we did when it first surfaced as an environmental concern a decade ago, but much remains to be done. Most of our reliable data are too recent to show conclusive trends. So there are still large gaps in our understanding of the relationship between the three vital pieces in the scientific puzzle: emissions, deposition, and ecosystem effects.

The degree of uncertainty, and the timeframe for resolving it, differ at different points on the research continuum. Preliminary clues may emerge rapidly through laboratory experiments, but the researcher's and the society's needs for answers about acid rain can do nothing to speed up long-term data collection in the field. Clouds will form, winds will blow, rain will fall, and the seasons will turn at their own pace. It will take another 5 years, 10 years, or even more of widespread field observations to establish conclusive trends in deposition patterns and environmental effects.

What can we do in the meantime? How can we forestall the possibility of environmental damage, assuming that 5 or 10 years could make a difference? Emissions control cost projections run very high, even for selective controls that target specific sources rather than try to blanket entire regions of the eastern United States. Because such huge sums are at stake and because the groups that must bear these costs may not be the same as the ones receiving the benefits, apparently simple control decisions are not simple at all.

Given a choice of control strategies to meet tighter standards, many utilities might switch to cleaner, low-sulfur coal rather than install expensive scrubbers. Such a course would moderate the rate increases that are likely to hit midwestern utility customers no matter what form of emissions control is chosen, since all of them are costly. But this switch would mean other shifts as well. Coal-switching would move jobs from the Midwest and the East, center of the high-sulfur coal industry, to Wyoming and other western states.

This is just one example of the chain of consequences that could follow from an apparently straightforward decision. The other options for emissions control are similary complicated when we look at the interests of all the groups and regions that may be involved. Retrofitting scrubbers, for instance, would allow the continued use of high-sulfur coal and minimize geographic shifts in coal production and jobs, but the total bill to be somehow divided—probably among groups within the northeastern quadrant—would be higher by billions of dollars per year.

The policy-oriented GAO federal report sums up the situation this way: The choice among alternative courses of action, it says, "involves an allocation of the burden of risk, among industry, consumer, labor, environmental, and other

interests, and the different regions of North America. It is, therefore, more than a scientific and economic decision; it involves political judgments on how our society's resources—and burdens—should be spread among these constituencies."

To break down such complex decisions into more-manageable segments, some policymakers are turning to a method known as decision analysis. The ADEPT (acid deposition decisions tree model) method developed for EPRI is one example of the decision analysis approach. It can help separate scientific issues from political issues where such a separation is necessary to clarify the nature of the problem.

EPRI's charter is to provide scientific information. Scientific information alone cannot determine policy decisions. But it is essential input to well-informed decision making and it is essential in evaluating those policies that the body politic creates.

Further reading

Richard E. Balzhiser, René H. Malès, and Kurt E. Yeager. *Acidic Deposition Briefing.* Prepared for Wiliam Ruckelshaus, Administrator, Environmental Protection Agency, and Jan W. Mares, Assistant Secretary, Fossil Energy, Department of Energy. June 1983. EPRI (unnumbered).

U.S. Environmental Protection Agency. *The acidic Deposition Phenomenon and its effects: Critical Assessment Review Papers,* Vol. 1, Atmospheric Sciences; Vol. 2, Effects Sciences. May 1983. EPA-600/8-83-016A and -016B.

National Research Council. *Acid Deposition: Atmospheric Processes in Eastern North America.* Washington, D.C.: National Academy Press, 1983.

Inventory of Acid Deposition: Research Projects Funded by the Private Sector. Final report for RP1910-3, prepared by General Research Corp., February 1983. EPRI EA-2889.

Acid Deposition: Decision Framework. Vol. 1. Final report for RP2156-1, prepared by Decision Focus Inc., August 1982. EPRI EA-2540.

This article was written by Mary Wayne, science writer. Technical background information was provided by Ralph Perhac and René Malès, Energy Analysis and Environment Division; Kurt Yeager, Coal Combustion Systems Division.

Acid Rain: Sources and Transport

B. Tracing the Pathways of Atmospheric Conversion

Weather, the largest, most-complex interactive system on Earth, decides where acidic deposition will finally be a problem. New tracer experiments should speed understanding of how atmospheric processes affect emissions transport and chemical conversion.

Characterizing the phenomenon of acidic deposition by the popular statement "What goes up must come down," although valid in a broad sense, is not very useful if one wants to know *how, where,* and *in what form* what goes up comes down. The sources of fossil fuel emissions, their chemical transformation in the environment, their routes of transport, and their return to Earth's surface involve the totality of physical and chemical processes that occur in the atmosphere, as well as many that occur on the ground.

The atmospheric processes involved in acidic deposition are complex and defy simple characterization. But tracking the evolution of the chemical pollutants provides a quick overview of the transformation cycle. Oxides of sulfur (SO_2) and of nitrogen (NO_x) are among the principal chemical products of fossil fuel combustion. Over half of the emissions of these gases fall to Earth relatively near their source without changing form. With the aid of sunlight and chemical oxidants also present in the atmosphere, some of the remaining sulfur and nitrogen is transformed into new compounds—sulfates (SO_4) and nitrates (NO_3). These new substances, together with the oxide forms, can be chemical precursors of the acid compounds H_2SO_4 and HNO_3 formed in the atmosphere. These acids are the ones of final concern in the acid rain question.

A key uncertainty in acidic deposition is whether the relationship between SO_2 and NO_x and the acidity of wet or dry deposition—the source-to-receptor relationship—is linear. That is, if society reduces the level of its pollutant emissions, can it expect an improvement of similar degree in the deposition of acidic materials and in the resultant environmental effects?

The best that science can tell us at this point is maybe, maybe not. We know that the governing equations describing atmospheric processes are not linear. But can linear equations approximate some of these processes well enough to be used in making rational decisions on how to control acidic deposition? On this, the scientific jury is still out.

Based on 17 years' observations at one site in eastern North America, the National Academy of Sciences could conclude only that the source-receptor relationship is not strongly nonlinear. Before science can venture a more confident guess, it must have a much better understanding of the many physical, chemical, and biogeochemical variables involved overhead and in the ground. A major element of EPRI's research related to acidic deposition spans the gamut from emissions through atmospheric transport and chemical transformation to deposition and terrestrial transport. Objectives include the identification and measurement of key pollutant emissions; a quantitative understanding of the transformation; better techniques for measuring the quality of air and precipitation; and methods for predicting the behavior and fate of pollutants. More than 30 individual research projects completed, underway, or planned address these objectives.

Emissions

The first step in shedding light on the relationship between pollutant emissions and their fate is to learn how much of each of the important chemicals is emitted and from where. But there are few inventories of either recent or long-term emissions that are detailed or accurate enough to be considered of much use in continental-scale atmospheric research. That is, in part, because of the scope and precision of emissions monitoring required and the inherent uncertainties in estimating area emissions on the basis of a few point measurements.

According to the National Academy of Sciences' 1983 report on acidic deposition, the most reliable data on key pollutant emissions affecting rural areas comes from EPRI's Sulfate Regional Experiment (SURE) conducted in 1977 and 1978. SURE covered the 31 eastern states of the United States and parts of southeastern Canada. Extensive checks and supplemental field monitoring have narrowed the uncertainty in measurement to 10-15% of the observed or estimated value.

EPRI's SURE inventory provides the daily rates and seasonal totals of all SO_2 and NO_x emissions, as well as particulates and hydrocarbon species. SURE also logged emissions of a key acid-precursor chemical, the anion sulfate, which, to a lesser degree, is produced by fossil fuel combustion.

Despite the limitations of existing emissions inventories, some conclusions regarding man-made pollutant emissions, based in large part on SURE data, can be made.

The available long-term emissions data indicate that from 1965 to 1978, sulfur emissions in the United States, except in the Southeast, declined an average of 8%; the reductions in key specific regions have been more pronounced: 20% for SO_2 in the Northeast and 40% in the Midwest. In the Southeast, however, SO_2 emissions have increased about 20%. Whatever ecological effects of precipitation acidity are now observed must be viewed within the context that overall sulfur emissions have unquestionably declined in rate in recent years.

Based on annual averages from the SURE inventory, sulfur emissions from all sources in the study areas total 85,000 t/d. Total nitrogen emissions average

32,000 t/d. And 52,000 t/d of various chemically reactive hydrocarbons are released to the atmosphere.

In general, industrial sources, including utilities, account for most of the SO_2 and particulates and about half of the NO_x. The utility contribution to total sulfur emissions in the eastern United States varies from about 55% in the winter to about 65% in the summer. Utilities and highway vehicles contribute most of the NO_x, and industrial and transportation sources contribute most of the hydrocarbons.

Sulfur and nitrogen are also present in the atmosphere from natural sources, such as decaying organic matter. These sources were formerly believed to play a significant role in acidic deposition. But an EPRI-sponsored assessment of biogenic sulfur emissions in the SURE region indicates an estimated total of about 150 t/d, which might approach 1-2.5% of the man-made sulfur contribution.

Much, however, remains to be learned about North American pollutant emissions. The distribution of emissions over time and area must be known with greater accuracy in order to improve the performance of models, or numerical constructs, with which the source-receptor relationships are studied.

To overcome some of the limitations in existing emissions inventories, EPRI is sponsoring an inventory of 1982 emissions, which is expected to be available at the end of 1984. The project will locate major point sources exactly; minor point sources will be grouped in an 80-km^2 (31-mi^2) grid map. The inventory data will be produced in a format that will accommodate updates; more extensive checks on the accuracy of the data will be made.

Atmospheric transport

More than half of the SO_2 emitted from fossil fuel combustion returns to Earth as either dry gas or particles before it can be chemically transformed. Typically, this occurs within 200-300 km (130-190 mi) of a source. Winds carry the remaining SO_2, along with sulfates and nitrates, over longer distances, sometimes as far as 1000 km (620 mi).

As is the case with all stages in the continuum from emissions to biologic effects, the boundaries used to distinguish one category from another are often more conceptual than real. A case in point is the role of a source parameter—for example, stack height—in the succeeding arena of activity and research interest: the transport of pollutants in the atmosphere.

In the 1960s and early 1970s, many industrial and commercial emissions sources, including fossil-fuel-fired power plants, replaced short (typically 60 m, or about 200 ft) stacks with taller ones (averaging 200 m, or 650 ft) as a way to reduce ground level pollutant concentrations in the surrounding areas. This solution to limitations on local concentrations was eminently successful, but it also set the stage for occasional longer-range transport.

Local meteorology and terrain around a source exert such a dominating influence on stack emissions from any height, however, that it is difficult to pinpoint the effect of stack heights on pollutant transport against the background of weather and other local features, such as hills and valleys. Measurements of plume rise made during EPRI's plume model validation studies are adding to a detailed understanding of the role of effective stack height, a key feature in plume mixing and transport.

Transport processes represent the greatest realm of uncertainty in acidic deposition because they include the largest, most complex interactive system on Earth—regional weather. Weather drives the two processes that are most important in determining whether pollutant emissions are deposited as dry particles and gas near a source or are carried into the lower atmosphere to form acidic compounds. These are advection, the movement of a volume of air and pollutants with the wind, and diffusion, the intermixing of this volume with surrounding air and other reactive pollutants.

The extent to which advection, diffusion, and chemical transformation occur is a function of the pollutant's atmospheric residence time, or the time between emission and deposition. Residence time is, in turn, governed by meteorology, terrain, and transport processes. Even though there is no debate that long-range transport of some pollutants occurs, a meaningful assessment of the relative scales of transport—from local to regional to continental to global—cannot yet be made.

Models of wind trajectories, pollutant residence times, and large-scale meteorologic processes would be greatly improved if better methods were available for measuring long-range transport. Chemical tracers hold great promise of filling in significant portions of the transport puzzle.

Chemically unique or radioactive tracer materials can be detected and measured in very dilute concentrations long distances from their point of release upwind. With enough tracer releases and subsequent measurement, a detailed pattern of air flow begins to emerge. Perfluorocarbon compounds are considered ideal for this purpose because of their nonreactive nature as low as one part per quadrillion (10^{15}), making it possible to track pollutant transport over distances exceeding 1000 km (620 mi).

EPRI is currently working in this pioneering area of atmospheric research through its participation in the cross-Appalachian tracer experiment (CAPTEX), cosponsored with the National Oceanographic and Atmospheric Administration (NOAA), the Department of Energy (DOE), the Environmental Protection Agency (EPA), and the Canadian Atmospheric Environment Service (AES).

Six releases of perfluorocarbon tracer were made this fall in Ohio and Ontario. A wide arc of measurement sites—including 80 surface samplers, 20 upper air sounding stations, and 4 aircraft—recorded data on the tracer material as it moved from the sources to the mid-Atlantic, New England, and southeastern Canada.

CAPTEX is a major step toward a better understanding of pollutant transport in the lower atmosphere. The observation data will help define the number of upper air measurements necessary to improve trajectory models and calculations. More important, however, CAPTEX will serve as a pilot for the design of a much more extensive measurement of upper air flow—the massive aerometric tracer experiment (MATEX).

A coauthor with EPA of the preliminary conceptual design of MATEX is Glenn Hilst, manager of EPRI's Environmental Physics and Chemistry Program. MATEX, proposed to begin after 1984, would involve numerous releases of tracer material, according to Hilst, and a network of about 600 monitoring stations, 100 aerometric sampling stations, and as many as 8 aircraft covering an area of 160 million km^2 (62 million mi^2).

A preliminary estimate of the cost of MATEX, including a full year of field work, is about $110 million. "The price tag for MATEX may appear to be steep," sayd Hilst, "but viewed in terms of emission control costs on the order of billions of dollars per year that are envisioned for electric utility power plants alone, even a modest fine-tuning of control strategies on the basis of MATEX results would make the project highly cost-effective."

MATEX will be designed so that the data from the measurements will have several uses: in evaluation of present methods for estimating source-receptor relationships, in development of an improved empirical relationship, in identification of the key pollutants involved in long-range transport, and in assessment of alternative control strategies.

Atmospheric Conversion

When pollutants encounter precipitation in clouds as they move through the atmosphere, typically within 3-5 days in eastern North America, chemical oxidation is enhanced and sulfuric and nitric acids can be formed. But the chemical reactions depend on a host of meteorologic variables, including the intensity of solar radiation, the number and type of clouds, relative humidity, and the presence of other pollutants in the air.

The last chemical link in the transformation of SO_x and NO_x to acids— hydrogen—is available in the atmosphere from hydrocarbons, a pollutant form that results mainly from automobile exhaust. Add sunlight and water and the ingredients for forming acidic compounds are all present. Depending on the concentration of hydrogen ions, the basis for calculations of pH value, the subsequent precipitation—rain, snow, or fog—is more or less acidic.

The basic chemical mechanisms by which SO_x and NO_x react with oxidants to forms acids and acid-producing compounds are fairly well known. A precise characterization of the reactions and associated cloud processes, however, has only recently begun to evolve.

The two main pathways of chemical conversion of SO_2 and NO_x to sulfates and nitrates and to sulfuric and nitric acid correspond to the two principal states of matter in the atmosphere—gas and liquid. These involve reactions in dry air (gas phase) and in liquid water drops, such as found in clouds and fogs, suspended in the air (liquid phase).

In the gas phase, pollutants are oxidized to acids, which are scavenged by cloud water and precipitated as rain or snow. According to Peter Mueller, who manages many of EPRI's atmospheric chemistry projects, the basic chemistry of this path is well understood. The reaction mechanisms and oxidation rates of SO_2 and the hydroxyl radicals (OH), hydroperoxyl radicals (OH_2), and ozone (O_3).

On the basis of cloud chemistry measurements, EPRI-sponsored researchers have concluded that most of the nitrate in precipitation gets there by the gas-phase path; but only about 20-25% of SO_2 emissions appearing as sulfate in precipitation are attributable to gas-phase oxidation in clouds, fogs, rain, and snow.

Because SO_2 and NO_x concentrations are highest near major emission sources, the gas-phase chemical reaction path is considered the more important conversion route in the vicinity (10-200 km or 6-130 mi) of sources.

In the liquid phase, SO_x and NO_x are incorporated in atmospheric hydrogen peroxide (H_2O_2), ozone, oxygen, and particles. Reaction rates, at least of SP_2 with hydrogen peroxide and ozone, are manganese oxide particles can accelerate the oxidation of SO_2 to sulfate.

The amount of sulfate formed in the liquid phase is dependent on atmospheric levels of SO_2 and hydrogen peroxide. In some areas, on some occasions the amount of hydrogen peroxide available to oxidize the SO_2 is very small, while at other times more than enough H_2O_2 is present to oxidize all the SO_2.

Based on current knowledge, according to Mueller, the liquid-phase reaction with hydrogen peroxide is believed to be the most important route in eastern North America for the conversion of SO_2 emissions to sulfate in precipitation. But partly because of few measurements of SO_2 and hydrogen peroxide in the gas and liuqid phases, a definitive assessment of how sulfate found in precipitation at specific sites is related to SO_2 emissions from known locations cannot be made. This must await considerably improved understanding of atmospheric chemistry.

A major factor in atmospheric chemistry that should not be discounted is the extent to which acids formed in the atmosphere are neutralized by naturally occurring base (alkaline) materials, including gaseous ammonia and aerosol particles containing metal oxides or carbonates. The principal products are ammonium sulfates and nitrates and metal sulfates and nitrates containing calcium, magnesium, iron, and other materials.

This continuous neutralization process is influenced by geography and variations in the concentrations of other copollutants, however. More alkaline substances are present in the atmosphere of the Midwest than in the rural North-east. So, given equal sulfate concentrations, precipitation would be less acidic in the Midwest than in the Northeast.

Deposition

Pollutant deposition, whether near the source or far away, is controlled by a number of mechanisms but can be discussed in terms of the two main pathways, dry and wet. Compared with many of the other atmospheric variables, processes, and uncertainties, deposition may appear to be straightforward, but it is, in fact, marked by much uncertainty as well. The unknowns, especialy with respect to dry deposition, relate to difficulties in obtaining accurate measurement of deposition rates.

Only a few sites in eastern North America have monitored and collected weekly precipitation chemistry data over more than four or five years and with consistent sampling methods. One such site is the Hubbard Brook Experimental Forest in New Hampshire. The 17-year deposition record from Hubbard Brook reportedly indicates that the molar ratios of sulfates to nitrates in precipitation are similar to the ratios of SO_x to NO_x that can be seen in eastern U.S. emissions.

For dry deposition, several factors illustrate the current uncertainties, the most important of which is that measurements of dry deposition have tended to be under experimental conditions that were not realistic. Adding to the uncertainty is the fact that several complex mechanisms convey dry particles or gas to the ground, and it is difficult to determine which of these dominate under different conditions. Near-surface mechanisms for particle deposition include gravitational settling, thermodynamic propagation by warmer air molecules, and molecular-level electro- and thermophoretic effects.

In addition, considerable uncertainty marks the role of different surfaces as dry deposition receptors. For example, cell-size openings (stomata) on leaf surfaces are known to influence gaseous deposition rates, and reemission of sulfur compounds from plant surfaces has been detected.

Given these uncertainties, estimates of the time and area distributions of dry deposition are subject to wide variations. In general, the process of dry deposition is believed to be about as effective as wet deposition in removing pollutants from the atmosphere. Because about one-third of all SO_2 emitted in the eastern United States is known to be transported out of the region, it is assumed that roughly one-third of northeastern North American emissions are dry-deposited on the continent and the remainder are deposited wet.

Wet deposition can be any form of precipitation—rain, snow, or sleet. It differs from the dry variety in that liquid-phase chemical reactions occur in clouds and below clouds as preciptation is falling. Wet deposition may be more important than dry processes in areas of high average precipitation far from emission sources.

The processes by which precipitation elements are formed are well understood. The most important of these is nucleation, the kinetic process by which water molecules condense on the surface of aerosol particles. Others include diffusional attachment (in which pollutant molecules or particles diffuse through the air to the surface of water molecules) and scavenging (in which falling water droplets or ice crystals collide with aerosols).

Several major deposition chemistry programs are under way in the United States to obtain more detailed and accurate measurements of these and other factors. One of the most detailed daily monitoring networks is being managed by EPRI for a group of 35 electric utility organizations in the eastern United States. The Utility Acid Precipitation Study Program (UAPSP), established in 1981 and now totaling 20 monitoring stations, incorporated five stations set up initially for EPRI's SURE project.

Preliminary analyses of data from the monitoring stations, some of which have been operating since late 1978, tend to confirm observations made in other networks, including SURE. Hydrogen, sulfate, and nitrate ion concentrations tend to be highest in the Midwest and occasionally in Pennsylvania and Massachusetts. Sulfate concentrations generally correlate with hydrogen ion concentrations, indicating that sulfate and, to a lesser extent, nitrate are the materials that cause the acidic properties of rain. Ammonium, nitrate, and calcium are sometimes—but not as consistently—associated with reduced hydrogen ion concentrations.

Sulfate concentrations are generally higher than nitrate except in snow, where nitrate is up to four times higher than sulfate. For as yet unknown reasons, snow accumulates nitrates more efficiently than rain, while accumulations of sulfates are about equal in rain and snow.

Terrestrial transport

Deposition of acids and acid-forming substances marks the end of their journey through the atmosphere, but it does not mean the end of chemical transformations. Biogeochemical processes take over where atmospheric processes leave off to produce or consume acids and other chemicals that shift the pH equilibrium in soil or water. Sulfates and nitrates that have not been transformed into acids in the atmosphere may still undergo chemical reactions back on Earth.

EPRI-sponsored research has begun to clarify many of the variables that influence terrestrial transport and transformation. These include bedrock geology, deep soil layer mineralogy, and precipitation flow paths.

The relevance of understanding atmospheric chemical processes, as well as those that occur on the ground, becomes apparent when considering different control strategies. If the strategy is simply to control or limit all sulfur emissions from all sources, much of the uncertainty surrounding atmospheric processes becomes moot, while transformations on the ground take on more importance. If, however, it can be determined that either local or long-range transport dominates the movement of pollutants, the effectiveness of selected control options demands a clearer understanding of atmospheric processes than we presently have.

Further reading

National Research Council. *Acid Deposition: Atmospheric Processes in Eastern North America.* Washington, D.C.: National Academy Press. 1983.
U.S. Environmental Protection Agency. *The Acidic Deposition Phenomenon and Its Effects: Critical Assessment Review Papers,* Vol. I., Atmospheric Sciences. Public revie draft prepared by EPA, May 1983. EPA-600/8-83-016A.
U.S. Environmental Protection Agency. *Atmospheric Sciences and Analysis: Work Group 2.* Final Report, No. 2F, November 15, 1982, to fulfill requirements of the Memorandum of Intent on Transboundary Air Pollution signed by Canada and the United States, August 5, 1980. Washington, D.C.: Office of Environmental Protection Processes and Effects Research.
EPRI Sulfate Regional Experiment: Results and Implications. Summary report, December 1981. EPRI EA-2165-SY-LD.
National Research Council. *Atmospheric-Biosphere Interactions: Toward a Better Understanding of the Ecological Consequences of Fossil Fuel Combustion.* Washington, D.C.: National Academy Press, 1981.

This article was written by Taylor Moore, Technical background information was provided by Mary Ann Allan, Robert Goldstein, Glenn Hilst, John Huckabee, Peter Mueller, Robet Patterson, and Ralph Perhac, Energy Analysis and Environmental Division.

Acid Rain: Environmental Effects

C. Discerning the Change in Waters and Woodlands

Lakes and forests are at the center of controversy over the effects of acid rain. Research on the response of these ecosystems to acidic deposition focuses on the questions of how, how much, and how fast.

Acid rain falls on lakes and ponds, streams and rivers, forests, crops, buildings, and statues in parts of eastern North America and Europe—that much is certain. But once it lands, the environmental effects of acid rain are anything but certain. Some claim that rain is acidifying lakes and killing fish in the northeastern United States, Canada, Sweden, and Norway; withering forests in West Germany and New England; and crumbling historic buildings. Others assert that these charges have yet to be proved. Most researchers agree, however, that it is hard to quantify the connections between acid rain and environmental effects.

Starting with lakes

Lakes and other surface waters are as good a place as any to begin to explore the complex effects of acid rain on the environment. Some of the most puzzling questions concerning lake are also some of the most basic questions: Is there an increasing trend toward acidification of lakes, and if so, what is causing it? How does acidification affect lakes? How extensive are the susceptible regions? How fast is change occurring, and is change reversible?

These questions are difficult to answer because of a shortage of scientifically credible, consistent, long-term data on trends in surface-water acidity, especially for the inaccessible, high-elevation lakes that are most likely to be affected by acid deposition. Not only are data limited, but there are differences in the way data were acquired. Surface-water acidity measurements made before 1960 used methods that tended to yield too-high pH readings because of chemicals added to the water in the testing process. Further, many of the reported acidity trends are based on comparisons made from only two points in time, several years apart. Often, data were colleted at different times of the day or on different days of the year, ignoring normal daily, monthly, or seasonal variations in water acidity levels caused by, for example, algae activity.

Sometimes, different data can even lead to seemingly different conclusions on acidification trends. For example, one data base collected from 1929 to 1937 for a sample of 217 high-elevation lakes in the Adirondacks showed that 4% of the sample had pH levels below 5.0. By 1975, 51% of those lakes reportedly had a pH below 5.0. According to another recent data base collected on 2273 high-elevation Adirondack lakes—which represent 80% of the lakes above 1000 ft (305 m)—only 4.7% of the lakes sampled had a pH less than 5.0. In both recent data bases, roughly 105 lakes were at a pH of less than 5.0, but the statistics—51% vs. 4.7%—sound markedly different.

Despite the shortage of good data, the conflicting reports that the available data sometimes produce, and the questions regarding long-term data comparisons, many researchers now agree that some lakes in different parts of the world have shown trends toward increased acidity over the past 50 years or so. Reports from the Adirondacks, Norway, Sweden, and Canada cite examples of such lakes. Similarly, some correlations between lake acidity and decreased fish population can be established.

Unfortunately, the true extent of lake acidification or of damage to fisheries is far from understood. In a recent report to Congress, the U.S. National Acid Precipitation Assessment Program notes, "Many of the reported cases that suggest acid deposition is the cause of some observed change in an aquatic ecosystem are based on circumstantial evidence and lack documentation of the mechanisms linking cause and effect. At present, the extent of actual acquatic damage is not well established."

Some examples of lake acidification may have been caused by man-made pollution. In Ontario, for example, recent acidification of the La Cloche lakes downwind from the Sudbury region has been well documented, and much of the damage has been attributed to atmospheric emissions of acids and metals from vast smelting operations near Sudbury and Wawa.

The causes behind acidification trends in northern New England, the Adirondacks, and other parts of Canada are still uncertain, and natural causes cannot be overlooked. In the Adirondacks, for example, many lake watersheds are underlain by noncalcareous bedrock and thin soils. Such watersheds are underlain by noncalcareous bedrock and thin soils. Such watersheds may lack the ability to neutralize acidity, whether the acidity is produced by organic decay of dead material in the watershed or by rain itself. Logging, forest fires, storm blowdown, and other conditions can also instigate or aggravate acidification by permitting normal or acid rain to run freely into lakes without neutralization by rock and soil.

Seeking explanations

Convinced that at least some of the acidified lakes were the result of acidic deposition, researchers needed to establish a firm relationship between the deposition of atmospheric acids and the acidity of surface waters before they could determine how many lakes were susceptible to acidification, how fast changes were occurring, and whether the changes were reversible. One of the most popular acidification hypotheses among scientists compared lake acidification by atmospheric input to the titration of a beaker of alkaline solution. A lake was assumed to have some finite alkalinity that was consumed as acidic deposition was added.

Another popular hypothesis held that the bedrock underlying a lake was the deciding factor in lake acidification. If this rock contained adequate amounts of calcite and other minerals to neutralize acidity; the lake would be unscathed; if the bedrock was nonreactive, the lake was likely to become acidified. But this hypothesis, too had flaws. Researchers noticed that lakes with the same bedrock did not become acidified at the same rate. Plainly, a better understanding of the acidification process was required.

EPRI entered research into the ecological aspects of acidification in 1977, with the initiation of the integrated lake watershed acidification study, or ILWAS. The study was designed to look at three neighboring lakes in New York State's Adirondack Mountains. Woods, Sagamore, and Panther lakes received the same amount of acidic deposition and shared the same nonreactive granitic bedrock. Yet their pH values were different: Woods Lake was highly acidic; Sagamore Lake was slightly acidic; Panther Lake was neutral. More than rain and bedrock was at work here, and EPRI reasoned that this was a good opportunity to investigate how acid rain and lakes interacted.

ILWAS was solidly based on intensive data collection. The lakes and their surrounding watersheds were divided into eight distinct data compartments: atmosphere, forest canopy, snowpack, soil, hydrologic catchments, bogs, streams, and lakes. Field studies painstakingly characterized the properties of each compartment. At selected locations measurements were made of the ambient air quality and of the quantity and quality of water that moved through each compartment. Data were collected monthly, weekly, or as otherwise required.

"Five years and some half a million data points later, we had found some very interesting things," notes Robert Brocksen, until recently manager of the Ecological Studies Program in EPRI's Energy Analysis and Environment Division and now director of the Wyoming Water Research Center of the University of Wyoming at Laramie. "We learned that bedrock geology alone was not adequate to define the sensitivity of lakes to acid rain and that you can't understand the relationship between atmospheric acid and acid in surface waters without considering the entire watershed."

Bedrock, soil, hydrology, vegetation—everything about a lake and its watershed had an effect on how that lake reacted to incoming acidity, according to Project Manager Robert Goldstein, who has directed ILWAS since its inception. Bedrock, as researchers earlier suspected, does indeed have an important—although not all-important—influence on acidification. ILWAS showed that soil is also a critical buffering agent and that most of the neutralization of incoming acidity, in fact, occurred in deep soils. The thicker the layers of neutralizing soil, the more neutralization could be accomplished. Loamy soils that held onto the water did a better job of buffering than impermeable soils, such as hard-packed clay, or soils that permitted too-rapid transfer of water, such as sand and silt.

Hydrology was also found to have important effects on incoming acidity. If the terrain surrounding a lake was extremely steep, incoming rain would tend to rush into the lake without an opportunity for adequate buffering in the soil layers. A more gentle terrain would tend to increase neutralization time. Snowpack, too, had an important influence on lake acidity. During the winter, precipitation acidity collected in the snowpack. In spring, when the snow melted, large volumes of this acidified water rushed over the soil or through only the topmost soil layers, greatly reducing buffering capacity and resulting in a springtime acidity surge.

Even watershed vegetation had an impact on lake acidity. Researchers found that the foilage of broadleaf deciduous trees tended to neutralize the acid in incoming rain, whereas the needles of conifers tended to increase the acidity of the rainwater that passed through them.

These and many more ILWAS findings were incorporated into a comprehensive model that permitted the researchers to simulate how the lakes would perform under different acidity inputs, factoring in not only the altered acidity but also the watershed's soil depth and type, bedrock, vegetation, and other characteristics. This year an expended version of ILWAS, called the regional integrated lake watershed acidification study (RILWAS), was begun to assess the ILWAS framework, as well as the acidification vulnerability of an entire region.

The first RILWAS objective involves applying the ILWAS model in both northern Wisconsin and the southern Appalachians. At two lakes in Wisconsin and a large stream watershed in North Carolina—each of which has unique hydrologic soil, and climatic features—researchers will try to refine the ILWAS model.

The second RILWAS objective, development of a regional assessment methodology, calls for applying a less-intensive ILWAS approach to 20 watersheds in the Adirondack Park area. The watersheds have been selected to represent a variety of mineralogies, bedrock types, hydrologies, and fisheries that exist in the region. Both parts of RILWAS are cofunded by local utilities, as well as by state and local governmental agencies.

Once completed, the ILWAS—RILWAS model will help explain the influences that different factors have on lake acidification. The completed model will be able to suggest how a given lake will behave under either increased or reduced acidic deposition input. It will also be able to indicate if a lake will be subject to acidification over a given period of time. It should also be able to indicate if the acidification of a given lake can be reversed.

Vanishing fish

While researchers are grasping for the connections between acid rain and acid lakes, they are also trying to define the connections between lake acidification and the disappearance of fish. A lack of fish and other biota has been observed in some lakes that had high levels of acidity, leading many researchers to conclude that acidic deposition—either man-made or natural—was responsible for the losses (there is no evidence to suggest that loss of fish is related to loss of their food). Because fish are the species most noticeably absent from acidified waters and because of their high public visibility as both food and sport, fish have been the focus of researchers' concerns.

High acidity levels in water can most definitely disrupt the bodily salt balance of a fish to the point where the fish is unable to function, and dies. Even if acidity levels are not high enough to kill mature fish, acidity may damage delicate fish eggs and fry, possibly during the sudden surges of acidity that can accompany spring snowmelt. Researchers also know that acidification of waters can mobilize toxic metals, such as aluminum, from soil and bedrock and that these metals can harm fish, possibly by disrupting gill structure, by causing oversecretion of mucous and subsequent suffocation, and through toxic accumulation in body tissues.

But despite the known effects of acidification of fish, circumstantial correlations between surface water acidity and the absence of fish do not establish a direct cause-and-effect relationship between the two. In only a few cases are data available showing that fish populations declined as water acidity gradually increased. Researchers have yet to establish a direct connection between acidifiation and fish losses, a task complicated by the fact that many other factors can cause fish loss and many fish losses of past years were reported without any attempt to document the possible causes.

For example, a lake may have been naturally acidic all along. Such a lake may never have supported fish, or if it was stocked with fish in recent times, the fish would have eventually died out. Even if fish were indigenous to a lake, they may have been killed by other causes besides acidification, such as by pesticides. Where acidity seems to be the cause of fish losses, the condition may be the result of either natural or man-made causes.

To sort out the fish loss mystery, researchers need more and better data on fish populations and the effects of acidity on fish. This information would ordinarily be gathered through field monitoring, but 10 or more years could be required to get a good data base, and by that time, more fish populations could be affected. EPRI has opted for a faster research approach that combines modeling, laboratory tests, and field tests.

This new study, beginning now and expected to run five years, will start by developing a model of fish populations in nonacidic but environmentally sensitive watersheds, such as those with thin soil layers and noncalcareous bedrock. According to Project Manager Jack Mattice, laboratory tests will then be conducted to assess the effects of acidification, such as decreased pH and increased aluminum, on fish at various stages of development. The results will be factored into the model to predict the relative population levels of the fish in an acidified environment. Model predictions will then be verified by comparison with field observations. Meanwhile, another EPRI project will examine how mobilized aluminum affects fish and other biota by studying data collected at some 17 different sites.

The final fish population model may ultimately be used in conjunction with the ILWAS model. The ILWAS model will characterize the chemistry to be expected for different watersheds; then the fish population model can be used to predict how the fish population in that watershed will respond. Through these two models, informed decisions can be made on how to protect fish in sensitive areas.

Where acidification has caused fish disappearance, there are several possibilities for restoring the fish. Liming acidified lakes is one technique. Large quantities of crushed limestone, slaked or hydrated lime, or unslaked lime or quicklime are added to the water or applied to the nearby ground or forest. The high alkalinity of the lime neutralizes the water, and the lake can be restocked as necessary. Sweden, which has no way of controlling emissions of neighboring countries that are blamed for its acid rain problems, has been carrying out an intensive liming program for several years; lakes are being limed in Norway, Canada, and the United States.

Although liming is being used to improve the pH of many lakes, a number of unanswered questions concerning this approach still remain, according to Project Manager Robert Kawaratani. Studies to date have not fully quantified how effective liming is in restoring lake habitat, how long it is effective, what the most economical and effective ways of applying the lime are, and what possible liming side effects might be. EPRI is now beginning to study to test-lime three acidified lakes in the Adirondacks—two without fish and one with fish—to try to come up with some of the answers.

Another way of dealing with acidified lakes may be to introduce fish that can survive in higher acidities. Such fish would be either individuals that have been deliberately and systematically acclimated to higher acidities or species that are naturally acclimated. Researchers still have questions about this strategy, too. Most acclimation experiments have investigated only the short-term response of adult fish to acid stresses, and no one knows if acclimation ensures long-term survival and successful reproduction. It may not be possible, for example, to acclimate newly fertilized eggs to acidified water. Further research will tell.

In the forests

Meanwhile, in the forests, another acid rain puzzle is confounding researchers. Trees, primarily conifers, have been damaged or are dying at unusually rapid rates in recent decades in certain areas of the northeastern United States and Europe. According to quantitative documentation, red spruce have declined atop summits in Vermont's Green Mountains, a decline that has also been observed in the upper elevations of New York and New Hampshire. Pine diebacks have been observed in New Jersey's Pine Barrens. And in West Germany, large areas of Norway spruce and fir have died or appear to be injured.

What troubles researchers is that all these areas receive large amounts of acid rain and other pollution. Certain researchers believe that there may be a connection betwen tree dieback and acidic deposition. Some think that acid rain leaches essential nutrients, such as calcium, from soil and tree leaves. Other hypothesize that the rain's acidity breaks down the protective waxy cuticle on the surfaces of leaves or needles, opening the way for infection or insect infestation. A more recent hypothesis holds that when acid rain mobilizes aluminum in the soil, the fine root hairs of trees absorb the phytotoxic metal, the metal disrupts the roots' delicate water equilibrium, the fine roots die, and so do the trees.

Despite all these hypotheses, there has never been any direct evidence that acid rain is the cause of declining tree growth in North America and Germany. What researchers have found, according to Project Manager John Huckabee, is a collection of conflicting research results suggesting that if acidic deposition is creating problems, it may be acting in combination with other factors (including droughts, which stress the tree population, or highly phytotoxic gaseous pollutants, such as ozone and sulfur dioxide). For example, researchers speculate that the spruce affected in New England may have been weakened by a long drought in the late 1950s and early 1960s; a similar drought may have debilitated the pine stands in New Jersey's Pine Barrens. And in West Germany, sulfur dioxide levels are reported to be relatively high because of heavy industrialization.

Two EPRI studies being conducted by the University of Washington at Seattle and Oak Ridge National Laboratory are examining the hypothesis that acid rain leaches calcium and other nutrients from forest soils. The Washington study is a control study, monitoring conifer and red alder growth in the relatively unpolluted northwest forests to assess normal soil nutrient levels. The Oak Ridge study is monitoring deciduous hardwoods in an eastern forest where there are acid rain and other pollutants. By comparing the two sets of data, researchers will get a better picture of what effect pollutants have on soil nutrients.

EPRI's above-mentioned study of aluminum mobilization in lakes and streams will also examine the effects of aluminum mobilization on forests. Data on precipitation, aluminum chemistry, soil chemistry, and tree growth will be collected at about 17 different sites, most in North America but some in Europe, in an effect to see if trees are really being harmed by aluminum mobilized by acid rain.

Acidity overwhelmed

Although the extent to which acid rain affects lakes and forests still baffles researchers, at least one acid rain question seems well on the way to resolution. That question is the effect of acid rain on agricultural crops and soils. Although early studies suggested that acidic deposition might cause crop damage, later studies on field-grown crops have resulted in the consensus that crops face no danger from acid rain. According to the 1982 Annual Report of the National Acid Precipitation Assessment Program, "The most consistent conclusion to be drawn from agricultural research at all scales and with all species has been 'no effect' at current average ambient pH levels of 4.0 to 4.2."

Although acidic deposition by itself appears to do no harm to crops, it may possibly work in combination with other atmospheric pollutants, such as ozone, to damage plants. Both the EPA and EPRI have research under way to study this possibility. Results are very preliminary, but it seems that the ozone and acid rain doses required to have an effect on crops are so large as to be unlikely to occur in nature.

As for agricultural soils, acid rain is apparently only a drop in the agricultural bucket. Routine applications of fertilizers appeared to dwarf any acid input that rain might provide, and the lime that farmers regularly added to their soils to counteract fertilizer-induced acidity also overwhelmed any rain-borne acid. Further, researchers found that acid rain depositions of sulfur and nitrogen compounds can even prevent sulfur and nitrogen deficiences in certain soils. All in all, researchers seem reasonably confident that the nation's agriculture industry is safe from acid rain.

Statues and cisterns

Two last areas of acid rain effects under investigation are materials and human health. Both are relative newcomers to the research agenda, so even fewer data are available for these areas that for lakes, fish, forests, and crops. The effects of acid rain on materials and human health may be even harder to quantify, so these subjects promise to be under research scrutiny for some time to come.

Reports are frequently made that acid rain is chemically corroding all kinds of materials, from the Statue of Liberty to the Washington Monument. Calcareous building materials, such as limestone, marble, cement, concrete, and masonry, would seem to be acid rain's particular prey, although metals, paints, plastics, and other substances may also be affected. Acidic deposition could certainly corrode these materials, but researchers are up against two problems when they attempt to quantify the link between acid rain and materials.

First, materials damage occurs naturally to some extent, a product of temperature, humidity, precipitation quality and frequency, wind velocity, and other everyday factors. Second, materials damage may be attributable to many other kinds of pollution, such as ozone and particulate matter. To link acid rain with materials damage, the researcher has to be able to distinguish materials effects caused by acid rain from effects caused by other pollutants and natural weathering.

Unfortunately, even though a sizable amount of industrial research has been done over the years to evaluate the susceptibilities of materials to airborne damaging agents and to develop more resistant materials, little has been done to identify which causes are having which effects. There is, in fact, no available method for distinguishing what causes any given instance of degradation.

To remedy the situation, the federal government's National Acid Precipitation Assessment Program is planning and beginning several studies to define the role of acid rain in materials degradation. These studies will include assessing which portion of materials damage can be attributed to acid rain, specific rates of damage, and amount of damage in economic terms. Based on this information, better-informed decisions can be made as to how to deal with the problem.

The effects of acid rain on human health are similarly unknown. Some researchers contend that acid rain can indirectly contaminate drinking water with toxic heavy metals. Some of these metals, such as lead and copper, may be leached out of plumbing systems by the rain; still others, such as aluminum, may be leached out of soil by acid rain.

No definite connections have yet been made between acid rain and human health, but a few data available suggest that if there is a problem, it would apply under limited conditions. Householders who draw their water unmonitored and untreated from rainwater cistern systems may be exposed to metals that the rain corrodes out of water pipes. Those who draw their water from shallow wells may also be at risk if aluminum or other metals leached from the soil by acid rain find their way into the groundwater supply. All these possibilities will require further study.

Answers yet to come

Further study is, of course, the key to quanitifying acid rain's environmental effects, whether they concern lakes, forests, crops, materials, or human health. The work of recent years was the first attempt of governments, research organizations, utilities, and universities to understand these effects. In what were once broad areas of uncertainty, the ongoing research effort has added, slowly and painstakingly, the bits and pieces that will eventually add up to definitive answers.

Researchers are not quite at the point where they can look at any lake and say whether or not it is susceptible to acid rain. They still cannot say for certain why fish are disappearing or what is decimating trees in West Germany and New England. They cannot point to a crumbling statue and pick out acid rain as the cause. They are beginning, however, to understand the complex interaction of precipitaiton, geology, and vegetation that predisposes lakes to acidification and are learning what fish populations are normal for a given lake. They are examining the forests for dieback clues. They are confident that crops will not suddenly wither from acid rain and reasonably certain that acid rain is not a large-scale public health problem. "We've narrowed the boundaries of what we don't know about acid rain's effects on lakes, forests, crops, materials, and human health," concludes Brocksen. And as the research continues the boundaries of the unknown will continue to narrow.

Further reading

An Assessment of the Relationship Among Acidifying Depositions, Surface Water Acidification, and Fish Populations in North America, Vol. 1. Final report for RP1910-2, prepared by Western Aquatics, Inc., June 1983, EPRI EA-3127.

U.S. Environmental Protection Agency. *The Acidic Deposition Phenomenon and Ifs Effects: Critical Assessment Review Papers,* Vol. 2, Effects Sciences, Public review draft prepared by EPA, May 1983. EPA-600/8-83-016B.

The Integrated Lake-Watershed Acidification Study: Proceedings of the ILWAS Annual Review Conference, Conference proceedings, prepared by Tetra Tech, Inc., January 1983. EPRI EA-2827.

Response of Agricultural Soils to Acid Deposition. Workshop proceedings, prepared by Battelle, Columbus Laboratories, July 1982. EPRI EA-2508.

Feasibility Study to Utilize Liming as a Technique to Mitigate Surface Water Acidification. Interim report for RP1109-14, prepared by General Research Corp., April 1982. EPRI EA-2362.

This article was written by Nadine Lihach. Technical background information was provided by Robert Brocksen, Robert Goldstein, John Huckabee, Robert Kawaratani, Jack Mattice, Ishwar Murarka, and Ralph Perhac, Energy Analysis and Environment Division.

Subject Index